Advances in Spatial Science

Springer
Berlin
Heidelberg
New York
Barcelona
Budapest
Hong Kong
London
Milan
Paris
Santa Clara
Singapore
Tokyo

Titles in the Series

Christoforo S. Bertuglia, Manfred M. Fischer and Giorgio Preto (Eds.)
Technological Change, Economic Development and Space
XVI, 354 pages. 1995

Harry Coccossis, Peter Nijkamp (Eds.)
Overcoming Isolation
Information and Transportion
Networks in Development Strategies for Peripheral Areas
VIII, 272 pages. 1995

Luc Anselin, Raymond J.G.M Florax (Eds.)
New Directions in Spatial Econometrics
XIX, 420 pages. 1995

H. Eskelinen, F. Snickars (Eds.)
Competitive European Peripheries
VIII, 271 pages. 1995

Advances in Spatial Science
Van den Bergh, Nijkamp and Rietveld (Eds.)
Recent Advances in Spatial Equilibrium Modelling
ISBN 3-540-60708-0

Errata

The last sentence of the preface on page V should read as follows:
"The importance of the models and their result is therefore not confined to the spatial economist's domain".

Further to this Figs. 17.1 - 17.3 and Tables 17.1 and 17.2 are missing in Chapter 17.

	Fixed Wage Cases	Competitive	Exogenous Labour Supply
Nominal Wage	0.00	-2.34	-23.08
Real Consumption Wage	0.00	-1.30	-13.43
Total Employment	0.31	2.91	35.86
Manufacturing	0.12	3.46	45.06
Non Man Traded	0.50	3.74	45.61
Non Traded	0.27	1.53	17.57
Unemployment Rate	14.62	13.20	0.00
Population	1.38	3.89	35.87
Value Added			
Manufacturing	0.12	3.33	42.82
Non Man Traded	0.50	3.58	42.81
Non Traded	0.27	1.41	15.89
Consumption Composites	0.98	3.25	32.77
Capital Stocks			
Manufacturing	0.12	2.80	34.72
Non Man Traded	0.50	3.15	36.33
Non Traded	0.27	1.06	11.32
Consumer Price Index	0.00	-1.16	-12.11

Table 17.2. The long run impact of an increase in amenity with a varial capital stock

Jeroen C.J.M. van den Bergh
Peter Nijkamp • Piet Rietveld (Eds.)

Recent Advances in Spatial Equilibrium Modelling

Methodology and Applications

With 26 Figures
and 34 Tables

Springer

Dr. Jeroen C.J.M. van den Bergh
Professor Dr. Peter Nijkamp
Professor Dr. Piet Rietveld

Free University
Faculty of Economics
Department of Spatial Economics
De Boelelaan 1105
NL-1081 HV Amsterdam, The Netherlands

Cataloging-in-Publication Data applied for

Die Deutsche Bibliothek - CIP-Einheitsaufnahme

Recent advances in spatial equilibrium modelling :
methodology and applications ; with 34 tables / Jeroen C. J. M.
van den Bergh ... - Berlin ; Heidelberg, New York ; Barcelona
; Budapest ; Hong Kong ; London ; Milan ; Paris ; Santa Clara
; Singapore ; Tokyo : Springer, 1996
(Advances in spatial science)
ISBN-13:978-3-642-80082-5 e-ISBN-13:978-3-642-80080-1
DOI: 10.1007/978-3-642-80080-1
NE: Bergh, Jeroen C. J. M. van den

ISBN-13:978-3-642-80082-5

Softcover reprint of the hardcover 1st edition 1996

SPIN: 10488585 42/2202-5 4 3 2 1 0 - Printed on acid-free paper

Preface

Prices and quantities of both stock and flow variables in an economic system are decisively influenced by their spatial coordinates. Any equilibrium state also mirrors the underlying spatial structure and a tâtonnement process also incorporates the spatial ramifications of consumer and producer behaviour. The recognition of the spatial element in the formation of a general equilibrium in a complex space-economy already dates back to early work of Lösch, Isard and Samuelson, but it reached a stage of maturity thanks to the new inroads made by T. Takayama. This book is devoted to spatial economic equilibrium (SPE) analysis and is meant to pay homage to the founding father of modern spatial economic thinking, Professor Takayama.

This book witnesses his great talents in clear and rigorous economic thinking regarding an area where for decades many economists have been groping in the dark. Everybody who wants to study the phenomenon of spatial economic equilibrium will necessarily come across Takayama's work, but this necessity is at the same time a great pleasure. Studying his work means a personal scientific enrichment in a field which is still not completely explored.

The present volume brings together recent contributions to spatial equilibrium analysis, written by friends and colleagues of Takayama. The structure of the book is based on four main uses of spatial equilibrium models:

(i) the imbedding of spatial flows in the economic environment, related to e.g. regions, trade, transport and labour markets;
(ii) the analysis of financial instruments for dealing with congestion and environmental externalities, including attention for (sub)optimality and interactions between instruments;
(iii) the combination of intersectoral and interregional or spatial interactions in a single framework;
(iv) the inclusion of spatial and economic dynamics in the description of spatial economies: dealing with different time periods.

Although the coverage of topics is never complete, many important issues in the field of spatial equilibrium analysis receive attention here. It is clear that the SPE and other spatial equilibrium models typically allow for an analysis of issues related to price-instruments, market-interactions, transport and trade. The importance of the models and their result is therefore not confined to the spatial

Especially in the fields of transportation science, international and environmental economics, concepts and frameworks adopted by regional and urban economists may prove useful. This book can thus inform a broad audience by giving an account of the established field of spatial equilibrium analysis, as well as its progress and extensions.

In composing this volume, much support was given by the Tinbergen Institute (Amsterdam) which hosted a meeting on this topic and which also encouraged the publication of this volume. The editors wish to thank Elfie Bonke and Miep Oomes for the professional support. Gwyneth Moore (Contact Europe) has to be thanked for her skilful editing. Finally, we wish to express our thanks to Marianne Bopp (Springer-Verlag) who puts the 'crown' on our work by publishing this volume.

Amsterdam, October 1995

Jeroen C.J.M. van den Bergh
Peter Nijkamp
Piet Rietveld

Contents

I

Spatial Equilibrium Theory and Models

1 Thirty Years with Spatial and Intertemporal Economics[1]

Takashi Takayama
University of Western Australia

1.1 Introduction

The decade of the 1950s could be called the most productive decade of applied economics. The nonlinear programming paper by Kuhn and Tucker (1951), announced the dawn of the decade, followed almost immediately by George Dantzig's *Maximisation of Linear Function of Variables subject to Linear Inequalities* (1951). Within the decade a number of promising applications of especially 'linear programming (LP) methods' appeared in the professional journals and bookstores all over the world. In the field of agriculture, the late Professor E. O. Heady, an outstanding educator and pioneer in applying neoclassical economic methods to economic problems (*Economics of Agricultural Production and Resource Use,* 1952), energetically applied linear programming methods to economic decision making problems which culminated in a seminar work (*Linear Programming Methods*, with W. Candler 1958). In the field of manufacturing industries, Alan S. Manne produced another seminar work (*Scheduling of Petroleum Refinery Operations*,) in 1956, which has since been used by most major oil companies in the world. A large number of transportation cost minimisation LP formulations and applications were published during this period (see Gass, *Linear Programming: Methods and Applications*, 1969).

In the field of *Economics Proper* S. Enke showed how the so-called spatial equilibrium can be solved by electric analogue (1951), followed by Paul A. Samuelson's 1952 paper in *American Economic Review.* In the fields of applied economics and operations research Karl Fox published his US livestock economy analysis in 1953. Charnes et al. (1952) showed how to use LP in arriving at optimal blending of aviation gasolines, and Alan Mann (1956) extended LP application to the whole oil refinery operation.

[1]This paper was first read at an international conference on 'Trade, Knowledge and the Network Economy' in Mallacoota, Victoria, Australia and was revised to present for the workshop entitled 'Advances in Spatial Equilibrium Modeling' held at the new Tinbergen Institute in Amsterdam, the Netherlands on January 24-25 1994.

Surrounded by all these exciting developments Japanese economists remained isolated from the outside world during the late 1940s and early 50s. Of course war-ravaged Japan was too preoccupied with its survival to look after the intellectual development of university staff and students.

Three years after my graduation from the Hokkaido University, a branch of the Rockefeller Foundation (the Council of Economic and Cultural Affairs) provided me a postgraduate fellowship, and I chose the Iowa State University, where I stayed for one academic year. What an eye-opening twelve months it was! How exciting it was to see neoclassical tools, LP and Quadratic Programming (QP) tools fully used for (farm) business decision making!

I was awe-struck, and then felt that something had to be done to change the environment of agricultural economics in Japan. Since 1956 my teaching duty at the Hokkaido University covered a wide range of subjects such as mathematical economics, economic statistics, production economics and agricultural policy. In 1959 my first book was published, in which modern management methods were then applied as useful and effective tools for improving the farm management of a large number of farms in Hokkaido, Japan. I mentioned about the tools being 'effective', as the applications of the tools enhanced the communication between the tool developers and tool users.

It is an embarassing realization on the part of economists that there is no 'perfect' model in this world. I plan to come back to this qualifier 'perfect' at the end of this paper.

The farm firm (micro) level models in their static framework did not automatically aggregate the micro firm production information to form a system of regional supply functions, which together with the market demand functions for the products involved in the regional supply functions, generated market clearing (equilibrium) prices for the products, and optimal (in the sense of spatial equilibrium) individual farm outputs consistent with the regional market demand quantities for all the products in the model system.

In the late 1950s the major challenge for me was to successfully formulate and practically solve the so-called 'Multi-Product Spatial Price Equilibrium Activity Analysis Models'.

This was my humble beginning in the STPA modelling field, and allow me to emphasise the fact that the motivation was 'practical' or 'pragmatic' rather than 'academic' or 'just theoretical'.

1.2 Theory and Practice of Spatial Economics

Looking back, I must confess that I started out with 'spatial economics' from the 'practical' side. By the time I finished my PhD thesis and was fortunate to be awarded a post doctoral fellowship by the Council on Economic and Cultural Affairs in 1961 to pursue my research in the USA, I had just about completed the activity analysis formulation of the spatial equilibrium model. However, at this stage computer facilities in Japan were hopelessly backward and outdated, and the algorithmic knowledge I had was only the gradient method. At Penn State I tried some versions of gradient methods for my quadratic programming formulation (due to the linear regional demand functions in the model), which hopelessly failed.

If I were an operations research specialist I would have known of the existence of P. M. Wolfe's 1959 *Econometrica* paper showing how to modify the simplex method to solve quadratic programming (QP) problems. It was my colleague, Professor Robert Hutton at Penn State who saved me from remaining in a state of ignorance and eventually solved my QP problem by using his LP computer program after appropriate modifications. What an excitement it was when one morning in the summer of 1962 I first saw the IBM solution printouts!

This was the first breakthrough in introducing demand functional information into activity analysis based location theory. Of course, one of the forerunners of the location theory, Professor Martin Beckman, had already a paper using activity analysis method to location theory in 1955, which paved the way to a number of publications such as *Allocation in Space* by L. Lefeber (1958) and *An Interregional Linear Programming Model* by B. H. Stevens (1959).

In the late summer of 1962 I attended the Annual Conference of the American Agricultural Economics in Storrs, Connecticut, where I became acquainted with my future colleague and a co-author of many papers and books, Professor George G. Judge, of the University of Illinois. George had already published a number of 'practical' papers related to such agricultural products as egg (1956) and livestock (1959 with T. D. Wallace). These papers followed the tradition of Karl A. Fox (1953).

While I was busy finishing up writing my 'activity analysis QP' paper at the Penn State, George Judge invited me to the University of Illinois for a few days. The research issue he raised on the first working day in his office was, "Is it possible to successfully reformulate and solve the Samuelson spatial price equilibrium LP model in my QP format?". I did not anticipate this challenge, and read the Samuelson paper seriously for the first time in my life.

It was not purely theoretical issues that motivated me to develop the activity analysis QP model, and, quite naturally, mentally I felt reluctant to tackle this issue. But while I was thinking about the way to formulate this Samuelson problem in the bed of a YMCA room on the campus, I suddenly realised the fascinating aspect of the primal-dual nature of this quadratic problem. In the case of LP duality the product market prices (or price ratio) once fixed, they do not interact with the levels of optimal outputs. In my QP formulation the quantity formulation naturally induces the price formulation, and vice versa. The mutual primal-dual formulations were presented the next day, and the rough draft was finished before I left for Penn State the following day.

In March 1963 I moved from Penn State to the University of Illinois to complete a few papers within that year with George. The first paper completed and published was the one in the Samuelsonian (theoretical) framework in *Journal of Farm Economics* (JFE) (1964a), immediately followed by an *Econometrica* paper (1964b).

Finally, however, the very paper I had devoted my time and energy to complete came out in print in the QP cum activity analysis format in late 1964, with George Judge as my co-author (1964c). I should have been extremely proud of this, and should have planned to apply the model to the real world situation in Hokkaido, Japan where the practical issue originated from. However, the academic research environment at the University of Illinois was carrying me away from such a pragmatic approach as I planned back in Japan.

An immediate challenge was waiting for me to apply my QP approach to the Samuelsonian (1957) intertemporal price equilibrium formulation. This task was quickly accomplished and the end result was published in JFE (1964).

It was made abundantly clear that mathematical programming, especially non-linear programming was a powerful tool in both theoretical and applied economics.

Due to visa irregularities, I had to leave the United States and return to Japan where I stayed for only two months. In February 1965 I moved to Australia. At the University of New England I had two major research goals. The first one was to supervise as many higher degree researchers who were willing to do serious research in both theoretical and applied work in my field. The second goal was to write a comprehensive research monograph in the STPA modelling field.

Before delving into serious theoretical questions such as the concept of 'net social payoff' I had at this stage, the application of the Takayama-Judge Spatial Equilibrium (especially Activity Analysis) models had spread all over the USA; (e.g. Plessner Y, Heady EO 1965; West DA, Brandow GE 1964; Hall HH, Heady EO, Plessner Y 1968).[2]

Up to 1963 the spatial equilibrium problems remained 'interregional' competitiveness problems. The post World War II deregulation of the interstate trade of agricultural products was considered to have played an important role in promoting the popularity of the Takayama-Judge models. This enthusiasm lasted till the end of the decade in the USA, and continued into the 1970s in other countries such as Japan. (See Sasaki K 1969 and 1970).

A natural extension of the Takayama-Judge models in application was in the field of international trade. This direction was taken up in 1966 in my inaugural lecture published as *Spatial Economic Analysis*. A revised version was published in 1967. The volume and value of the international trade had been steadily rising since the 1950s. The time was ripe to evaluate a number of national and regional (say EEC) policies on international prices of wheat and other farm products and other issues such as impact of trade on the GNPs of trading countries.

The national governments were seriously engaging in modelling for policy evaluations. In the US Department of Agriculture, for instance, A. S. Rojko's group was working on the Grain-Oilseeds-Livestock (GOL) model of the world (see Rojko AS, Urban FS 1971; Naive JJ 1971). Pinhas Zusman's group in Israel was evaluating EEC trade policies (see Zusman, Pinhas, Melamed, and Katzir 1969).

Back in Australia I partially supervised several PhD candidates who completed their theses in the field of 'Spatial-Temporal Equilibrium' after I returned to the University of Illinois in 1968. From the theoretical point of view, Alan Woodland's (presently Professor of Econometrics at the Sydney University) 1969 PhD thesis was a milestone in this difficult field of research.

---

[2]I quoted just a few publications here, with full knowledge that there were a large number of important empirical works in the real world that I am unable to refer to here due to space constraint of this short paper.

Incompleteness of a model formulation plagues the model creator once any defect is revealed. In the case of the 1964 *Econometrica* paper some irregular cases such as the regional self-supply quantity ($\bar{x}_{ii}$ for some i) turning out to be negative were found. This problem required an almost complete overhaul of the previous formulation, but the end result of our salvage work was extremely satisfactory and appeared in *Econometrica* (see Takayama T, Woodland AD 1970). This paper finally established firmly the self-dualism of the price and quantity formulations. Our scheme was to separate the regional consumption, y_i (i = 1, 2, ..., n) from the total regional inflows $\sum_{i=1}^{m} x_{ji}$, setting $y_i \leq \sum_{j=1}^{m} x_{ji}$, and, similarly, separating the regional supply x_j (j = 1, 2, ..., m) from the total regional outflows $\sum_{i=1}^{n} x_{ji}$, setting $x_j \geq \sum_{i=1}^{n} x_{ji}$.

With the self-dualism issue resolved I started to work on my research monograph on STPA models as soon as I returned to the University of Illinois campus in 1968. The monograph entitled *Spatial and Temporal Price and Allocation Models* was completed at the end of 1970 and published from the North-Holland Publishing Company in 1971. In retrospect, it was extremely tough going at the end as I entered Part VII Pricing and Allocation over Time. Dynamics intrigued me and still does. There were a number of unresolved problems buried in this part to which I return in section 1.5 of this paper.

Besides many painful memories there were a few pleasing experiences too. In writing Chapter 11, Single Product Monopoly Spatial Equilibrium Models, I found the results extremely clear and satisfactory. If the reader carefully examines the solutions for the spatial monopolist and for the precautionary monopolist, he/she will find the regional supply quantities for the two cases are exactly the same (even though the consumption quantities are different). For the answer one is referred to Joan Robinson's *The Economics of Imperfect Competition* (1933). (This observation was made to me by Professor D. Levhari, Hebrew University of Jerusalem).

With the publication of the research monograph, George felt that we needed to publish some research results in close relation with the book. Another two years effort finally saw the publication of an international compendium of theoretical and applied works in the field of STPA (see Judge GG, Takayama T 1973).

1.3 Turbulences in Commodity and Oil Markets and the STPA Modelling

The publication of the edited book cited above coincided with the two consecutive turbulences in the world commodity markets. The first was started with almost simultaneous crop failures in the USSR, China and India, and skyrocketed the price of wheat in the 1972-73 crop year. The interdependency amongst the agricultural product market produced chain reactions, which prompted President Nixon to declare an embargo of US soybean to Japan. The second turbulence occurred in the world oil market in October 1973 as OPEC declared the oil embargo to the USA and the Netherlands and cut the supply of the crude oil by one third (about ten million barrels/day). The first edition of the book was sold out almost as soon as it appeared and the second edition was published in a paperback form.

My research proposal submitted to the Ford Foundation - the World Bank sponsored 'International Competition on International Economic Order' was funded and I became involved in carrying out a large scale modelling project related to the agricultural commodities in 1974. In the same year I was asked by the Venezuelan Ministry of Mines and Hydrocarbon to assist their modelling effort in collaboration with OPEC.

The time finally arrived for the full-fledged application of the STPA models to the agricultural commodities and natural resources trade policy analysis in a worldwide scale. Some of the results were reported in two recent publications (see Labys W, Takayama T, Uri N (Eds) 1989; Gurenen O, Labys WC, Lesourd JB (Eds) 1991).

These represent only the tip of the iceberg of world-wide applications of the STPA models.

OPEC was forced to develop their own pricing mechanism for their 47 API differentiated crudes since 1975. The Queen Mary College group led by Professor Deam was contracted to do the job using the BP data and some new information from OPEC. The group naturally used the Alan Manne type refinery LP model for each country refinery group along with the LP transport network model to solve the price differentials among the 47 crudes. The dimension of the LP model was said to be over 10,000 x 20,000. Both British and French high powered computers failed to solve the initial problem. Finally in St Paul-Minneapolis, the initial solution was generated (of course with the approval of the US Government!).

At the University of Illinois my project team developed a multi-country (region) energy price and allocation model in the 1975-76 period and produced a number of interesting results. The constraint was the smallness of the dimension of the QP software to handle a potentially gigantic dimension of the problem at hand.

A sketch of my energy modelling scheme was belatedly published in the *Journal of Regional Science* (see Takayama T 1979).

My agricultural commodity trade policy research project group steadily produced a series of reports, some of which were recorded in the Labys-Takayama-Uri edited book (1989).

At the final stage of the project we had nine commodities in our system; they were beef, pork, chicken, wheat, maize, other grains, soybeans, soybean meal and soybean oil. Regionally we had eight countries and one region called 'rest of the world'. The eight countries were USA, Japan, Canada, Australia, USSR, China, India and Argentine.

The larger the model, the costlier it became. Other disadvantages go along with academic research projects such as the ones I pursued in the late 1970s. Another lesson I had to learn, was that without close collaboration with the petrochemical engineers, transportation specialists, international trade policy specialists, agronomists and others, this kind of project could not be run successfully. I maintain that I conducted my research projects as conscientiously as an academic scholar could, but I am not sure, to be honest, whether the projects were successful or not.

One exciting sub-project in the world agricultural commodity trade project was the one carried out by Dr H. Hashimoto. This work concentrated on establishing the US price stabilisation policy based on a five year dynamic model with eight commodities and several scenarios for pre-set price stabilisation bands (policies). The dual solution for the bands for each year generate government purchase (or withdrawal) quantities of the products concerned.

The whole project was carried out on the Linear Complementarity Programming (LCP) framework. The Hashimoto's model reached the matrix size of about (550 x 550).[3]

There was good reason to believe that the so-called 'buffer stock' policy for price stabilisation worked in the manner in principle as we modelled. However, how to implement the policy was a completely different matter as the commodity markets were (and still are) run by several (handful of) international companies cum government enterprises.

Other theoretical (practical?) issues one encounters in this field are (a) 'future uncertainties' in crop yields in various regions in the world and, (b) the (5th) terminal year stock carryover quantities (terminal boundary conditions).

It was becoming steadily clear to me that the governments and business institutions, small or large, strenuously avoided 'uncertainties' in their policy decision-making (just like 'nature abhors a vacuum' (Pascal)), only to use the term to defend the failure of their policy (strategy) ex-post. In carrying out large scale modelling projects such as my projects during the 1974-78 period at the University of Illinois, the OPEC crude oil pricing project, the US Department of Energy projects[4], etc, there was no room for uncertainty modelling schemes. This was due mainly to non-existence of efficient computer software to solve stochastic and adaptive programming problems of the scale mentioned above, and this situation is unfortunately still with us. In Section 1.6 of this paper I present dynamic models with stochastic element in its structure from a purely theoretical point of view.

1.4 Model Validation and Social Evaluation

Model validation practice has been with the STPA modelling exercises since the beginning, Robert Thompson (1989) called for rigorous validation of agricultural commodity models, and W. C. Labys (1984, 1985, 1987, 1991) did it for energy and mineral resources STPA models. Theoretical purists may not be interested in

---

[3]For detail see Hashimoto (1977), and for a compact summary, Chapter 10 of Labys, Takayama and Uri (1989).

[4]Interested readers are referred to a series of 'energy model validation' publications from The Energy Information Administration, the US Department of Energy, Washington, D.C.

model validation at all, but empirical economists have conscientiously applied certain criteria to judge the validity of their models.[5]

In the case of the Spatial Equilibrium Model with K commodities, m_k $(k = 1, 2, \ldots, k$ supply regions and n_k $(k = 1, 2, \ldots, K$ demand regions, we know that the maximal number of optimal flows among regions do not exceed $K \pi_{k=1}^{k} (m_k \cdot n_k - 1)$, while in the real world trade we find trade flows well over the above number. This may be due to a particular international trade policy of a national government that tries to diversify the importing sources to avoid excessive dependency on a specific producer nation for both economics and international politics. I have never considered this aspect of the model a defect. From the policy point of view the optimal network solutions reveal the most rational way to carry out the trade, while the actual statistics are the product of compromises.

The aspect of the STPA models applied to international trade of agricultural products that has concerned me throughout my career, is the existence of the Northern and Southern Hemispheres and the necessity to incorporate this fact into our modelling practice. This means that the annual model must cover eighteen months with two overlapping periods of twelve months each (1st - 12th month for the Northern Hemisphere and 6th - 18th month for the Southern Hemisphere, for instance).

I anticipate that this modelling scheme may introduce a number of new and more realistic features into the basic STPA models. And I realistically hope that someone will succeed in modelling this aspect one day in the near future.

During the Carter Presidency, the US Department of Energy (DOE), developed tremendous modelling capabilities especially in the Energy Information Administration (EIA), the Solar and Conservation (SC) Division, and the Policy Evaluation (PE) Office. The EIA used the now rather infamous PIES (Project Independence Evaluation System - a modelling system developed in the last year of the Nixon Administration to evaluate the possibility of the USA getting free from the yoke of the importation of oil from the Persian Gulf states) fully expanded to accommodate the ten (Department of Energy classification) regions. The integrated energy system contained the down stream energy demand information, upstream energy supply information, process technology (production functions in process analysis format) of the latter into the former forms of energy, and policy components. Such a gigantic system can be solved with interactive procedures for each year and then recursively solved forward in time. Sometimes, even the Government high power computer near Washington, D.C. could not complete a run during the weekend (two full days)!!

There were a number of energy (projection or analysis) models run for the US DOE. A large number of University researchers and the US National Laboratory

---

[5]Model testing has been done extensively and intensively in the applied econometrics field and the tentative conclusion is that anyone of the statistics such as the mean squared error, Chi-square, U-statistics, Wald-test, etc is uniformly better than the others. Since the *true* model is unknowable in our profession, such econometric exercises may prove to be unproductive after all.

researchers collaborated with us. There had never been such a large scale support of energy-related research project on university campuses and in the National Laboratories in the US history. During the second half of the Carter Administration, Congressional and public criticisms of the model-based projections produced by the US DOE[6], forced the DOE to validate all the models used by the Department.

As President Reagan entered the White House in 1981, Dr David Stockman, the then Director of the Office of Budget and Management issued his stark criticism of economists and economic forecasters; 'Economists have been consistently wrong'. In terms of world energy modelling both OPEC and the US DOE stepped over the data and even computer capability (and therefore modelling common sense) and ended up discrediting themselves in the end.

As to the STPA modelling, I did not feel guilty of overextending our modelling thus misleading the Governments and the public.

In 1981 I moved from the US DOE to the World Bank as Economic Adviser, and in 1984 moved to the University of Western Australia to enjoy my cherished academic freedom again.

1.5 STPA Model House-Cleaning

In the 1980s the STPA models turned out to be a set of common professional tools among agricultural economists, general economists, geographers, regional economists, energy economists, etc. It is impossible for a single researcher such as myself to fathom the depth of sophistication the models attained during this period.

In my mind, one remarkable progress made in this decade was the computer software development for large scale (more than 500 inequality constraints according to George Dantzig) general non-linear programming problems and their applications. I reported this in 'Recent Development in Spatial (Temporal) Equilibrium Models: Non-Lineality, Existence and Other Issues', (with Dr. Gordon MacAuley) a paper presented at the 1988 XXVth International Conference on International Commodity Market Modelling, World Bank, Washington, D.C. In this work we used the MINOS[7] and successfully solved an STPA model with non-linear (non-quadratic) objective function and non-linear terms contained in the constraint functions.

The above experience expelled my bias toward QP formulation of STPA models to some extent. Some rather uncomfortable features still remain unresolved in the

---

[6]Due to the Public Disclosure Law applied to Federal Government information (except for National Securities matters), the model validation of the US DOE-EIA models was legislated.

[7]The packaged used was MINOS 5 developed by Murtagh and Saunders (1983) and incorporated in the GAMS (Generalised Algebraic Modelling System) developed by Melrous in the World Bank during the 1981-83 period.

MINOS type approach. One outstanding problem, in my mind, is that a solution may not be found in a finite number of iterations.

The most recent developments reported by Professor Anna Nagurney of non-linear programming software experimentations on a super computer in the USA promised a great hope for wider use of large dimensional programming problems including my STPA models (this aspect must be covered in her paper, 'A Dynamic Approach to Network Oligopolies and Variational Inequalities' in Network Dynamic Session of this Conference).

Despite a number of problems surrounding the dynamic version of the STPA models, a great deal of effort has been expended by a few dozen researchers in this field whose publications were cited in Judge-Takayama (1973) and Labys-Takayama-Uri (1989) to quote a few.

Shortly after I finished the 1971 Takayama-Judge book, I started to explore further into the dynamics in the STPA modelling. In the remainder of this paper I would like to present a trace of my thoughts (not application) in that direction.

However, before entering the technical aspect of dynamic spatial equilibrium modelling, let me make my personal evaluation of the STPA models vis-a-vis real world environment.

At the beginning of this paper I mentioned my own motivation in stepping into the STPA modelling which was 'practical' and 'useful' and *not* 'theoretical'. In this sense I have never claimed that my STPA modelling was a theoretical breakthrough. If there was any positive contribution, I can point out without hesitation that the 'purified duality' of quantity and price formulations has been firmly established in the STPA models. Therefore, I would like to point out unambiguously that *my STPA models are both Marshallian and Walrasian* (see Takayama and Woodland 1970; Takayama and Judge 1971). Compare this with Samuelson's 'full-duality' in his 1965 *Econometrica* paper, and I hope this dispels misunderstanding by some researchers that the STPA models are 'Marshallian' to emphasise their models are 'Walrasian'.

Even though introduction of consumers' surplus (utility) was firmly in my activity analysis based model (1964), P. Samuelson as the theoretical forerunner must be recognised. The supply generation using the activity analysis formulation is due to W. Leontief, T. Koopmans and G. Dantzig (via E.O. Heady 1952). Operationally, the contribution of Dantzig and Wolfe (1959) - Cottle (1968) must be duly acknowledged.

From the beginning user identification for the STPA models required no special effort. In the 1960s agricultural economists, geographers and some general economists turned out to be the major users. Naturally, the Federal Government, State and local government, research institutions and private companies used the models in one form and purpose or the other.

During the 1970s the user groups spread over agriculture, oil, minerals and some manufactured products. As I mentioned before, the US Government Departments, the Japanese Government Ministries, other National Governments, the FAO, the World Bank, the OPEC, the Multinational companies, consulting firms and University researchers were deeply involved in running international resource allocation models. The STPA models provided the very basis of policy evaluation of the day.

The researchers in this field turned more critical eyes to the STPA models and their results vis-a-vis realities in the 1980s.

By then (and from the beginning) it became clear that the STPA models are not forecast models or prediction models. They contain two distinct elements that go against politico-economic realities of the existing world. The first element is economic rationality clearly expressed in their optimisation principles employed. The real world realities consist of political compromises among national governments and even among domestic regional units (agents).

Another factor that creates model deviations from the historical data is the stochastic characteristics prevalent in econometrically estimated demand and/or supply functions embedded in the STPA models.

As I mentioned before, one group of researchers in this field carried out the so-called 'model validation' and pointed out that the international pattern of commodity movements among countries is much more complicated than the STPA model solutions indicate (Thompson R 1989).

Another group studied the commodity characteristics that made the STPA models approximately useful in the policy decision making situation, and stated that goods with low value per unit weight such as farm products, oil and minerals are more suitable for the STPA models than high valued goods such as gold and hi-tech light weight merchandises (Batten D, Westin L 1990 for instance).

Both findings mentioned above are justifiable once the two elements of the STPA models are taken into consideration in modelling real world situations. There is always a hope that if one is allowed (fund-wise and data-wise) to study in more detail policy interventions by national governments and multinational firms, the finer classification of the commodity, regions (sub-regions), production cost differences (oligopolist firms in manufacturing sectors), production seasons between the northern and southern hemispheres for farm products, etc., then the above findings may prove to be less important.

However, one important lesson a researcher in this field needs to learn is the price to pay for making the model more and more detailed and comprehensive, thus inviting the 'curse of dimensionality'.

No economic model has ever been perfect in relation to politico-economic realities we have faced in the history of mankind. But the basic urge to make it perfect is always with us.

Within the past decade the Swedish group led by A. Andersson, F. Snickars, B. Johansson, D. Batten, L. Westin, along with collaborative colleagues such as John Roy from Australia, have pushed the frontier of the STPA modelling in a number of directions. It is quite natural that they have gradually departed from the main framework of the STPA modelling strategies as the politico-economic environments in the world have changed drastically in these years. Along with the changes have come the drastic shift in issues and problems the policy decision-makers in both public and private sectors face in the world.

Through my thirty years experience with research and modelling in academia, and public life in Washington, D.C. I have maintained the belief that a successful modeller should be an effective communicator with his/her policy decision-maker counterparts(s). The longevity of LP models in general and the Leontief I/O models reveals that they possess that admirable quality of 'simplicity' and therefore 'self-communicability'.

My STPA models have been very easily understood by policy decision-makers, and therefore have been absorbed in both public and private sector modelling activities. However, research must be pushed forward constantly.

In the next section I plan to present my search for establishing the basic framework for dynamic STPA modelling.

1.6 Intertemporal-Spatial Price Equilibrium Models

In the following, an intertemporal price equilibrium model is presented in the calculus of variation format (section 5.1), which is then reformulated in the optimal control framework (section 5.2). In section 5.3 two examples are solved in their quadratic-linear form, which is then extended to include space (two country trade-stock operation model) in section 5.4. In section 5.5 discrete time dynamic-spatial equilibrium models are presented in deterministic and stochastic formats.

The subjects treated in this section are not new, but they show step by step, how much complication is introduced as we wade into this dynamic direction. At the same time this process acutely points out how much more information is required to successfully formulate dynamic (stochastic) models.

The basic dynamic SPTA models were already presented in the 1971 Takayama-Judge book. In the following, let me step right into the Samuelsonian intertemporal price equilibrium model.

1.6.1 A Simple Deterministic Intertemporal Price Equilibrium Model in Classical Calculus of Variation Form

Abstracted from complexities in the present day price and commodity movements on the floor of the commodity exchange board, a commodity such as wheat or rice in the temperate zone is harvested at a certain time of the year and, if the producing region or country is closed from the outside world, it may conceptually be anticipated to show some smooth patterns of price and consumption quantity movements over the year and stationary periodic movements over years. This Holbrook-inspired Samuelson (1957) conceptualisation of *intertemporal* price and quantity movements may be given a proper mathematical formulation.

Assume the production takes place only at $t = 0$, and the total production is K. Define the relationship among the consumption intensity (of felicity) at the moment of time t, $y(t)$, total stock of the storable commodity, $Q(t)$, and price of the commodity, $p(t)$, as:

$$\frac{dQ}{dt} = \dot{Q} = -y(t) = -f\,(p(t)), \tag{1.1a}$$

where $f(p(t))$ is the indirect demand (felicity) function.

Let us assume that f in (1.1a) is invertible, and express this as:

$$p(t) = F(y(t)) \ (= F(-\dot{Q})), \tag{1.1b}$$

which is called the demand function in the Marshallian sense.

Define the social payoff function as the integral of the demand function over $y(t)$:

$$SP(y(t)) + SP(-\dot{Q}(t)) = \int_a^{y(t)} F(c)dc, \tag{1.2}$$

and the total storage cost as:

$$R(Q(t)) = \int_b^{Q(t)} T(q)dq, \tag{1.3}$$

where $T(.)$ is the storage cost function and is assumed to depend only on $Q(t)$, and a and b are appropriate constants.

This problem may be formulated as the following calculus of variation problem and the solution may be obtained rather easily if the functional forms of SP, R, and f are simple enough as will be shown in the example problems at the end of this section.

The Samuelsonian Maximum Problem: Find the optimal stock profile, $Q(t)$ for $t \in [0,1]$ that maximises:

$$NSP = \int_0^1 \{SP(-\dot{Q}\ (t)) - R(Q(t))\}dt \tag{1.4}$$

subject to

$$\begin{cases} \text{(a)} \quad \dfrac{dQ}{dt} = -f(F(-\dot{Q})), \\ \text{(b)} \quad Q(0) = K, \text{ and} \\ \text{(c)} \quad Q(1) = 0. \end{cases} \tag{1.5}$$

Here, assumptions such as the total supply is consumed completely at the end of the year (1.5c), and time horizon is taken to be one year or, more accurately, one harvest year are incorporated. This is a classical two point boundary value (TPBV) problem and the Euler differential equation as a necessary condition is given as

$$\frac{d}{dt}\frac{d[SP(-\dot{Q}(t))]}{d\dot{Q}(t)} + \frac{dR(Q(t))}{dQ(t)} = 0 \tag{1.6a}$$

or, by (1.2) and (1.3)

$$-\frac{d}{dt}F(-\dot{Q}(t)) + T(Q(t)) = 0 \tag{1.6b}$$

(or by using (1.1b) we can write (1.6b) as $\frac{dp(t)}{dt} = T(Q(t))$).

In essence (1.6b) states that the optimal profiles of price and stock can be obtained if the instantaneous price change is exactly the same as the storage cost at that time. Another necessary condition, Legendre condition can be spelled out as:

$$\frac{d^2SP(-\dot{Q})}{d\dot{Q}^2} \le 0, \tag{1.7}$$

which can be rewritten by using (1.1b) as

$$\frac{dp(t)}{dy(t)} \le 0. \tag{1.7a}$$

By assuming that the demand function is downward sloping in the Marshallian sense, this condition is guaranteed with strict inequality.

The temporal equilibrium path $\bar{Q}(t)$ can be obtained by solving the following system of necessary conditions;

$$\begin{cases} \text{(a)} \quad \frac{d}{dt}F(-\dot{Q}(t)) = T(Q(t)) \\ \text{(b)} \quad \frac{dQ}{dt} + \dot{Q}(t) = -f(F(-\dot{Q}(t))) \\ \text{(c)} \quad Q(0) = K \text{ and } Q(1) = 0. \end{cases} \tag{1.8}$$

This system is written in its classical calculus of variation form. In the following section the optimum control formulation of the problem and the necessary conditions will be developed, and the equivalence of these two formulations will be shown. Two example problems will then be presented and the solutions will be derived and discussed.

1.6.2 A Simple Deterministic Optimal Control Formulation of Intertemporal Price Equilibrium Model

In the previous calculus of variation formulation, we entertained variables such as Q, $\dot{Q}$, and t. As is customary in optimal control theoretical framework, we call $Q(t)$ a state variable and in this case $-\dot{Q}(t) = y(t)$ as a control variable as we may be able to assume that the consumers can control the intensity of consumption somewhat in our economy and define our optimal control path as follows:
Find the optimal state path, $\bar{Q}(t)$, and the optimal control path, $\bar{y}(t)$ that maximise

$$NSP = \int_0^1 (SP(y(t)) - R(Q(t))dt \tag{1.9}$$

subject to

$$\begin{cases} (a) \quad \dot{Q}(t) = -y(t) \\ \text{and} \\ (b) \quad Q(0) = K \text{ and } Q(1) = 0. \end{cases} \tag{1.10}$$

It now remains to be shown that the above optimal control formulation is equivalent to the calculus of variation formulation given in Section 1.6.1.

The Hamiltonian for this problem is:

$$H(Q(t), y(t), \rho(t), t) = SP(y(t)) - R(Q(t)) + \rho(t)\,(-y(t)), \tag{1.11}$$

and the optimal paths may be obtained by solving the following system of necessary conditions (see Athans and Falb 1966 for details or Chapter 14 of Intrilligator, 1971):

$$\begin{cases} \text{(a)} \quad \dfrac{\partial H}{\partial Q(t)} = -T\,(Q(t)) = -\dot{\rho}\,(t) \\ \text{(b)} \quad \dfrac{\partial H}{\partial y(t)} = \dfrac{dSP(y(t))}{dy(t)} - \rho(t) = F(y(t)) - \rho(t) = p(t) - \rho(t) = 0 \\ \text{(c)} \quad \dfrac{\partial H}{\partial \rho(t)} = -y(t) = \dot{Q}(t) \\ \text{(d)} \quad Q(0) = K \text{ and } Q(1) = 0. \end{cases}$$

(1.12)

It is almost obvious that the systems (1.8) and (1.12) are equivalent. In fact, (1.12b) is to identify $\rho(t)$ as $p(t) = F(-\dot{Q}(t))$. Then (1.12a) is equivalent to (1.8a). (1.12c) is equivalent to (1.8b). And finally (1.12d) is the same as (1.8c). Thus, this simple intertemporal price equilibrium problem can be solved by using either one of these formulations.

In the following section two examples using linear demand functions and a constant storage cost for the first example and a linear cost function for the second example are used to clarify the nature and the properties of intertemporal price equilibrium.

1.6.3 Two Examples of Intertemporal Price Equilibrium

In both examples to follow, we employ the linear demand function of the following form:

$$p(t) = \lambda - \omega y(t) = \lambda + \omega \dot{Q}(t), \tag{1.13}$$

where $\lambda > 0$, $\omega > 0$ are fixed constants for all $t \in [0, 1]$. This demand function can be written in the following indirect demand functional form:

$$y(t) = -\dot{Q}(t) = \alpha - \beta\, p(t) \tag{1.14}$$

where obviously $\alpha = \dfrac{\lambda}{\omega}$ and $\beta = \dfrac{1}{\omega}$.

Example 1

In this example the storage cost is assumed as δ, a constant for all $t \in [0, 1]$, that is:

$$T(Q(t)) = \delta > 0 \text{ for all } t. \tag{1.15}$$

By using (1.12b), (1.12a), and (1.12c) in that order, one can derive the following second order differential equation of $Q(t)$:

$$-T(Q(t)) \overset{(1.15)}{=} -\delta \overset{(1.12)}{=} -\dot{p}(t) \overset{(1.12)}{=} -\dot{p}(t) \overset{(1.13)}{=} -w\, \ddot{Q}\,(t), \tag{1.16}$$

that is,

$$\omega\, \ddot{Q}\,(t) = \delta \tag{1.17}$$

or

$$\ddot{Q}\,(t) = \beta\delta \tag{1.17a}$$

From (1.17a) we get the following expression containing unknown constants κ and ξ.

$$Q(t) = \kappa + \xi t + \tfrac{1}{2}\, \beta\delta t^2. \tag{1.18}$$

By using the TPBV condition (1.12d), we get

$$Q(t) = K - (K + \tfrac{1}{2}\, \beta\delta)t + \tfrac{1}{2}\, \beta\delta t^2 \tag{1.19}$$

as the equilibrium stock profile, and the equilibrium price profile is given by (1.13) or

$$p(t) = \lambda - \omega K - \frac{\delta}{2} + \delta t \tag{1.20}$$

an upward sloping linear function with the slope exactly equal to the storage cost. The stock profile is a quadratic function starting at K when $t = 0$ and ending at 0 when $t = 1$. However, $Q(t)$, the stock and price, $p(t)$, cannot be negative at any $t \in [0, 1]$.

Thus we have the following restriction on the parameters λ, ω(or β), and δ:

$$(0<)\ \beta \leq \frac{2K}{\delta}, \tag{1.21}$$

$$\lambda - \omega K - \frac{\delta}{2} > 0 \tag{1.22}$$

and, using (1.21),

$$0 < \delta < \lambda\ . \tag{1.23}$$

Example 2

In this example, let us assume that the storage cost is a linear function of the stock, that is

$$T(Q(t)) = \delta + sQ(t) \tag{1.24}$$

where $\delta > 0$ and $s > 0$ and are constant over $t \in [0, 1]$.
For this situation we get the following second order differential equation for (1.12);

$$\omega \dot{Q}(t) = \delta + sQ(t). \tag{1.25}$$

The particular solution of (1.25) is:

$$z = -\frac{\delta}{s} \tag{1.26}$$

and the two characteristic values of the complementary solutions are:

$$\rho = \pm \sqrt{\beta s}\ . \tag{1.27}$$

Therefore the general form of the solution is given by:

$$Q(t) = ae^{\sqrt{\beta s}\,t} + be^{-\sqrt{\beta s}\,t} - \left(\frac{\delta}{s}\right). \tag{1.28}$$

By taking the TPBV condition (1.12d) into consideration in (1.28), we finally get the optimal stock profile as:

$$Q(t) = \frac{1}{e^{\sqrt{\beta s}} - e^{-\sqrt{\beta s}}} \Big\{ \Big[\left(\frac{\delta}{s}\right)\left(1 - e^{-\sqrt{\beta s}}\right) - Ke^{-\sqrt{\beta s}}\Big] e^{\sqrt{\beta s}\,t} + \Big[He^{\sqrt{\beta s}} + \left(\frac{\delta}{s}\right)\left(e^{\sqrt{\beta s}} - 1\right)\Big] e^{-\sqrt{\beta s}\,t} \Big\} - \left(\frac{\delta}{s}\right). \tag{1.29}$$

and the optimal price profile is given by

$$p(t) = \lambda + \frac{\sqrt{\omega s}}{e^{\sqrt{\beta s}} - e^{-\sqrt{\beta s}}} \left\{ \left[\left(\frac{\delta}{s}\right)\left(1 - e^{\sqrt{\beta s}}\right) - Ke^{-\sqrt{\beta s}} \right] e^{\sqrt{\beta s}\,t} - \left[Ke^{\sqrt{\beta s}} + \left(\frac{\delta}{s}\right)\left(e^{\sqrt{\beta s}} - 1\right) \right] e^{\sqrt{\beta s}\,t} \right\}. \tag{1.30}$$

The non-negativity condition of $Q(t)$ is guaranteed if $\dot{Q}\ (1) \leq 0$, that is

$$2\left(K + \left(\frac{\delta}{s}\right)\right) \geq \left(\frac{\delta}{s}\right)\left(e^{\sqrt{\beta s}} + e^{-\sqrt{\beta s}}\right), \tag{1.31}$$

and $p(t) > 0$ for all $t \in [0, 1]$ if $p(0) > 0$.

In both examples one can conclude that if the storage cost δ is larger, the more the consumers intend to consume at the early part of the [0, 1] time horizon. The effect of a change in s can be similarly concluded. There are a few more interesting comparative dynamics case studies left in these examples but due to the page limitation they are left as exercises to the reader.

In the next section temporal-spatial equilibrium models will be treated in their linear-quadratic form to clearly grasp the characteristics of dynamic spatial equilibrium models and the equilibrium solutions.

1.6.4 Intertemporal-Spatial Price Equilibrium Model. Linear-Quadratic Formulation.

As a natural extension of the models in the previous section we can express our continuous optimal control model for a two country situation as follows:

Find optimal control consumption profiles of country 1, and 2, $(\bar{y}_1, \bar{y}_2)$, an intra-temporal optimal export (import) profile, $\bar{u}$ and optimal state stock quantities of the commodity, $\bar{Q}_1$ (t), $\bar{Q}_2$ (t), in countries 1 and 2 that maximise

$$NSP = \int_0^1 [\lambda_1 y_1(t) - \tfrac{1}{2}\,\omega_1 y_1^2(t) + \lambda_2 y_2(t) - \tfrac{1}{2}\,\omega_2 y_2^2(t) - \tfrac{1}{2}\,\theta u^2(t) - \delta_1 Q_1(t) - \\ -\delta_2 Q_2(t)\;]\,dt \tag{1.32}$$

subject to

(a) $\dot{Q}_1(t) = -y_1(t) - u(t)$

(b) $\dot{Q}_2(t) = -y_2(t) + u(t)$

which implies the storage quantity change at t in countries 1 and 2 is equal to consumption minus (or plus) the international movement of the commodity, $u(t)$ and

(c) $Q_1(0) = K_1$ constant, $Q_1\,(1) = 0$
$Q_2(0) = K_2$ constant, $Q_2(1) = 0$.

(1.33)

The Hamiltonian may be defined as

$$H = \lambda_1 y_1(t) - \tfrac{1}{2}\,\omega_1 y_1^2(t) + \lambda_2 y_2(t) - \tfrac{1}{2}\omega_2 y_2^2(t) - \tfrac{1}{2}\,\theta u^2(t) \\ - \delta_1 Q_1(t) - \delta_2 Q_2(t) + \rho_1(t)\,[(-y_1(t) - u(t))] + \rho_2(t)\,[(-y_2(t) + u(t))] \tag{1.34}$$

The canonical form optimality conditions are given as follows

$$
\begin{cases}
\text{(a)} & \dfrac{\partial H}{\partial y_1(t)} = \lambda_1 - \omega_1 y_1(t) - \rho_1(t) = 0, \\[2ex]
\text{(b)} & \dfrac{\partial H}{\partial y_2(t)} = \lambda_2 - \omega_2 y_2(t) - \rho_2(t) = 0, \\[2ex]
\text{(c)} & \dfrac{\partial H}{\partial u(t)} = -\theta u(t) - \rho_1(t) + \rho_2(t) = 0, \\[2ex]
\text{(d)} & \dfrac{\partial H}{\partial Q_1(t)} = -\delta_1 = -\dot{\rho}_1(t), \\[2ex]
\text{(e)} & \dfrac{\partial H}{\partial Q_2(t)} = -\delta_2 = -\dot{\rho}_2(t), \\[2ex]
\text{(f)} & \dfrac{\partial H}{\partial \rho_1(t)} = -y_1(t) - u(t) = \dot{Q}_1, \\[2ex]
\text{(g)} & \dfrac{\partial H}{\partial \rho_2(t)} = -y_2(t) + u(t) = \dot{Q}_2 \ .
\end{cases}
\tag{1.35}
$$

The boundary values are given by (1.33c).

From (1.35a and b) we get

$$
\begin{cases}
\text{(a)} & \dot{\rho}_1(t) = -\omega_1 \dot{y}_1(t) \\
\text{(b)} & \dot{\rho}_2(t) = -\omega_2 \dot{y}_2(t) \ .
\end{cases}
\tag{1.36}
$$

From (1.35f and g) we get

$$
\begin{cases}
\text{(a)} & \ddot{Q}_1(t) = -\dot{y}_1(t) - \dot{u}(t), \\[2ex]
\text{(b)} & \ddot{Q}_2(t) = -\dot{y}_2(t) + \dot{u}(t)
\end{cases}
\tag{1.37}
$$

From (1.35c), we have

$$
\dot{\rho}_2(t) - \dot{\rho}_1(t) = \theta \dot{u}(t). \tag{1.38}
$$

From (1.36) and (1.37) we get

$$\begin{cases} \text{(a)} \quad \dot{\rho}_1 = -\omega_1 \left(-\dot{u}(t) - \ddot{Q}_1(t) \right), \\ \\ \text{(b)} \quad \dot{\rho}_2 = -\omega_2 \left(-\ddot{Q}_2(t) + \dot{u}(t) \right). \end{cases} \tag{1.39}$$

From (1.38) and (1.39), we get

$$\begin{cases} \text{(a)} \quad \dot{\rho}_1(t) = \dfrac{\omega_1}{\theta} \left(\dot{\rho}_2(t) - \dot{\rho}_1(t) \right) + \omega_1 \ddot{Q}_1(t), \\ \\ \text{(b)} \quad \dot{\rho}_2(t) = -\dfrac{\omega_2}{\theta} \left(\dot{\rho}_2(t) - \dot{\rho}_1(t) \right) + \omega_2 \ddot{Q}_2(t). \end{cases} \tag{1.40}$$

Equation (1.40) may be written as

$$\begin{bmatrix} \dfrac{\omega_1+\theta}{\theta} & \dfrac{\omega_1}{\theta} \\ -\dfrac{\omega_2}{\theta} & \dfrac{\omega_2+\theta}{\theta} \end{bmatrix} \begin{bmatrix} \dot{p}_1(t) \\ \dot{p}_2(t) \end{bmatrix} = \begin{bmatrix} \omega_1 & o \\ 0 & \omega_2 \end{bmatrix} \begin{bmatrix} \ddot{Q}_1(t) \\ \\ \ddot{Q}_2(t) \end{bmatrix},$$

which yields

$$\begin{bmatrix} \dot{\rho}_1(t) \\ \dot{\rho}_2(t) \end{bmatrix} = \begin{bmatrix} \dfrac{\omega_1+\theta}{\theta} & -\dfrac{\omega_1}{\theta} \\ -\dfrac{\omega_2}{0} & \dfrac{\omega_2+\theta}{0} \end{bmatrix}^{-1} \begin{bmatrix} \omega_1 & 0 \\ 0 & \omega_2 \end{bmatrix} \begin{bmatrix} \ddot{Q}(t) \\ \\ \ddot{Q}(t) \end{bmatrix} \overset{(1.35)\&e)}{=} \begin{bmatrix} \delta_1 \\ \delta_2 \end{bmatrix},$$

Finally

$$\begin{bmatrix} \ddot{Q}_1(t) \\ \ddot{Q}_2(t) \end{bmatrix} = \begin{bmatrix} \frac{1}{\omega_1} & 0 \\ 0 & \frac{1}{\omega_2} \end{bmatrix} \begin{bmatrix} \frac{\omega_1+\theta}{\theta} & -\frac{\omega_1}{\theta} \\ -\frac{\omega_2}{\theta} & \frac{\omega_2+\theta}{\theta} \end{bmatrix} \begin{bmatrix} \delta_1 \\ \delta_2 \end{bmatrix}$$

$$= \begin{bmatrix} \frac{-\delta_2\omega_1+\delta_1(\theta+\omega_1)}{\theta\omega_1} \\ \frac{-\delta_1\omega_2+\delta_2(\theta+\omega_2)}{\theta\omega_2} \end{bmatrix} = \begin{bmatrix} \mu_1 \\ \mu_2 \end{bmatrix} \quad \text{, a constant vector.} \tag{1.41}$$

Therefore, we get

$$\begin{bmatrix} \dot{Q}_1(t) \\ \dot{Q}_2(t) \end{bmatrix} = \begin{bmatrix} a_1+\mu_1 t \\ a_2+\mu_2 t \end{bmatrix}$$

and

$$\begin{bmatrix} Q_1(t) \\ Q_2(t) \end{bmatrix} = \begin{bmatrix} b_1+a_1 t+\frac{1}{2}\mu_1 t^2 \\ b_2+a_2 t+\frac{1}{2}\mu_2 t^2 \end{bmatrix} . \tag{1.42}$$

By taking (1.33c) into consideration, we get

$$\begin{bmatrix} Q_1(t) \\ Q_2(t) \end{bmatrix} = \begin{bmatrix} K_1-(K_1+\frac{1}{2}\mu_1)t+\frac{1}{2}\mu_1 t^2 \\ K_2-(K_2+\frac{1}{2}\mu_2)t+\frac{1}{2}\mu_2 t^2 \end{bmatrix} , \tag{1.42a}$$

which is an optimal storage quantity vector in typical quadratic form.

From (1.35c) we get $u(t) = \dfrac{\rho_2(t) - \rho_1(t)}{\theta}$, and by substituting this into (1.35a and b), we get

$$\begin{cases} \text{(a)} \quad \lambda_1-\omega_1 y_1=\lambda_1-\omega_1\left[K_1+\tfrac{1}{2}\mu_1-\mu_1 t-\dfrac{\rho_2(t)-\rho_1(t)}{\theta}\right]=\rho_1(t), \\ \text{(b)} \quad \lambda_2-\omega_2 y_2=\lambda_2-\omega_2\left[K_2+\tfrac{1}{2}\mu_2-\mu_2 t+\dfrac{\rho_2(t)-\rho_1(t)}{\theta}\right]=\rho_2(t). \end{cases} \tag{1.43}$$

Solving (1.43) for $\rho 1$(t), and $\rho 2$(t), yields

$$\begin{bmatrix}\rho_1(t)\\ \rho_2(t)\end{bmatrix}=\begin{bmatrix}\dfrac{\theta+\omega_1}{\theta} & \dfrac{-\omega_1}{\theta}\\ \dfrac{-\omega_2}{\theta} & \dfrac{\theta+\omega_2}{\theta}\end{bmatrix}^{-1}\begin{bmatrix}\lambda_1-\omega_1(K_1+\tfrac{1}{2}\mu_1-\mu_1(t))\\ \lambda_2-\omega_2(K_2+\tfrac{1}{2}\mu_2-\mu_2(t))\end{bmatrix}. \tag{1.44}$$

Therefore, by using (1.35c) again, we get the optimal shipment (impulse)

$$u(t)=\frac{1}{\theta}\left(\rho_1(t)\ \rho_2(t)\right)\begin{pmatrix}-1\\ 1\end{pmatrix} \tag{1.45}$$

By using (1.35f and g), we get the optimal consumption profile

$$\begin{bmatrix}y_1(t)\\ y_2(t)\end{bmatrix}=\begin{bmatrix}-\dot{Q}_1(t)-u(t)\\ -\dot{Q}_2(t)+u(t)\end{bmatrix}. \tag{1.46}$$

When the storage costs are

$$\begin{cases}\delta_1+s_1Q_1(t)\\ \delta_2+s_2Q_2(t)\end{cases} \tag{1.47}$$

for countries 1 and 2 respectively we can solve problems similar to the one above for optimal control and state variable, but we will not develop the solution procedures here in this paper[8].

In application, these continuous time (dynamic) optimal control models encounter almost insurmountable difficulty that almost no economic observation is taken on the continuous time basis. Thus, in reality we are forced to formulate our models

---

[8]Aoki in his unpublished paper developed a theoretical model similar to the above and derived optimal control solutions.

in a discrete time form to which we will turn our attention in the following sections.

1.6.5 Temporal and Spatial Price Equilibrium Models; Deterministic and Stochastic Models

In the first part of this section we develop deterministic temporal-spatial price equilibrium models, in line with the Modigliani-Hohn (1955), Simon (1956) and Theil (1964) models, for production and inventory planning over time. This reveals a mathematical equivalence of two models that are distinctly different from an economic-theoretical point of view. In the second part of this section we follow Theil and develop a stochastic temporal-spatial price equilibrium model and establish the certainty equivalence first period decision rule. In the final part of this section we investigate various ways of formulating stochastic models and point out possible ways of attacking and solving these problems.

a. Temporal and Spatial Price Equilibrium Models and Over Time Production-Inventory Planning Models

The market oriented temporal price equilibrium model in its linear-quadratic formulation for a single commodity will be formulated below in a discrete deterministic optimal control format. This formulation reveals an obvious equivalence of this model with those developed by Modigliani-Hohn, Simon and Theil mentioned above.

Let us first formulate our temporal price equilibrium model in a typical quadratic programming form. First, let us assume that the demand function for each time period is known to be

$$p_{dt} = \lambda_t - \omega_t y_t \text{ for all } t \in [0, T-1]. \tag{1.48}$$

Similarly, the supply function is also assumed to be known as

$$p_{st} = \nu_t + \eta_t x_t \text{ for all } t \in [0, T-1] \tag{1.49}$$

Finally, for theoretical convenience let us assume that the storage cost, s_t, of the commodity from one period, t, to the next period, t+1, is assumed to be a linear function of the storage quantity during the period, z_t,

$$s_t = b_t + d_t z_t \text{ for all } t \in [1, T]. \tag{1.50}$$

In (1.48) - (1.50) we assume λ_t, $\omega_t(>0)$, ν_t, $\eta_t(>0)$, b_t and $d_t(>0)$ are assumed to be known constants unless otherwise stated.

We now state our optimal control problem as follows: Find optimal control vectors for market supply quantity, $\bar{x}$, and market demand quantity, $\bar{y}$, and state vector carry over quantity, $\bar{z}$, that maximises

$$NSP = -b'z - \tfrac{1}{2} z'Dz - \nu'x - \tfrac{1}{2} x'Fx + \lambda'y - \tfrac{1}{2} y'\Omega y \tag{1.51}$$

subject to

$$Rz \leq x - y + z_0 e_1 \tag{1.52}$$

$$x \geqq 0, \quad y \geqq 0, \quad z \geqq 0 \, , \tag{1.53}$$

where

$$b = \begin{bmatrix} b_1 \\ \vdots \\ b_T \end{bmatrix}, \quad z = \begin{bmatrix} z_1 \\ \vdots \\ z_T \end{bmatrix}$$ storage quantity vector and z_t is carry-over quantity from $t-1$ to t, $1 \leq t \leq T$,

$$\nu = \begin{bmatrix} \nu_o \\ \vdots \\ \nu_{T-1} \end{bmatrix}, \quad x = \begin{bmatrix} x_0 \\ \vdots \\ x_{T-1} \end{bmatrix}$$ supply quantity vector,

$$\lambda = \begin{bmatrix} \lambda_0 \\ \vdots \\ \lambda_{T-1} \end{bmatrix}, \quad y = \begin{bmatrix} y_0 \\ \vdots \\ y_{T-1} \end{bmatrix}$$ demand quantity vector,

$$D = \begin{bmatrix} d_1 & & 0 \\ & \ddots & \\ O & & d_T \end{bmatrix}, \qquad F = \begin{bmatrix} \eta_0 & & 0 \\ & \ddots & \\ 0 & & \eta_{T-1} \end{bmatrix}.$$

$$\Omega = \begin{bmatrix} \omega_0 & & 0 \\ & \ddots & \\ 0 & & \omega_{T-1} \end{bmatrix}, \quad R = \begin{bmatrix} 1 & & & \\ -1 & 1_{\ddots} & & \\ & -1_{\ddots} & 1_{\ddots} & \\ & & -1 & 1 \end{bmatrix} \text{ and } e_1 = \begin{bmatrix} 1 \\ 0 \\ \vdots \\ 0 \end{bmatrix}$$

Equation (1.52) may be written in the following alternative forms:

$$\begin{cases} z_1 & \leqq z_0 & +x_0 & -y_0 \\ z_2 & \leqq z_1 & +x_1 & -y_1 \\ \vdots & \vdots & \vdots & \vdots \\ z_T & \leqq z_{T-1} & -x_{T-1} & -y_{T-1} \end{cases} \tag{1.52a}$$

or

$$z = \begin{bmatrix} z_1 \\ \vdots \\ z_T \end{bmatrix} \leqq R^{-1}\{(x-y)+z_0 e_1\} = \left(\begin{bmatrix} R^{-1}, -R^{-1} \end{bmatrix} \begin{bmatrix} x \\ y \end{bmatrix} + z_0 l \right)$$

$$= \begin{bmatrix} 1 & & & 0 \\ 1 & 1 & & \\ \vdots & \vdots & \ddots & \\ 1 & 1 & \cdots & 1 \end{bmatrix} (x+y) + z_0 l \ , \tag{1.52b}$$

$$= \begin{bmatrix} x_0 \\ x_0+x_1 \\ \vdots \\ x_0+x_1+\cdots+x_{T-1} \end{bmatrix} - \begin{bmatrix} y_0 \\ y_0+y_1 \\ \vdots \\ y_0+y_1+\cdots+y_{T-1} \end{bmatrix} + z_0 \begin{bmatrix} 1 \\ 1 \\ \vdots \\ 1 \end{bmatrix} \ ,$$

where $e_1 = (1\ 0\ \dots\ 0)'$ and $l = (1\ 1\ \dots\ 1)'$.

This is exactly an extension of the Simon, Modigliani-Hohn, Theil type production-inventory planning models; an extension since demand variables are explicitly introduced in (1.52) and note also that the net social payoff is used instead of the net revenue or cost concept.

Thus we can rewrite (1.51), by assuming that the equality instead of inequality holds in (1.52b), as

$$NSP = z_0^2 \iota D \iota + \begin{bmatrix} -R^{-1}b - \nu - z_0 \iota' D R^{-1} \\ R^{-1}b + \lambda + z_0 \iota' D R^{-1} \end{bmatrix}' \begin{bmatrix} x \\ y \end{bmatrix} - \tfrac{1}{2} \begin{bmatrix} x \\ y \end{bmatrix}' \begin{bmatrix} R^{-1\prime} D r^{-1} + F R^{-1\prime} D R^{-1} \\ R^{-1\prime} D R^{-1} R^{-1} D R^{-1} + \Omega \end{bmatrix} \begin{bmatrix} x \\ y \end{bmatrix} . \tag{1.54}$$

From this we can solve for the optimal control: by differentiating (1.54) with respect to $\begin{bmatrix} x \\ y \end{bmatrix}$, we get

$$\frac{\partial NSP}{\partial(x,y)} = \begin{bmatrix} -R^{-1}b - \nu - z_0 \iota' DR^{-1} \\ R^{-1}b + \lambda + z_0 \iota' DR^{-1} \end{bmatrix} - \begin{bmatrix} R^{-1\prime}DR^{-1} + FR^{-1\prime}DR^{-1} \\ R^{-1\prime}DR^{-1}R^{-1}DR^{-1} + \Omega \end{bmatrix} \begin{bmatrix} x \\ y \end{bmatrix} . \tag{1.55}$$

and setting this equal to 0 and solving for $\begin{bmatrix} \bar{x} \\ \bar{y} \end{bmatrix}$, we get

$$\begin{bmatrix} \bar{x} \\ \bar{y} \end{bmatrix} = \begin{bmatrix} R^{-1\prime}DR^{-1} + FR^{-1\prime}DR^{-1} \\ R^{-1\prime}DR^{-1} \; R^{-1\prime}DR^{-1} + \Omega \end{bmatrix}^{1} \begin{bmatrix} -R^{-1}b - \nu - z_0 \iota' DR^{-1} \\ R^{-1}b + \lambda + z_0 \iota' Dr^{-1} \end{bmatrix} , \tag{1.56}$$

which is a deterministic optimal control for this problem.

Example 3

As an example let us assume that there are three periods in our discrete time horizon, and the quantity carried in from period 0 to 1 and that carried out from period 3 to 4 are both known (atypical) two point boundary value problem) to be $z_0 = 5$ and $z_3 = 5$.

The demand functions are supposed to shift upward in the following fashion.

$$\begin{aligned} p_1 &= 100 - 10y_1 \\ p_2 &= 110 - 10y_2 \\ p_3 &= 120 - 10y_3 . \end{aligned}$$

We assume the supply quantity in each period is known to be

$$\bar{x}_1 - 50 - \bar{x}_2 - \bar{x}_3 .$$

The control variables in this problem are y_1, y_2 and y_3, consumption quantity in each period, and the state variables are z_1 and z_2, carry-over quantity from period 1 to 2 and from period 2 to 3 respectively. Since z_3 is constant, we have to modify (1.52b) (R becomes a (3x2) matrix and $\bar{R}^{(-1)}$ cannot be taken).

The solution for this problem can be confirmed to be

$$\bar{p}_1 = 5{,}5,\ \bar{p}_2 = 6.0,\ \bar{p}_3 = 6.5$$

$$\bar{y}_1 = 45,\ \bar{y}_2 = 50,\ \bar{y}_3 = 55$$

$$\bar{z}_1 = 10,\ \bar{z}_2 = 10$$

In this example the method developed above is as effective as the Lagrangian method in non-linear programming as the solutions satisfy (1.52a) with equality. In general application the Lagrangian approach presented in Takayama-Judge (1971) proved to be superior for the reason already advanced by Modigliani and Hohn.

At this stage let us introduce the spatial dimension into this model. We assume as before a well-behaved social net pay-off function for each of the n regions in the economy. These n regions are connected by transportation networks and the total regional transportation cost for each time period is given as

$$TC_t = \sum_{i=1}^{n} \sum_{j=1}^{n} (t_{ij}\,\xi_{ij}(t) + \tfrac{1}{2} s_{ij}(t)\,\xi_{ij}(t)^2) \tag{1.57}$$

$$= t_t'\xi(t) + \tfrac{1}{2}\xi(t)' S_t\,\xi(t)$$

where $t_t = (t_{12}(t) \ldots t_{1n}(t) \ldots t_{nl}(t) \ldots t_{nn-1}(t))$ *and* $\xi(t) = (\xi_{12}(t) \ldots \xi_{1n}(t) \ldots \xi_{nl}(t) \ldots \xi_{nn-1}(t))$

$$\text{and } S_t = \begin{bmatrix} s_{12}(t) & & & & & & \\ & \ddots & & & & & \\ & & s_{1n}(t) & & & & \\ & & & \ddots & & & \\ & & & & s_{nl}(t) & & \\ & & & & & \ddots & \\ & & & & & & s_{nn-1}(t) \end{bmatrix}_{((n^2 - n)\times(n^2 - n))}$$

The economy wide net social payoff is defined as:

$$\mathrm{NSP} = \begin{bmatrix} -b_1 \\ \vdots \\ -b_T \end{bmatrix}' \begin{bmatrix} z_1 \\ \vdots \\ z_T \end{bmatrix} - \tfrac{1}{2} \begin{bmatrix} z_1 \\ \vdots \\ z_T \end{bmatrix}' \begin{bmatrix} D_1 & & \\ & \ddots & \\ & & D_T \end{bmatrix} \begin{bmatrix} z_1 \\ \vdots \\ z_T \end{bmatrix} - \begin{bmatrix} \nu_0 \\ \vdots \\ \nu_{T-1} \end{bmatrix}' \begin{bmatrix} x_0 \\ \vdots \\ x_{T-1} \end{bmatrix}$$

$$- \tfrac{1}{2} \begin{bmatrix} x_0 \\ \vdots \\ x_{T-1} \end{bmatrix}' \begin{bmatrix} F_0 & & \\ & \ddots & \\ & & F_T \end{bmatrix} \begin{bmatrix} x_0 \\ \vdots \\ x_{t-1} \end{bmatrix} + \begin{bmatrix} \lambda_0 \\ \vdots \\ \lambda_{T-1} \end{bmatrix}' \begin{bmatrix} \Omega_0 & & \\ & \ddots & \\ & & \Omega_{T-1} \end{bmatrix} \begin{bmatrix} y_0 \\ \vdots \\ y_{T-1} \end{bmatrix}$$

$$- \begin{bmatrix} t_0 \\ \vdots \\ t_{T-1} \end{bmatrix}' \begin{bmatrix} \xi_0 \\ \vdots \\ \xi_{T-1} \end{bmatrix} - \tfrac{1}{2} \begin{bmatrix} \xi_0 \\ \vdots \\ \xi_{T-1} \end{bmatrix}' \begin{bmatrix} S_0 & & \\ & \ddots & \\ & & S_{T-1} \end{bmatrix} \begin{bmatrix} \xi_0 \\ \vdots \\ \xi_{T-1} \end{bmatrix} . \tag{1.58}$$

The dynamic system relations is expressed by the following:

$$\begin{array}{lllllllllll}
z_{11} & \leqq & z_{10} & + & x_{10} & - & y_{10} & + & \sum_{i \neq 1} \xi_{i1}(0) & - & \sum_{j \neq 1} \xi_{1j}(0) \\
\vdots & & \vdots & & & & & \vdots & & \vdots & \\
z_{n1} & \leqq & z_{n0} & + & x_{n0} & - & y_{n0} & + & \sum_{i \neq n} \xi_{in}(0) & - & \sum_{j \neq n} \xi_{nj}(0) \\
\vdots & & \vdots & & & & & \vdots & & \vdots & \\
z_{1T} & \leqq & z_{nT-1} & + & x_{1T-1} & - & y_{1T-1} & + & \sum_{i \neq 1} \xi_{i1}(T-1) & - & \sum_{j \neq 1} \xi_{1j}(T-1) \\
\vdots & & \vdots & & & & & \vdots & & \vdots & \\
z_{nT} & \leqq & z_{nT-1} & + & x_{nT-1} & - & y_{nT-1} & + & \sum_{i \neq n} \xi_{in}(T-1) & - & \sum_{j \neq n} \xi_{nj}(T-1)
\end{array} \tag{1.59}$$

or equivalently:

$$\begin{bmatrix} z \\ \vdots \\ z_T \end{bmatrix} \leqq \begin{bmatrix} I & & & 0 \\ -I & I & & \\ & & \ddots & \\ 0 & & -I & I \end{bmatrix} - \left[\begin{bmatrix} I & -I & & & & \\ & I & -I & & & \\ & & & \ddots & & \\ & & & & I & -I \end{bmatrix} \begin{bmatrix} x_0 \\ y_0 \\ \vdots \\ x_{T-1} \\ y_{T-1} \end{bmatrix} + \right.$$

$$\left. + \begin{bmatrix} G(0) & & & \\ & G(1) & & \\ & & \ddots & \\ & & & G(T-1) \end{bmatrix} \begin{bmatrix} \xi_0 \\ \vdots \\ \xi_{T-1} \end{bmatrix} + \begin{bmatrix} z_0 \\ 0 \\ \vdots \\ 0 \end{bmatrix} \right]$$

$$= \begin{bmatrix} I & & & 0 \\ I & I & & \\ \vdots & \vdots & \ddots & \\ I & I & \dots & I \end{bmatrix} \left[\begin{bmatrix} I & -IG(0) & & & & & 0 \\ & & I & -IG(1) & & & \\ & & \dots & \vdots & & & \\ & & & \dots & \vdots & & \\ & & & & \dots & \ddots & \\ 0 & & & & & I & -IG(T-1) \end{bmatrix} \begin{bmatrix} x_0 \\ y_0 \\ \xi_0 \\ x_1 \\ y_1 \\ \xi_1 \\ \vdots \\ x_{T-1} \\ y_{T-1} \\ \xi_{T-1} \end{bmatrix} + \begin{bmatrix} z_0 \\ 0 \\ \vdots \\ 0 \end{bmatrix} \right]$$

$$= \begin{bmatrix} I & -IG(0) & & & & & 0 \\ I & -IG(0) & I & -IG(1) & & & \\ \cdot & \dots & \cdot & \dots & \cdot & & \\ \cdot & \dots & \cdot & \dots & & \cdot & \\ \cdot & \dots & \cdot & \dots & & & \cdot \\ I & -IG(0) & I & -IG(1) & \dots & I & -IG(T-1) \end{bmatrix} \begin{bmatrix} x_0 \\ y_0 \\ \xi_0 \\ x_1 \\ y_1 \\ \xi_1 \\ \vdots \\ x_{T-1} \\ y_{T-1} \\ \xi_{T-1} \end{bmatrix} + \begin{bmatrix} z_0 \\ z_0 \\ z_0 \\ \vdots \\ z_0 \end{bmatrix} \tag{1.59a}$$

where

$$G(t) = \begin{bmatrix} -1 & -1 & \dots & -1 & 1 & & & & 1 & & & \\ 1 & & & & -1 & -1 & & \vdots & & 1 & & \\ & 1 & & & & 1 & & & & & \ddots & \\ & & \ddots & & & & \ddots & \vdots & & & & 1 \\ & & & 1 & & & 1 & \vdots & -1 & -1 & \dots & -1 \end{bmatrix}_{(nx(n^2 - n))}$$

By substituting (1.59a) assuming equality holds everywhere in (1.59a), into (1.38) and solving (1.58) for the optimal control solution, we get

$$\begin{bmatrix} x_0 \\ y_0 \\ \xi_0 \\ \vdots \\ x_{T-1} \\ y_{T-1} \\ \xi_{T-1} \end{bmatrix} = [A'DA+Q]^{-1} \left[A' \begin{bmatrix} b_1 \\ \vdots \\ b_T \end{bmatrix} + \begin{bmatrix} -\nu_0 \\ \lambda_0 \\ \vdots \\ -\nu_{T-1} \\ \lambda_{T-1} \end{bmatrix} \right] \tag{1.60}$$

where

$$A = \begin{bmatrix} I - IG(0) & & & \\ I - IG(0) & I - IG(1) & & \\ I - IG(0) & I - IG(1) & & \\ \vdots \quad \vdots & \vdots & \vdots & \ddots \\ I - IG(0) & I - IG(1) & \dots & I - IG(t-1) \end{bmatrix}$$

$$Q = \begin{bmatrix} F_0 & & & & & & \\ & \Omega_0 & & & & & \\ & & S_0 & & & & \\ & & & \ddots & & & \\ & & & & F_{T-1} & & \\ & & & & & \Omega_{T-1} & \\ & & & & & & S_{T-1} \end{bmatrix}, \; and \quad D = \begin{bmatrix} D_0 & & & \\ & D_1 & & \\ & & \ddots & \\ & & & D_{T-1} \end{bmatrix}.$$

When a usual two point boundary value problem with z_0 and z_T both known vectors of constants is to be solved, the classical Lagrangean approach with an additional constraint:

$$z_T = k_T \text{ (constant vector)} \tag{1.61}$$

may be found to be most effective as was already shown in Example 3 above.

Thus, a deterministic temporal and spatial price equilibrium model is seen to be almost one to one to the Modigliani-Hohn, Simon, and Theil production-inventory planning models except for the following differences:

2) The former assumes the existence of market demand and supply functions (at most), but the latter usually assumes known demand quantities for each period.

3) The former has been developed based on mathematical programming considerations such as non-negativity of all the primal variables (control and state variables) and inequality constraints in (1.59a) but the latter is based on equality in (1.59a) and no non-negative constraints on the variables involved.

The last characteristic difference is rather crucial in applied works in relation to temporal and spatial price equilibrium analysis. For example, during a period of fifty-two weeks carry-over quantity of such commodity as banana from one weeks to the next turn out to be positive for some consecutive weeks, the longest sequence is 14 weeks, and then zero for some other weeks. If this takes place, the above analytical approach may fail, and, moreover, there is no infinite horizon argument applicable to this type of problem.[9]

Comparison between the model developed here and that in the 1971 Takayama-Judge book is also important. In the book the total storage cost (and/or transportation cost) are assumed to be a linear function; but in this model they are in positive definite quadratic form. From a programming point of view this difference may not mean much, but from an estimation point of view this may turn out to a be fairly heavy burden (in many real cases a larger haul (or quantity stored of a commodity may not cause sharply increased transport (storage) cost).

b Stochastic Temporal-Spatial Price Equilibrium Models

In this subsection we introduce stochastic components into our temporal-spatial price equilibrium models. There are many and varied ways of accomplishing this without damaging the usefulness of the models in application.

b1 Optimal Control of Stochastic Systems with Known Coefficients

Although the stochastic components in the objective function or performance function will be the central theme at the end of this section, let us first assume that the market demand functions, storage cost functions, and transport cost functions are not subject to stochastic disturbances by themselves.

The third and fourth terms on the right hand side of (1.58) are dropped due to the following assumption:

Supply quantity in each period in each country is assumed as a random element with mean $\bar{x}_i(t)$ and variance $\sigma_i^2(t)$ and also as an independent process over space and time and other state and control variables; $E(x(t)) = \bar{x}(t)$, $E(x(t)x'(t)) =$

[9]For further clarification, see Modigliani-Hohn (1955) and Theil (1964).

Supply quantity in each period in each country is assumed as a random element with mean $\bar{x}_i(t)$ and variance $\sigma_i^2(t)$ and also as an independent process over space and time and other state and control variables; $E(x(t)) = \bar{x}(t)$, $E(x(t)x'(t)) =$

$$\begin{bmatrix} \sigma_1^2(t) & & 0 \\ & \ddots & \\ 0 & & \sigma_n^2(t) \end{bmatrix}$$ for all $t = 0, \ldots, T\text{-}1$, $E(x(t)x'(t')) = 0$ for all $t \neq t'$ or

$$E(x\,x') = \begin{bmatrix} \Sigma_0 & & \\ & \ddots & \\ & & \Sigma_{T-1} \end{bmatrix}, \text{ and}$$

$$E[x(t)\xi'(t')] = E[x(t)Z'(t')] = E[x(t)y'(t')] = 0 \textit{ for all } t \leq t'.$$

Thus, due to this stochastic component in the model, we assume that the performance of the system is based on the following expected net payoff:

$$\mathrm{E(NSP)} = \left\{ \begin{bmatrix} -b_1 \\ \vdots \\ -b_T \end{bmatrix}' \begin{bmatrix} z_1 \\ \vdots \\ z_T \end{bmatrix} - \tfrac{1}{2} \begin{bmatrix} z_1 \\ \vdots \\ z_T \end{bmatrix}' \begin{bmatrix} D_1 & & \\ & \ddots & \\ & & D_T \end{bmatrix} \begin{bmatrix} z_1 \\ \vdots \\ z_T \end{bmatrix} \right. \tag{1.62}$$

$$+ \begin{bmatrix} \lambda_0 \\ \vdots \\ \lambda_{T-1} \end{bmatrix}' \begin{bmatrix} y_0 \\ \vdots \\ y_{T-1} \end{bmatrix} - \tfrac{1}{2} \begin{bmatrix} y_0 \\ \vdots \\ y_{T-1} \end{bmatrix} \begin{bmatrix} \Omega_0 & & \\ & \ddots & \\ & & \Omega_{T-1} \end{bmatrix} \begin{bmatrix} y_0 \\ \vdots \\ t_{t-1} \end{bmatrix}$$

$$\left. - \begin{bmatrix} t_0 \\ \vdots \\ t_{T-1} \end{bmatrix}' \begin{bmatrix} \xi_0 \\ \vdots \\ \xi_{T-1} \end{bmatrix} - \tfrac{1}{2} \begin{bmatrix} \xi_o \\ \vdots \\ \xi_{T-1} \end{bmatrix} \begin{bmatrix} S_0 & & \\ & \ddots & \\ & & S_{t-1} \end{bmatrix} \begin{bmatrix} \xi_0 \\ \vdots \\ \xi_{T-1} \end{bmatrix} \right\}$$

The system dynamics may be expressed by[10]:

$$
\begin{bmatrix} z_1 \\ \vdots \\ z_T \end{bmatrix}
\begin{bmatrix} -IG(0) & & & 0 \\ -IG(0) & -IG(1) & & \\ \vdots & \vdots & \ddots & \\ -IG(0) & -IG(1) & \cdots & -IG(T-1) \end{bmatrix}
\begin{bmatrix} y_0 \\ \xi_0 \\ y_1 \\ \xi_1 \\ \vdots \\ y_{T-1} \\ \xi_{T-1} \end{bmatrix} +
$$

$$
+ \begin{bmatrix} z_0 + x_0 & & & \\ z_0 + x_0 + x_1 & & & \\ \vdots \quad \vdots \quad \vdots & \ddots & & \\ z_0 + x_0 + x_1 + \cdots + x_{T-1} & & & \end{bmatrix}
$$

$$
\equiv A_x \begin{bmatrix} y_0 \\ \xi_0 \\ \vdots \\ y_{T-1} \\ \xi_{T-1} \end{bmatrix} + s \tag{1.63}
$$

In this formulation and those to follow we assume away measurement errors in the state variables z_t, $t=1, \ldots, T$. We assume that z_0 and x_0 are completely known to the society, and our immediate problem is to determine an optimal strategy for y_0 and ξ_0 in such a way that (1.62) is maximised subject to (1.63) which is, except for the first component, a stochastic dynamic system.

First, our optimal strategy for consumption and interregional shipments of the commodity can be given by:

[10]We assume that we know which regions are importing regions and which are exporting regions for each period.

$$E\begin{bmatrix}\bar{y}_0\\ \bar{\xi}_0\\ \vdots\\ \bar{y}_{T-1}\\ \xi_{T-1}\end{bmatrix} = [A_x' DA_x + Q_x]^{-1}\left[\begin{bmatrix}\lambda_0\\ \iota_0\\ \vdots\\ \lambda_{T-1}\\ \iota_{T-1}\end{bmatrix} - A_x' b - A_x' DEs\right] \tag{1.64}$$

where

$$Q_x = \begin{bmatrix}\Omega_0 & & & & \\ & S_0 & & & \\ & & \ddots & & \\ & & & \Omega_{T-1} & \\ & & & & S_{T-1}\end{bmatrix}$$

The first period certainty equivalence is now clear; that is the initial period decision $(\bar{y}_0, \bar{\xi}_0)$ is identical with the first period subvector of the decision vector which maximises (1.62b) without the expectation subject to (1.63) with s replaced by its expectation, Es.

Our basic scheme in revising the content of information and the optimal strategies while the time horizon keeps on moving forward by one period is based on this first period certainty equivalence rule (Takayama and Judge 1971, Chapter 19), and is called 'adaptive intertemporal spatial equilibrium models.'

Derivations of this certainty equivalence rule through stochastic dynamic programming procedures is fully explained in Aoki (1957), and the equivalence of these two approaches may be easily proved for this case.

For instance, when the objective function is (1.62) and the system dynamics is expressed as

$$\begin{bmatrix}z_1\\ \vdots\\ z_T\end{bmatrix} = \begin{bmatrix}z_0\\ \vdots\\ z_{T-1}\end{bmatrix} + \begin{bmatrix}-IG(0) & & \\ & \ddots & \\ & & IG(T-1)\end{bmatrix}\begin{bmatrix}y_0\\ \xi_0\\ y_{T-1}\\ \xi_{T-1}\end{bmatrix} + \begin{bmatrix}x_0\\ x_1\\ \vdots\\ x_{T-1}\end{bmatrix}, \tag{1.63a}$$

the $(T\text{-}1)$st period stochastic optimal control vector can be uniquely determined as:

$$E\begin{bmatrix}\bar{y}_{T-1}\\ \bar{\xi}_{T-1}\end{bmatrix}=\left[\begin{bmatrix}\Omega_{T-1} & \\ & S_{T-1}\end{bmatrix}+[-IG(T-1)]'D_T[-IG(T-1)]\right]^{-1}$$

$$\left[\begin{bmatrix}\lambda_{T-1}\\ -t_{T-1}\end{bmatrix}-[-IG(T-1)]'b_T-[-IG(T-1)]'D_T[Ez_{T-1}+Ex_{T-1}]\right]$$

$$=\left[\begin{bmatrix}\Omega_{T-1} & \\ & S_{T-1}\end{bmatrix}+[-IG(T-1)]'D_T[-IG(T-1)]\right]^{-1}[-IG(T-1)]'D_TEz_{T-1}$$

$$+\left[\begin{bmatrix}\Omega_{T-1} & \\ & S_{T-1}\end{bmatrix}+[-IG(T-1)]'D_T[-IG(T-1)]\right]^{-1}\left[\begin{bmatrix}\lambda_{T-1}\\ -t_{T-1}\end{bmatrix}-\right.$$
$$\left.-[-IG(T-1)]'[b_T+D_TEx_{T-1}]\right] \tag{1.65}$$

This is as expected, the sum of the two components; (1) linear homogeneous feed back component dependent on Ez_{T-1} and (2) other component formed by linear term vectors in the performance function and the expected values of supply quantities.

The kth stage feed back control can then be expressed as follows:

$$E\begin{bmatrix}\bar{y}(k)\\ \bar{\xi}(k)\end{bmatrix}=\left[\begin{bmatrix}\Omega_k & \\ & S_k\end{bmatrix}+[-IG(k)]_{k+1}[-IG(k)]\right]^{-1}[-IG(k)]'K_{k+1}Ez_k+g_k\,, \tag{1.66}$$

where K_k is determined recursively by

$$K_k=D_k+\left[K_{k+1}-K_{k+1}[-IG(k)]\left[\begin{bmatrix}\Omega_k & \\ & S_k\end{bmatrix}+[-IG(k)]'K_{k+1}[-IG(k)]\right]^{-1}[-IG(k)]'K_{k+1}\right] \tag{1.67}$$

and $K_T = D_T > 0$, and g_k will also be determined recursively.

This is a general optimal feed back control for a non-homogeneous linear plant and general quadratic performance function case (see Aoki 1967; Popovic, 1972 for other cases).

It is quite natural that this recursive optimal feed back control (1.66) leads us to exactly the same formula and conclusion as those derived by Theil 1964; (1.63) and the first period certainty equivalence rule. For detailed discussion of this point, see Popovic (1972).

b2 Optimal Control of Stochastic Systems with Unknown Coefficients in the System and/or Performance Functions

In this subsection, we discuss our models in relation to some of the known results in the area of stochastic optimal control theory.

Due to the nature of the model structure, conceptually there doesn't seem to be any room for coefficient matrix $[I\text{-}IG(t)]$ to become stochastic since we ignore state or control variable measurement errors in our model. Of course, we can make them stochastic, but the result may be meaningless. However, if we consider temporal and spatial equilibrium models that utilise an activity analysis mechanism to generate supply quantities for various products (see Chapters 17 and 18 of Takayama and Judge 1971), and consider the input-output coefficients used in the model as stochastic variables whose estimation would take into account the observations generated as the process advances, then we will be able to develop a fairly meaningful open loop feedback optimal control model along the lines of Zellner (1971), Prescott (1967), and Popovic (1972).

The model structure will take the following form:

The performance function is defined as $E(NSP$ - Loss) where NSP is defined by (1.62) and expected Loss is given by

$$E(Loss) = E\sum_{t=1}^{T} \theta_t(v_t - a_t)'(v_t - a_t) \ or \ E\sum_{t=1}^{T} \theta_t(v_t - a_t)'W_t(v_t - a_t) \tag{1.68}$$

where v_t is a vector of (n) resources to be employed at the time period t (=1, ...,T), a_t is the planned or target resource vector at t, and θ_t is a positive scalar for each t (W_t is an $(n \times n)$ positive semi-definite weight matrix).

The production process takes the following form:

$$v_{t+1} = B_t x_t + \eta_t, \ t = 0,1, \ldots, T-1, \tag{1.69}$$

where B_t may be considered to be an $(n \times n)$ matrix and $[B_t \ v_t]$ is of rank n, and η_t a random vector.

If we further assume that $B(t)$ and $B(t+1)$ are linearly related as

$$B(t+1) = M(t)\ B(t) + \gamma(t) \tag{1.70}$$

where $M(t)$ is a known $(n \times n)$ matrix, and $\gamma(t)$ is an $(n \times n)$ random matrix, then we can formulate the following optimal control problem:

Find an optimal control sequence $\left(\bar{y}(t),\ x(\bar{t}),\ \xi(\bar{t})\ \middle|\ t = 0, \ldots, T-1\right)$ that maximises $E(NSP$ - Loss$)$ = (1.62) - (1.68) subject to (1.69), (1.70), and (1.63).

With proper transformation of the system dynamics (5.22), (5.23) and (5.16) and with proper distribution assumptions on random variables and coefficients (see pages 68-69 and 71-72 of Popovic 1972), we can develop an analytical sub optimal open loop feedback optimal (OLFO) control rule (see Dreyfus 1954, Tse and Athans 1966; Popovic 1972).

An interesting and most challenging problem arises when we place the stochastic nature of our social payoff function under close scrutiny. The social payoff function at time t, in our models, is expressed as

$$\text{Social Payoff } (t) = \int_0^{y_t} (\text{demand function in } (\eta_t))\, d\eta_t, \tag{1.71}$$

and we may write the demand function as $p_t = f(y,\ u \mid t)$, or in its linear form

$$p_t = \lambda_t - \omega_t y_t + u_t \tag{1.72}$$

where λ_t and ω_t are distributed with $\left(\hat{\lambda}_t,\ s^2_{\lambda t}\right)$ and $\left(\hat{\omega}_t,\ s^2_{\omega t}\right)$ respectively, and u_t with $\left(0,\ \sigma^2_t\right)$ and intertemporally independent.

From (1.71) and (1.72) we get the social payoff as

$$\begin{aligned}\text{Social Payoff } (t) &= \int_0^{y_t} \left(\lambda_t - \omega_t \eta_t + u_t\right) d\eta_t \\ &= \lambda_t y_t - ½\ \omega_t y_t^2 + u_t y_t\ .\end{aligned}$$

Evaluating the expected social payoff yields,

$$\text{E(Social Payoff (t))} = E\left(\lambda_t y_t - ½\ w_t\ y_t^2 + u_t y_t\right) = \hat{\lambda}_t\ y_t - ½\ \hat{\omega}_t\ y_t^2\ . \tag{1.73}$$

Thus, we avoid complication of optimal strategies in our models by evaluation of the expected value of social payoff in the way developed above (for contrast, see Zellner 1971 pp 325-326).

However, in our formulation and applied work, generally, (1.72) parameters, λ_t and ω_t, are extrapolations of the same parameters estimated on the basis of the past data.

Of course, we can solve the (5.2a) type problem for the certainty equivalent rule by simply assuming that the expected social payoff in the future will be in the form of (1.73) and u_t is independent of the random variables in the system dynamics for each t.

Nevertheless, as the time progresses, the extrapolation of λ_1, ω_1 for, say $t = T$, $T + 1, \ldots, 2T$ based on the past data ($t = -1, -2, \ldots, -\tau$) may turn out to be rather unreliable, and require updating. Here a technique similar to the 'sequential updating' technique proposed by Zellner (1971 pp 338-340) may be effectively employed for the revision of the parameters. One problem is how to determine the target values in the loss function.[11]

A study of the Sydney banana market by Guise, Aggrey-Mensah (1973) revealed that the longest inventory period is fourteen weeks and the one week or two week parameter updating procedures mention above may face difficulties (Prescott 1967).

A desirable way to resolve the problem of updating the parameters in the objective function and possibly in the system dynamics (say matrix, for instance) is to realise some adaptive updating *simultaneously*. However, an easy formulation to attain this objective doesn't seem to exist at this stage.

1.6.6 Further on Dynamics

Since the time I was finishing up the 1971 book, a strong urge to explore deeper into systems dynamics drove me into a closer collaboration with a group of optimal control practitioners, mainly electrical engineering researchers. In the process of interdisciplinary interactions, Marwan Simaan, now Professor of Electrical Engineering at the University of Pittsburgh, and I co-authored several papers in Automatica and other engineering journals (1976, 1978, 1980, 1984).

Deep in my mind I had planned to find a breakthrough that can contribute to the analytical and/or programming solution of the dynamic STPA models through the optimal control and differential game paths. Naturally, the optimal control approach used in (1.1) and (1.3) (bang bang control) paved the way toward multi agent (multi-region) differential game approach employed in (1.2) and (1.4). Beside the theoretical contribution of the differential game paper (1.2) above to open up the door to some of the theoretical work by various economists (see Fershtman C, Kamien MI (1987); Driskill RA, McCafferty S (1989); Tsukui S, Mino K (1990)) the path both Professor Simaan and I opened up a decade and a half ago has not found its way effectively into the operational STPA modelling field.

---

[11] One suggestion is to use the certainty equivalent solution paths as the target paths. However, the implication of employing this strategy is rather involved and I do not want to go into it in this paper.

By the time these optimal control and differential game related papers were published, I was out of academia and deeply involved in the US Federal bureaucracy (1979-1981) and then in the World Bank (1981-1983), where most future uncertainties were handled in the form of scenario modelling (a crude form of certainty equivalence modelling).

After this short period of my bureaucratic service I felt that I saw enough of the so-called 'top-down' - 'secondary data - based' large scale modelling practices in the US Federal and international organisations. I talked about the operational problem of the US DOE's PIES (and later 'MEFS') model in the early section of this paper, which is just a tip of a huge ice-berg.

I believed that my personal mission in the bureaucracy was to oversee the modelling work as conscientiously (defensibly) as humanely possible. Almost indifferent to my effort the bureaucratic machines in Washington, D.C. ground their own agenda forward!

A consolation during this period was a successful completion of PhD thesis supervision of Jeff Sohl who applied complementarity programming to the national policy issue of deregulation of natural gas (1983), in a discrete time dynamic model framework.

The brighter side of my experience in the Washington bureaucracy was that I enjoyed close working relationships with a number of prominent economists and other professionals such as Professors K. Arrow, W. Leontief, R. Dorfman, David Freedman (statistics), David Rose (Nuclear Physics), George Dantzig, L. Klein, J. Hausman, D. Aigner, Arnold Zellner, J. Gibbons (then Director, the Congressional Office of Technology Assessment and presently, Science and Technology Adviser to the President of the United States), D. Jorgenson, J. Sweeney, H. Chenery, late B. Barassa and many others.

With the change of the US Administration from President Carter to President Reagan, the large scale modelling fever quickly died away. That feeling of uneasiness brooded during this bureaucratic period of my life started to dissipate. The time had come for me to move out of that most fiercely political city in the world.

In Perth, a hidden secret of the world, I have had a number of opportunities to exchange my views of STPA modelling and beyond with young and bright researchers in the field; Drs and/or Professors B. Johansson, L. Westin, John Roy, David Batten, A. Nagurney, R. Betterham, B. Fisher, T.G. MacAuley, and others. Time was ripe for me to come to terms with 'fully non-linear' STPA modelling. Finally in 1991 recent developments in spatial temporal equilibrium models, especially in the issues of non-linearity and existence was published (with T.G. MacAuley). This has finally settled a rather operational issue that the STPA models are basically fully non-linear models (the quadratic programming presentation was practically and operationally necessary but not absolutely essential).

1.7 Conclusion

Looking back I have to admit that I have lived a rather variegated life for a university scholar. In the period of 1978-79, the Oak Ridge National Laboratory asked me to head the Economic Analysis Section of the Energy Division. The task of the section was to develop, update, and supply information to the US Energy Department. At that time, I heard the expression 'RD and D'. In academia 'R and D' was the usually accepted phraseology. The RD and D, as was explained to me then, turned out to be 'Research, Development and Deployment'. There are a number of cases in which R and D was successful, but a large scale commercialisation or commercially successful delivery failed. For manufactured and or hi-tech products, entrepreneurial spirit, availability of financial facilities, and minimum critical domestic consumption basis (per capita income and population) seem to be three necessary factors for a success of new product deployment.

In the case of STPA model development a similar reasoning may apply. The first two factors aside, ever increasing interregional and international trade in commodities and knowledge since the end of World War II has been a fertile soil to accept, nurture and deliver the STPA models to the consumers (the model users and future model developers), and this trend will be steadily expanded since the successful conclusion of the Uruguay Round GATT negotiations and establishment of the World Trade Organisation.

So far in this paper I have recounted my own experiences of thirty years with STPA modeling in a topical (if not chronological sequence). There remain a number of important developments in this field that I did not mention so far.

While I was busily engaging in practical policy-oriented problems in the period of late 1970s to early 1980s, I did not have enought time to look into purely theoretical, academic papers or books. However, back in Academia at the University of Western Australia I have been gently led to realize how ignorant I was about the progress that had been made in my own field in the regional science domain. Without refering to individual papers, I would like to mention just the names of some of the younger generation researchers who have positively contributed to the progress of this field during the past decade: J. Aronson, M. Carey, S. Dafermos, M. Florian, T.L. Friesz, D. Gabay, P.T. Harker, H. Hashimoto, J.L. Horowitz, C.D. Kolstad, H. Moulin, R. Roson, M.J. Smith, T.E. Smith, R. Stough and A. Weskamp.

It was made abundantly clear to me that younger researchers had been trying hard to make a breakthrough in the directions of Takayama and Judge (1971) generalisation of spatial and intertemporal models to embrace general equilibrium modeling strategies (see C.T.M. Elbers (1992) and the bibliography therein), and Simaan and Takayama (1978) innovative ways of solving their problems (see Anna Nagurney (1993) and her extensive compilation of references.)

There may be many other omissions and commissions of the important issues and problems for which I alone am responsible. There is one most important element that has kept me active in this field. That is the warm friendship extended to me by such outstanding scholars in this field as Professors Martin Beckmann (see his pioneering work with Marschak (1955)), Peter Nijkamp, Jeff Hewings, Walter Isard, and H. Kohno to name only a few. I would like to express my gratitude to them at this special occasion, and also to a large number of those who have worked with me in the past thirty some years.

Acknowledgement

The author would like to express his gratitude to Professor Bojan Popovic for his careful reading and suggestion on Section 1.6 of this paper. The remaining errors are my own.

References

Aoki M (1967) Optimisation of Stochastic Systems. Academic Press, New York

Aoki M (1976) A Stochastic Optimal Control Model of International Trade. Memoed, U.C.L.A.

Athans M, Falb PL (1966) Optimal Control. McGraw-Hill

Batten D, Westin L (1990) Modelling Commodity Flows on Trade Networks, Retrospect and Prospect, New Frontiers in Regional Science. Chatterji et al. (Eds) Macmillan, London

Beckmann M, Marschak T (1956) An Activity Analysis Approach to Location Theory, Proceedings of the Second Symposium in Linear Programming, National Bureau of Standards and Directorate of Management Analyses, Washington, D.C.

Charnes A, Cooper WW, Mellon B (1952) Blending Aviation Gasolines. Econometrica 20:135-159

Cottle RW, Dantzig GB (1968) Complementary Pivot Theory of Mathematical Programming, Linear Algebra and Its Applicators pp 103-25

Dantzig GB (1951) Maximization of Linear Function of Variables Subject to Linear Inequalities. In: Koopmaris TC (Ed) Activity Analysis of Production and Allocation. Cowles Commission Monograph 13:339-347

Dreyfus SE (1954) Some Types of Optimal Control of Stochastic Systems. Journal of Society of Industrial and Applied Mathematics

Driskill RA, McCafferty S (1989) Dynamic Duopoly with Adjustment Costs. A Differential Game Approach. J Econ Theory 49:324-338

Elbers CIM (1992) Spatial Disaggregation in General Equilibrium Models, with an application to the Nepalese Economy, VU Press, Amsterdam

Enke S (1951) Equilibrium among Spatially Separated Markets Solution by Electric Analogue. Econometrica 19:40-48

Fershtman C, Kamien MI (1987) Dynamic Duopolistic Competition with Sticky Prices. Econometrica 55:1151-1164

Fox KA (1953) A Spatial Equilibrium Model of the Livestock Feed Economy. Econometrica 21:547-566

Gass SI (1969) Linear Programming. Methods and Applications, 3rd edition. McGraw-Hill, New York

Guise JWB, Aggrey-Mensah W (1973) An Evaluation of Policy Alternatives facing Australian Banana Producers. Studies in Economic Planning over Space and Time. Judge GG, Takayama T (Eds) North-Holland

Guvenen O, Labys WC, Lesourd JB (Eds) (1991) International Commodity Market Models. Chapman and Hall, New York

Hall HH, Heady EO, Plessner Y (1968) Quadratic Programming Solution of Competitive Equilibrium for US Agriculture, Am J Agricult Econ 50:536-555

Hashimoto H (1976) World Food Projection Models, Projections, and Policy Evaluation. PhD Thesis. Department of Economics, University of Illinois (unpublished)

Heady EO (1952) Economies of Agricultural Production and Resource Use. Prentice Hall, New York

Heady EO, Candler W (1958) Linear Programming Methods, Iowa State University Press

Intrilligator MD (1971) Mathematical Optimisation and Economic Theory. Prentice-Hall. Inc. Englewood Cliffs. N.J.

Judge GG (1956) Spatial Equilibrium Model for Eggs, Bulletin 318. Storrs, Connecticut Agricultural Experiment Station

Judge GG, Takayama T (Eds) (1973) Studies in Economic Planning over Space and Time. North-Holland, Amsterdam

Judge GG, Wallace TD (1959) Spatial Equilibrium Analysis of the Livestock Economy, Technical Bulletin TB-78, Department of Agricultural Economics, Oklahoma State University

Kennedy M (1974) Optimum Utilization of Transportation System. In Proceedings of the International Statistics Conferences, Washington, DC, 5

Kuhn HW, Tucker AW (1951) Nonlinear Programming. In: Neyman J (Ed) Proceedings of the Second Berkely Symposium on Mathematical Statistics and Probability. University of California Press, Berkeley, pp 481-492

Labys WC, Polak PK (1984) Commodity Models for Forecasting and Policy Analysis. Croom Helm, London

Labys WC, Wood DO (1985) Energy Modelling. Economics of Mineral Industry. In: Vogel W (Ed) American Institute of Mining Engineers, New York

Labys WC (1987) Commodity Markets and Models. An International Bibliography, Gower Publishing Co

Labys WC, Takayama T, Uri N (1989) Quantitative Methods for Market-Oriented Economic Analysis over Space and Time, Gower, London

Labys WC, Yang CW (1991) Advances in Spatial Equilibrium Modelling of Mineral and Energy. International Regional Science Review 14:61-94 (A most recent review of works in this subject)

Lefeber L (1958) Allocation in Space, North-Holland, Amsterdam

Mann AS (1956) Scheduling of Petroleum Refinery Operations, Harvard University Press, Cambridge

Modigliani F, Hohn F (1955) Production Planning over Time and the Nature of the Expectation and Planning Horizon. Econometrica 23

Nagurney A (1993) Network Economics, Kluwer, Dordrecht

Plessner Y, Heady EO (1965) Competitive Equilibrium Solutions with Quadratic Programming. Metroeconomica 17:117-130

Popovic B (1972) Analytic Solutions to Some Stochastic, Adaptive, and Dual Optimal Control Problems for Economic Decision Making. PhD Thesis (unpublished). Department of Economics. University of Illinois. Also available in the form of Report T-4 (September 1972) UILU-ENG 72-2245. Coordinated Science laboratory, University of Illinois, Urbana, Illiois

Prescott EC (1967) Adaptive Decision Rules for Macro Economic Planning PhD Thesis (unpublished) Graduate School of Industrial Administration. Carnegie-Mellon University

Robinson J (1933) The Economics of Imperfect Competition, MacMillan, London

Rojko AS, Urban FS, Naive JJ (1971) World Demand Prospects for Grains in 1980 with Emphasis on Less Development Countries, FAER-75, Economic Research Service, USDA, Washington, DC

Samuelson PA (1952) Spatial Price Equilibrium and Linear Programming. Am Econ Rev 42:283-303

Samuelson PA (1957) Intertemporal Price Equilibrium. A Prologue to the Theory of Speculation. Weltwirtschaftliches Archiv. 79:181-221

Samuleson PA (1959) Using Full Duality to Show that Simultaneously Additive Direct and Indirect Utilities Implies Unitary Price Elasticities of Demand. Econometrica. 33:731-796

Sasaki K (1969) Spatial Equilibrium in Eastern Japan's Milk Market (in Japanese), J Rural Econ 41

Sasaki K (1970) Spatial Equilibrium in Eastern Japan's Port Industry (in Japanese), J Rural Econ 42

Simaan M, Takayama T (1976) Optimal Monopolist Control in a Dynamic Market. IEEE Trans Syst Mon and Cybern 6:799-807

Simaan M, Takayama T (1978) Game Theory Applied to Dynamic Duopoly Problems with Production Constraints. Automatica 14:161-166

Simaan M, Takayama T (1980) Optimal Harvesting Policies and Regulations for Renewable Resources, Int J Syst Sci 11:281-290

Simon HA (1956). Dynamic Programming under Uncertainty with a Quadratic Criterion Function. Econometrica 24

Stevens BH (1959) An Interregional Linear Programming Model, J Reg Sci 1:60-98

Takayama T (1959) Production Economics for Agriculture (Nogyo no Seisan Keizaigaku) Tenshi-in Press, Sapporo

Takayama T, Judge GG (1964a) Spatial Equilibrium and Quadratic Programming. Journal of Farm Economic (JFE) 44:67-93

Takayama T, Judge GG (1964b) Equilibrium among Separated Markets; A Reformulation. Econometrica 32:510-524

Takayama T, Judge GG (1964c) An Interregional Activity Analysis Model of the Agricultural Sector. JFE 46:349-365

Takayama T, Judge GG (1964d) Intertemporal Price Equilibrium Model. JFE 46:477-484

Takayama T, Judge GG (1966) Spatial Economic Analysis. Inaugural Lecture Publication. University of New England, Armidale, NSW, Australia

Takayama T (1967) International Trade and Mathematical Programming. Australian J Agricult Econ 11:36-48
Takayama T, Judge GG (1971) Spatial and Temporal Price and Allocation Models. North-Holland
Takayama T, Hashimoto H (1976) World Food Projection Models. 1972 and 1974, Illinois Agricultural Economics. 16
Takayama T, Woodland A (1970) Equivalence of Price and Quantity Formulations of Spatial Equilibrium. Purified Duality in Quadratic and Concave Programming. Econometrica 38:889-906
Takayama T, Simaan M (1984) Differential Game Theoretic Policies for Consumption Regulations of Renewable Resources, IEEE Trans Syst Man Cybern. pp 764-766
Theil H (1964) Optimal Decision Rules for Government and Industry. North-Holland
Thompson R (1989) Spatial and Temporal Price Equilibrium Models. Quantitative Methods for Market-Oriented Economic Analysts over Space and Time. Labys Takayama Uri (Eds). Avebury pp 49-65
Tse E, Athans M (1970) Optimal Minimal-Order Observer-Estimators for Discrete Linear Time-Varying Systems. I.E.E.E. Trans. on Automatic Control. AC 15:416-426
Tsukui S, Mino K (1990) Nonlinear Strategies in Dynamic Duopolistic Competition with Sticky Prices. J Econ Theory 52:136-161
Uri N (1975) Toward an Efficient Allocation of Electric Energy, Health Lexington Books. Lexington, MA
West DA, Brandow GE (1964) Space-product Equilibrium in the Dairy Industry in the North Eastern and North Central Regions, J Farm Econ 46:719-731
Westin L (1990) Vintage Models of Spatial Structural Change, Umea Economic Studies 227. University of Umea, Sweden
Wolfe PM (1959) The Simplex Method for Quadratic Programming. Econometrica 27:382-398
Woodland A (1969) Competitive Equilibrium in International Trade, unpublished PhD thesis. University of New England, Armidale, NSW, Australia
Zellner A (1971) An Introduction to Bayesian Inference in Econometrics. Wiley, New York
Zusman P, Melamed A, Katzin I (1969) Possible Trade and Welfare Effects of EEC Tariff and Reference Price Policy on the European-Mediterranean Market for Winter Oranges. Monograph 24, Gianini Foundation, UC Berkeley

2 Spatial Equilibrium Models: A Survey with Special Emphasis on Transportation

Jeroen C.J.M. van den Bergh, Peter Nijkamp and Piet Rietveld
Vrije Universiteit

2.1 Introduction

Although both transportation science and economics have generated a wealth of formal models to address compelling policy questions, there seems to be a no-man's land of transport-economy interactions that has seen relatively little maturation in terms of modelling. The aim of this paper is to focus on suitable models for estimating the spatial and economic effects - economy-wide, regional or national, but preferably multisector - caused by transport policies as well as by (autonomous) changes in the transport sector. A main motivation for the focus here on specific approaches - oriented towards notions of market mechanisms, rational behaviour and equilibrium - is the belief that the scope for sectoral interactions depends, among other things, upon transportation costs, which are variable and affected by a number of factors, such as production technology, input structure, density factors and congestion phenomena. This means that any policy aimed at influencing transportation conditions may have a non-negligible economy-wide effect, in terms of impacts on national income, prices, sectoral distribution of factors and production as well as on spatial commodity and passenger flows. The call nowadays for curbing the negative external effects of road, rail and air traffic, causing environmental damage, health risks, accidents and congestion (time losses), indicates that we may have to face significant alterations with respect to transportation. It is evident then that tools are required for analysing economy-wide effects of transportation (scenarios) and transport policies, in particular those connected to commodity flows, sectoral performance, employment, and public transport. Surprisingly, this issue has not been tackled in a satisfactory formal-analytical way up till now. Clearly, this may be partly due to the interdisciplinary nature of the research questions, where economic phenomena interact with technical-physical ones, such as market, congestion and network processes.

There are of course several models available for dealing with some specific aspects of the broad issue of economy-wide effects of transportation. They may be classified as multisectoral, multiregional, endogenous-price, general equilibrium, and network models. In the class of endogenous-price models, spatial price equilibrium (SPE) models are attractive from a spatial commodity flow perspective. However, from a general economy-wide perspective one may use a more refined tool, viz. the general equilibrium (GE) model. This model is based on a microeconomic general equilibrium framework that allows for substitution possibilities at the production as well as the demand side of the economy, via an endogenous-price system. It thus takes account of structural relationships in an economy and is hence a more suitable tool for obtaining insight - rather than prediction of numbers - into economy-wide, direct and indirect, consequences of transportation policy.

A main deficiency of most economic equilibrium analyses is that they refrain from the spatial character of economic relationships, be it spatial flows, regional disparities, or network phenomena. Therefore, in the paper attention will also be devoted to comparing various approaches that can be used for analysing economic effects of transportation policy. The focus will be on integrating spatial and economic concepts of transport, with a special focus on an equilibrium framework. This does not mean an entirely new approach. One may look for spatial elements in SPE and network models, while endogenous spatial and sectoral price mechanisms can be based on both SPE and general equilibrium models.

In the remainder of this paper an overview is given of models that can be used for estimating spatial and economic effects of transport and transport policy. Relevant effects include changes in prices, costs, input structure, profits, commodity flows, modal shift, transport demand and supply, congestion and intersectoral interactions, as well as traditional indicators of sectoral and regional/national economic performance. Several types of models are discussed, most of which have a clear economic orientation. It is investigated in particular whether and how they incorporate transport features such as production, markets, infrastructure and congestion. The discussion is focused on essential characteristics of basic models and interesting extensions, as well as on ideas about how to enrich these existing models for the above stated purpose. Advantages and disadvantages of the various approaches are dealt with, and a systematic evaluation is provided as a conclusion.

2.2 Types of General Economic, Spatial and Transportation Equilibria

The economic foundation of the paper is that of a market, in which supply and demand interact on the basis of price signals. Markets are assumed to clear. Besides markets for factors, intermediate and final commodities, also those for transportation services will be considered here. In the case of freight transport this means that spatially segregated markets for commodities are assumed. The main distinct types of equilibria can then be formulated as follows.

(A) When space is assumed away (and implicitly no cost of transport is assumed), we have:

general equilibrium (competitive spaceless equilibrium) (GE)[1]:

(a) market equilibrium:
  (a.1) non-negative, homogeneous and unique prices;
  (a.2) no excess demand;
  (a.3) efficient market pricing (i.e. price times excess supply is zero);

(b) consumer equilibrium (consumption efficiency): maximization of utility subject to income constraint;

(c) producer equilibrium (production efficiency): maximization of profits subject to technological and institutional constraints.

(d) individual balance of payment.

(B) When space is considered from a discrete, multiregional perspective and when transport costs are fixed and demand and supply functions for a single product are given, then we have:

spatial (locational) price equilibrium (SPE):

(a) non-negative, homogeneous and unique market demand and supply prices;

(b) no excess demand or supply in all regions;

(c) on each interregional route the delivered price in the importing region must at least equal the producer price in the exporting region plus transportation costs; when this is not the case, there is no positive interregional transportation/trade flow.

SPE may be extended to multiple products: when independent transportation of each good is assumed, then all equilibria may be treated independently; when products interact via the transportation system (e.g. combined transportation, congestion, density effects), then an interrelated price system will result.

(C) When transportation is included as one of the economic profit-making sectors:

general spatial equilibrium (competitive spatial equilibrium) (GSE): the differences with respect to GE are:

(a) GE market equilibrium conditions (a.1) to (a.3) pertain to each region; in addition, we have:
  (a.4) non-negative, homogeneous and unique transportation costs on any route;
  (a.5) no excess demand for transportation services;
  (a.6) efficient pricing for transportation (see GE (a.3));
  (a.7) locational price equilibrium: SPE condition (c) for each product; GE producer and consumer equilibrium conditions (b) and (c) for each region;

(b) balance of payments equilibrium, which includes transportation costs and revenues.

---

[1]This is a somewhat adapted version of the classification given in Takayama and Judge (1971, p 26).

Other possible relevant equilibria are:

- a transportation equilibrium based on GSE conditions (a.5), (a.6), (a.7) and (b).
- a multimodal transport equilibrium where the same conditions hold for each mode separately; important is the condition that delivered prices of an imported commodity should be equal along different modes as well originating from different sources, as long as they are used.
- network (freight) equilibrium: transportation costs over all paths that are used from each origin to each destination are equal, and no unused path has a lower transportation cost (see Florian and Los 1982; Friesz et al. 1985; Harker 1987); important in network models are flow conservation conditions to ensure that over every subset of arcs of a network in- and outgoing flows are balanced.
- locational equilibrium in continuus space: a well known example is the monocentric land use model of a city (Alonso 1964; Mills 1967).

In the model types described above, equilibria are analyzed at a rather high level of spatial aggregation. The markets at which equilibrium exists are commodity and factor markets. The land market is not taken into account, however. For an appropriate analysis of land markets however, the analysis has to be carried out at a much lower spatial level, i.e. at the level of multizonal systems in metropolitan areas. For the transport component this would imply that passenger transport receives a much more prominent role compared to the spatial equilibrium models mentioned above. Reviews of such integrated transport land-use models can be found in Anas (1982), Anas and Duann (1986), Berechman and Gordon (1986), Henderson (1988), Berechman and Small (1988), Webster et al. (1988), and Rietveld (1994). Although these models address relevant transport aspects of spatial equilibrium (SE) and general equilibrium (GE) models, they will not be treated here due to lack of space. It is clear that models with this spatial orientation are complementary rather than substitutes to the SE and GE models referred to above. A multilevel approach is most likely a natural way to connect the two model types.

2.3 Transport Analysis

Irrespective of any specific type of equilibrium models, one can envisage three main directions for dealing with transport issues in a general economic context.

First, one may use descriptive models of commodity flows or passenger transport demand, and next derive transport and simultaneous economic effects. Transport parameters are here exogenous, i.e. they are not influenced by economic variables. Therefore, an estimation of effects may be biased, with regard to both economic and transport indicators. Examples are various types of multi-regional models such as input-output and purely econometric models, and endogenous-price models.

A second approach uses also fixed transportation parameters, but it searches for optimal transportation service levels and distribution (e.g. over a network or a multiregional system); optimality may mean least cost or maximum social welfare (consumer and producer surplus). Such planning or social-welfare oriented models can be used, for instance, to trace the optimal policy in the case of congestion and environmental externalities arising from transport. A particular type of model in this class is the SPE model.

A third approach is new. It allows for endogenous[2] transportation costs and prices, and is based on assumptions of individual rational behaviour and optimization of welfare, utility, profit or cost. The values of all price variables have then to be determined sequentially or simultaneously. The type of models in this class are of a general equilibrium and economy-wide character, since more markets are taken together in one frame. A limited and more partial approach - from an economic perspective - is the network equilibrium model, which is suitable for the analysis of a detailed network where individual agents choose optimal routes based on indiviually rational behaviour, and where congestion is important. There are three main reasons to allow for variable transportation costs. First, transportation is produced by a certain activity, which we may generally refer to as a transportation firm. A common terminology is a carrier. This production involves a cost, which provides a lower bound to the price of transportation services (profit maximization). Second, when we consider a significant shift in the demand for transportation, or in transportation infrastructure, then this may seriously affect the price of transportation. And thirdly, this in turn may give rise to supply and demand reactions which affect the level of transportation services used. It is clear that these reasons provoke the need for an analysis in a general equilibrium framework with production.

These various model approaches and their potential applications to the relevant issues will now be discussed in more detail in Sections 2.4 to 2.6, respectively. This can help improving the modelling of economy-transport relationships.

2.4 Transport in Space Using Multi-regional Models

2.4.1 Introduction

In this section we first discuss the input-output related approaches to multiregional modelling. This is followed by a concise review of econometric modelling approaches. We conclude with a specification of what a complete multiregional economic-transport approach would imply.

A complete multiregional economic-transport approach requires in general the following elements: consistency of imports and exports between region A and B; inclusion of both mobile and immobile resources; and transportation costs. Finally, multiregional is used here to encompass relationships between regions, because we are clearly not interested in treating independent regions (such as in the case of budget allocation or decentralization issues). Our starting point for the multiregional approach towards transportation is the pattern of regional interactions.

---

[2] Sometimes the term adjoint 'variable' is used. We prefer endogenous, because it indicates clearly that the variability is arranged within the model.

A multiregional structure can be regarded as a first approximation to a network. The data of such a multiregional structure, for instance those indicating distances between two points or regions, may be based on an approximate minimum transportation or minimum distance route, as is usual in for instance spatial price equilibrium models[3].

If transport is included in a multiregional model, then a simple network structure is adopted, including direct links only between each region. The main disadvantages of this structure is that congestion has a limited interpretative content. The same limitation holds with respect to density factors in transportation. Finally, optimal routes are independent from activity and demand levels in each region, since in a multiregional framework it is assumed that each interregional link represents somehow the optimal connection, i.e. the least transportation cost connection.

2.4.2 Multi-regional Input-Output in Modelling and Transport

Input-output (IO) models have a long tradition in regional economics, starting with the early contribution by Isard (1951a), the trade analysis of Moses (1955), and the work by Leontief and Strout (1963). Examples of empirical work in this field can be found in Polenske (1980). The IO models are mostly used for studies that wish to incorporate structural economic relationships, i.e. interindustry linkages by way of intermediate deliveries of goods and services. By calculating direct and indirect effects of exogenous variables (expenditures), one may obtain multipliers and effects on employment, and use of resources. Estimation of parameters requires essentially only one input-output table for a given reference time point.

Transport and IO models have usually been approached from the interregional trade perspective (economic-base, Chenery-Moses, balanced Leontief, Isard).[4] Here interregional gravity models are sometimes combined with the intersectoral IO models to arrive at completely variable interregional intersectoral models (Leontief-Strout). More limited is the approach to use fixed import-output coefficients, which can be regarded as the straightforward extension of general IO models[5]. Here one may repeat a given sectoral disaggregation for each region.

---

[3] Among the various extensions of the 'standard Samuelson/Takayama-Judge SPE models' presented in a very interesting article by Florian and Los (1982), the authors offer the perspective of a network for dealing with non-constant unit transport costs in an SPE framework.

[4] It is interesting to note that Krutilla (1989) shows that analysis on the basis of trade policy models which ignore or exogenize the transportation sector may give rise to estimates of trade policy actions and optimal tariff calculations that are biased. Based on monopolistic transportation supply, he finds that if the transportation sector is a resident of the importing country the optimal trade policy is a subsidy to stimulate trade beyond the monopolistic optimum.

[5] Isard (1951, p. 324) remarks that "The assumption of fixed production coefficients for inter-area industry relations is more restrictive than for inter-industry relations within a closed economy ... this is, in general, well compensated by the greater degree of accuracy obtainable from working with inter-area industry coefficients."

The fixed interregional trade coefficients are to be interpreted as constant patterns of supply areas or channels (see Isard 1951b, p 323). This is an important assumption, which holds more strongly for commodities which are associated with significant transportation costs (bricks, cement, glass, etc.) than for those with negligible transportation costs. Another class of models is formed by multicountry models using trade-theoretic concepts (Moses 1955 and 1960).[6]

Stepping aside from the theoretically less satisfying combined IO-gravity models approach, the disadvantages of the multiregional IO models are clear and depend mainly on their fixed character (see Nijkamp and Paelinck 1981): fixed input composition of output, fixed interindustry interdependencies, constant returns to scale, fixed import and export relationships, independence of prices and supply (i.e. analysis of only real variables), and the supply orientation combined with exogenous demand.

An extension of IO to linear programming (LP) models by including a mechanism for choosing among alternative IO-feasible options can take away some of the above deficiencies (see Dervis et al. 1982; and Lundqvist 1981). Such an extension involves three main elements: (i) conversion of IO equalities to generalized inequality relationships; the usual materials balance equations are adjusted to require total supply (import and domestic production) to be larger than or equal to total demand (intermediate, investment, final consumption and export); (ii) adding an objective function, e.g. representing a cost-benefit oriented social welfare function; (iii) inclusion of additional constraints: capacity constraints on capital and labour (e.g. using fixed capital-output rations); a foreign exchange constraint (value of imports does not exceed the sum of the value of exports plus net capital inflow); and non-negativity conditions on all real variables.

An especially interesting differentiating characteristic of IO extended LP versus IO models is that they generate a set of implicit, so-called shadow prices. These are the solution to a dual price variable model of the primal quantity (real or material) variable model formulation. This allows one to compare the approach to GE, since both of them give information on quantities and prices. In the process from GE via LP to IO, prices become less flexible, and production adjustments larger, for given exogenous changes. In LP the prices are not endogenous like in a GE model, since that would give rise to a nonlinear rather than a linear programming formulation (e.g. including expressions with variable prices times variable quantities). Therefore, price instruments cannot be included. Furthermore, the resource and production structure is not necessarily consistent with incomes and budgets implied by the solution of an LP formulated program and its dual (see Dervis et al. 1982, p 132). However, an iterative procedure using shadow price solutions to formulate income constraints and calculate new solutions has been tried by some authors (see Ginsburgh and Waelbroeck 1982).

GE and IO models both incorporate interindustry relationships. Moreover, both can be extended to include various other structural relationships among economic actors, such as household-industry and interregional linkages, and between variables such as prices and costs. One way of looking at a GE model is to consider it as an IO model with flexible (variable and endogenous) coefficients.

---

[6] A familiar terminology is Chenery-Moses models.

GE models can however include also the structural link between demands and incomes, which is not possible in IO models per se. The main differences between GE and IO models are with respect to income/budget constraints, prices and production functions. The budget constraints such as formulated on level of actors in a GE model are not included in an aggregated IO or LP extended model; here only an overall budget constraint is satisfied, namely that of the single maximizing central planner. Price relationships can be derived from IO models. The unit cost equation in matrix terms is $P=PA+RK+WL$, with P a commodity price vector, A an IO coefficients matrix, R the profit rate, W the wage rate, K the capital (value) vector and L the labour vector. Prices are relative since W and R are exogenous. However, they are separated from the quantity relationships because of the fixed character of production input-output relationships (although using the same data). This means that prices are fixed with respect to changes in quantities and vice versa. Exogenous primary factor prices determine all other prices. In GE models prices are endogenous and influenced by demand and supply, i.e. related to (equilibrium) quantities. Furthermore, production input-output relationships are variable as a result of so-called neoclassical production functions allowing for substitution between inputs. This means that input prices are linked to input combinations in production by way of cost minimization.

2.4.3 Multi-regional Econometric Modelling and Transport

Although a distinction between econometric and input-output models is the first one usually made with regard to (multi) regional modelling, combinations of the two are very common (see the overview by Rietveld 1982)[7]. This may be to some extent due to the objective of incorporating intersectoral/structural relationships. Econometric models have the power to efficiently summarize information implicit in large amounts of data, and consequently may under certain circumstances give more accurate predictions than any of the other types of model mentioned here (see Adams and Glickman 1980; Bolton 1985). An obvious condition is that the model is robust, i.e. that the structural relationships have not changed much.[8] Essentially it means that it is mainly suitable for marginal and trend analysis.

Now we will briefly pay attention to the link between econometric type of models and the treatment of transportation issues. They use accessibility and potential concepts as proxies for the relationship between the region and its input and output markets (see Snickars 1982). Few of these types of model are of serious interest for our line of discussion. This is the result of the impossibility to link transportation in a structural sense to variables in an econometric model. This

[7] Econometric equations may be suitable for dealing with consumption demand and labour markets, while IO modules can deal with interindustry relationships. These complement one another.

[8] Of course, econometric methods may have been used to estimate (part of) GE models. The essential point of difference in this context is that the econometric models mentioned here are not based on a systematic specification of the behaviour of the relevant actors on the relevant markets. The econometric models mentioned here use ad hoc specifications or reduced forms which make them less attractive from a theoretical point of view.

is a consequence of the lack of intersectoral relationships as mentioned above. In a way, one can say that econometric models depend too much on statistical regularities, while we are more interested in irregularities caused by changes in sectoral interdependencies and various types of substitution processes, such as production inputs, origins of transport, destinations of transport, consumed commodities[9].

Econometric models may include endogenous prices. In comparison with GE models, however, they are less able to deal with large shifts in quantities and relative prices, because they lack structural relationships. In a predictive sense many model types will not be very powerful with regard to economy-transport relationships. However, as opposed to econometric approaches structural models (IO, GE, SPE) can be useful for qualitative scenario analysis, finding plausible directions of change as a response to rather trend-breaking exogenous phenomena or policy measures.

2.5 Transport in Space: Transportation Network Models

An alternative approach which can address aspects of economic-transport interactions is made up by the class of transport (equilibrium) network models. These stem from engineering and operations research fields (see, e.g. Friesz et al 1983; Friesz and Harker 1985; Harker 1987; Boyce et al 1988; Batten and Westin 1990). These models are quite micro-oriented and detailed, and especially useful in the context of single commodity analysis. This means that they share a partial equilibrium approach - from the economy-wide perspective - so that they are very close in orientation to SPE models (see also the integrative study by Harker 1987). The advantage of transportation network models is that they can deal more accurately with freight transport, where the real transportation infrastructure consists of a complex network, e.g. with ports and cargo terminals. It seems especially adequate for dealing with multi-modal and multi-product questions, at least conceptually.

Florian and Los (1982) provide a formal general representation of extensions of variable transportation cost and multiproduct SPE models with network models, where network modelling concepts are translated to the SPE framework. Harker (1987) includes analytical, computational and applicational results of such integration. He also discusses the possibility of integrating GE with network equilibrium models, i.e. providing for simultaneous solutions. Harker describes the purpose of his work as "...formally treating the generation, distribution, modal split and assignment of freight movements in a consistent, general equilibrium framework" (Harker 1987, p 2).

---

[9] The econometric model is anyway much less popular in regional economics than in macro-economics. One reason is that regional relationships are less stable over time than their national counterparts.

Networks consist of arcs (links) and nodes. A path is a specific sequential combination of arcs from an origin of supply to a destination of demand. Some nodes represent regions, others transportation infrastructure elements such as ports, terminals, etc. Other important notions of network models are 'equilibrium flows', 'origin', 'destination', 'path', 'carrier', 'shipper' (see Friesz et al 1985; or Harker 1987). A shipper[10] is a decision-making entity desiring a particular commodity at a particular destination[11]. A carrier is a decision-making entity that actually executes the transport of freight. Therefore shippers are the demanders of transportation and carriers the suppliers of it. In transportation network analysis an important research area seems to be the simultaneous treatment of shippers and carriers decisions (see Harker 1987). Up till now most models using submodels for both actors have formulated them sequentially - as well as computing equilibrium solutions sequentially. In that case they are actually not equilibrium solutions. A GE approach can integrate the shippers' and carriers' behaviour, as well as integrate it with other actors in the economy. Problems usually occur when non-linear cost curves and imperfect markets for transportation are introduced.

With regard to the network characteristics of combined shipper-carrier models it is usual to employ a shipper's perceived network which is interlinked at certain nodes (transhipment, interline and other key nodes) with more detailed carrier-specific networks. In relation to shippers, carriers and networks we should mention here also the two Wardrop's principles (Wardrop 1952), that state conditions for user optimum and system optimum network flows, respectively.[12] Shippers may be modelled by the first Wardropian principle that can be stated as the minimization of delivered commodity prices (see Harker 1987). A difference between a spatial and a general equilibrium model is that the first adopts the second Wardropian system optimum principle (using programming model formulations), while the second uses decentralized optimization formulations and thus starts from the user optimum. However, under certain conditions the two equilibrium concepts can be shown to lead to the same outcome. That is why programming and decentralized formulations may sometimes be substituted for one another.

---

[10] Notice that the notion of a shipper does only apply to freight transport and not so much to passenger traffic (by private car), since freight flow patterns are determined by both transportation system users (shippers) and suppliers/owners (carriers) (see Friesz and Harker, 1985, p. 162).

[11] In network models the shipper operates in an environment where demand and supplies at various nodes are fixed. This may be extended to endogenous demand and supply, such as in SPE and CGE models. In both of the two latter cases trip generation becomes endogenous.

[12] A user-optimized equilibrium is reached when no user has an incentive to lower his transportation cost through unilateral action. A system-optimized equilibrium is reached when the marginal total costs of transportation alternatives are equal (see Friesz and Harker, 1985, p. 162).

2.6 Spatial Price Equilibrium (SPE) Models

Our interest in this overview is in multiregional economy-wide models. Thus one may wonder whether SPE models belong to this category[13]. Usually, they take a partial view, i.e. consider a single good. However, in principle generalizations to multiple goods are possible. Besides, the approach shows a clear view on spatially separated markets and transport from an equilibrium point of view. Therefore, it may certainly shed some light on the integration of transportation in GE models. Anyway, many steps have already been taken to fill the gaps between various alternative approaches, such as variable transportation costs, network models, dynamic models, and multi-commodity models. The two most important contributions in that respect are the very extensive treatment by Takayama and Judge (1964 1971) and the conceptually very rich approach by Florian and Los (1982)[14]. Harker (1987) integrates elements of network and SPE models.

The basic SPE formulation aims to solve an interregional trade or spatial price equilibrium, where demand prices equal the sum of supply costs and transportation costs. Samuelson (1952), starting from this equilibrium concept, was the first to derive a formal representation of the SPE problem, as a linear programming model. The problem can be cast as finding - for given transportation costs and domestic demand and supply curves at regional or spatial markets - the spatial competitive equilibrium. The latter consists of a set of prices in all the spatial markets, the demands and supplies at each location, and all exports and imports[15]. Samuelson has shown that some kind of consumers surplus concept is relevant in trying to convert the problem into a maximum or minimum formulation. He defines a 'net pay-off' (NSP) function as the sum of the social pay-off's in each region (spatial market) minus the transport cost. The social pay-off in a region is defined then as "the algebraic area under its excess-demand curve" (Samuelson 1952, p 288). Using excess-supply functions s_i for region i, transportation cost function t_{od} (covering both directions) and letting E_{od} denote the exports from region o to region d, then in the case of two regions only the NSP is as follows (see Samuelson 1952, p 290):

---

[13] Roson and Westin (1990) distinguish between input-output, transport assignment and gravity models. Transport assignment models probably include a wider class than SPE models, most of which however are less general and therefore of little interest for our discussion. Of more interest may be the combination of gravity and SPE models.

[14] It is also interesting to note that on the other end of the spectrum of spatial equilbrium models there are the more restricted, partial transportation planning problems, where demand and supply of commodities are fixed in amount (see the classical study by Beckmann et al., 1956).

[15] The difference with the simpler 'classical' transportation problem is that there the demands and supplies at each place are given (fixed), instead of their dependence on prices (demand/supply functions).

$$NSP = -\int_0^{E_{12}} s_1(x)dx - \int_0^{-E_{12}} s_2(x)dx - t_{12}(E_{12}). \tag{2.1}$$

This follows from the fact that $E_{21}+E_{12}=0$ for the case of only two regions. The conditions for an optimum imply that, assuming constant per unit transportation cost T_{od}[16], $-T_{21} \leq s_2(E_{21})-s_1(E_{12}) \leq T_{12}$. If E_{12} is positive, then in $s_2(E_{21})-s_1(E_{12}) \leq T_{12}$ the equality sign should hold, or, taking prices instead of excess demands, $P_2=P_1+T_{12}$. This is the fundamental result, namely that prices for the same good should in equilibrium (i.e. optimum) for different markets be either equal to the per unit transportation cost, or there should be no trade (i.e. neither imports nor exports).

A general multiregional one-product formulation of an SPE model is as follows:

$$\text{minimize} \sum_o \int_0^{x_o} S_o(z)dz - \sum_d \int_0^{y_d} D_d(z)dz + TC. \tag{2.2}$$

where S_o are the inverse supply functions and D_d the inverse demand functions, and TC the total transportation cost. The latter is a function of all link-specific shipments (t_o^d stands for the shipment of goods from source o to destination d). These shipments have to satisfy non-negativity restrictions. Supplies x_o at origins o are equal to the sum of all shipments out of o:

$$x_o = \sum_d t_o^d, \tag{2.3}$$

and demands y_d at destinations d are equal to the sum of all shipments arriving in region d:

$$y_d = \sum_o t_o^d. \tag{2.4}$$

Florian and Los (1982) note that the constraints in (2.3) and (2.4) may be interpreted as conservation of flows on a bipartite network, for which reason it may be regarded as a generalization of the classical transportation problem.

A different look at the basic SPE optimization formulation is proposed by Florian and Los (1982) by substituting (2.3) and (2.4) in (2.2), so that a non-linear programming problem in non-negative link-specific shipments variables t_o^d is obtained:

[16] To be more precise (following Samuelson, 1952), generally transportation costs may differ in two directions.

$$minimize \sum_o \int_0^{\sum_d t_o^d} S_o(z)dz - \sum_d \int_0^{\sum_o t_o^d} D_d(z)dz + TC. \qquad (2.5)$$

According to Florian and Los this is 'a relatively easy non-linear programming problem', for which various solution methods are available.

Now we are especially interested in the composition of total transportation costs TC (in (2.2) or (2.5)). Variation of these is anyway the most that can be done while still remaining in the SPE framework and the class of solvable mathematical constrained optimization problems. Our special interest is to allow for variable cost structures based on assumptions with respect to market structure (number of modes or firms), density effects on links, network characteristics and transportation of multiple commodities[17].

First, assume that the cost function belonging to an interregional link is to be associated with a single unique transportation firm, implying as many firms as links[18]. With one additional assumption, namely that c_o^d is the fixed cost per unit of the good to traverse link o-d, one obtains the most common cost structure used:

$$TC = \sum_o \sum_d c_o^d * t_o^d. \qquad (2.6)$$

An important extension for our survey is the one in which the cost of shipping one unit of product over a certain link is a non-linear function $c_o^d(t_o^d)$ of the total amount of transport on that link[19], in which case (2.6) is replaced by (2.7):

$$TC = \sum_o \sum_d \int_0^{t_o^d} c_o^d(z)dz. \qquad (2.7)$$

Two other market assumptions are of interest. The first one represents the other extreme case of a single firm serving all links. In such a case, the transportation price per unit of product and distance may be related to the total volume of transportation services; this requires that data be available on τ_o^d, the amount of transportation services required to ship a unit of product on a certain link, for each

---

[17] Notice that we are here, necessarily in the SPE framework, only considering non-constant transportation costs with respect to congestion and some inverse given supply function of transportation services. It is important to realize that such a formulation is not as rich as that of production of transportation services in a CGE format where inputs are made explicit. However, both lead to endogenous transportation prices, stressing different aspects.

[18] They add that the price of transportation services should at least vary with in proportion with the costs of supplying them. Though this property is not imposed by the structure of the model, it is interesting to note the closeness in mind to the CGE frame, where this typically is enforced.

[19] As Florian and Los suggest, congestion effects imply a convex shape and quantity discounts (density effects) lead to a concave shape.

link. The total transportation cost now depends on an inverse transportation supply or price curve p, as shown in (2.8):

$$TC = \int_0^{\sum_o \sum_d \tau_o^d * t_o^d} p(z)dz. \tag{2.8}$$

Note that the transportation cost per unit commodity on link o-d is formulated as $\tau_o^d * p\left(\sum_o \sum_d \tau_o^d * t_o^d\right)$.

The second market situation is between the two extremes, namely with a small number of firms (m) each controlling a disjoint subset of the routes (V_i), and each one having an inverse demand function related to the routes it serves[20]:

$$TC = \sum_{i=1}^{m} \int_0^{\sum_{(o,d) \in V_i} \tau_o^d * t_o^d} p_i(z)dz. \tag{2.9}$$

Further extensions of the latter case can be made in a network approach. The final extension discussed here is a situation of multiple products. For given inverse supply and demand functions for each good g and all regions, the optimization problem of equation (2.2) is generalized as follows (as above subject to non-negativity restriction on the shipments):

$$minimize \sum_g \left[\sum_o \int_0^{x_{g,o}} S_{g,o}(z)dz - \sum_d \int_0^{y_{g,d}} D_{g,d}(z)dz + TC \right]. \tag{2.10}$$

The most important restriction here is that the inverse demand functions are a function of only the respective product, i.e. it assumes separable demand functions in product prices. When cross-price elasticities of demand are non-zero, the inverse cannot be taken. See Florian and Los (1982) for a further discussion. In the present paper we are interested in the relevant transportation cost function belonging to the multiple goods problem (2.10). The specifications (2.7) and (2.9) for the single product case discussed above can be generalized. Next, it is easy to see that (2.11) is a generalization of (2.7):

[20] Note that the treatment of this market situation does not require any specific assumptions on oligopolistic price behaviour, since the firms do not compete with each other on the same interregional links. This implies that a next extension would include oligopolistic competion. See Harker (1987) for a further discussion on this issue in the context of linked network-SPE models. These issues - interregional link competition and oligopolistic assumptions - are probably more consistently dealt with in a general equilibrium context (although the second one causes many difficulties there as well). Finally, this issue should then be dealt with in a multi-product situation typical of CGE, as will be discussed hereafter in the context of SPE.

$$TC = \sum_{o} \sum_{d} \int_{0}^{\sum_{s} \mu_s * t_{s,o}^{d}} c_o^d(z) dz. \tag{2.11}$$

where it is assumed that all products transported over link o-d use the same services, where the difference determined by volume, weight, and service considerations is taken care of via the μ coefficient (transforming to comparable units of transportation units). Finally, (2.12) is a generalization of (2.9):

$$TC = \sum_{i=1}^{m} \int_{0}^{\sum_{(o,d) \in V_i} \tau_o^d * \left(\sum_{s} \mu_s * t_{s,o}^{d}\right)} p_i(z) dz. \tag{2.12}$$

The main complications here are of course of a computational character, in addition to the basic separability complication mentioned above. Thompson (1989, p 58) notes that a serious specification error may occur from estimating export supply and import demand as a function of only its own price, as is most common in applications. Cross-price effects are thus omitted. The theoretically sound way is to derive exports and imports as residuals of supply and demand equations. Finally, Thompson (1989) remarks that the matrix of transportation costs, a key input in SPE models, is difficult to estimate, since reliable data on freight rates between regions for specific commodities are hard to find.

Typical solution problems and proposed solution procedures show many similarities to those related to GE modelling (see Nagurney 1987). The most used form is the quadratic programming model resulting from linear supply and demand functions, as proposed by Takayama and Judge (1971). A fixed point problem view was chosen by MacKinnon (1975 and 1976). Rowse (1981) and Florian and Los (1982) consider non-linear programming solution algorithms. Solutions via variational inequality formulations are discussed in Nagurney (1993).

Though the multi-commodity SPE problem[21] shares several elements with GE models the most important difference is that the demand and supply are all mutually independent, and as a result the commodities only 'interact' in the transportation system. From a general equilibrium point of view this means in fact that even the most extended SPE formulation still remains very much in the tradition of partial transportation and interregional trade optimization. SPE versus GE models is therefore to some extent analogous to partial planning versus general behavioural/structural models. However, a similar result is a social welfare equilibrium, although with different interpretations (because of the different components of welfare). Furthermore, important to SPE models (see the above discussion) is the assumption of separable demand (and supply) functions (i.e. cross-price elasticities zero). This is not required for GE formulations. This

---

[21] Applications of multi-commodity SPE studies are rare. Interaction between commodities occurs on the transportation network, or in terms of substitution in demand functions. See for examples Blakeslee et al. (1973) and Takayama et al. (1989). The latter includes a model for the world food economy.

assumption underlines the independent treatment of products in the SPE model. In conclusion, the interesting contribution of SPE is the spatial price character, which results from linking a multiregional system to a transportation system.

2.7 Endogenous-Price Multi-Sector (EPM) Models

The linear models dealt with here are extensions of multiregional input-output models towards endogenous-price systems. They assume fixed transportation costs, and were - unlike the standard (non-linear) GE models to be dealt with later - specifically developed to allow for easily computable solutions to large problems, e.g. in multiregional-multisectoral analyses[22]. Important questions that may be studied with these type of models are economic, multiregional effects of changes in the transportation system, reflected as a change in the costs. Such questions were already studied by Amano and Fujita (1970) in the context of multiregional input-output models.

In a number of papers, Liew and Liew (1984 1985 1991) have derived linearized endogenous-price models, where prices of primary factors are however fixed, and which had - more importantly for the discussion here - also fixed positive transportation costs (see also Roson 1991). With the exception of one paper (Liew and Liew 1991), these models are quite far away from the general equilibrium framework: no primary factory markets are included, households are not considered, and consequently model closure based on household income formation is not required. The exception mentioned is the most interesting paper, as this includes both households and production[23].

Because of the similarity of the various models employed by Liew and Liew it is sufficient to consider their most extensive model (Liew and Liew 1991). The main extension with respect to earlier papers is with regard to endogenous household demand. It is noted by Liew and Liew (1991) that the output effect of changes in transportation costs is expected to be larger in the case where households can react in addition to producers. An interesting aspect not dealt with in that model but included in another one is the consideration of multiple modes (Liew and Liew 1984)[24].

Liew and Liew (1991) consider an economy with multiple regions and industries. Producers supply outputs and demand inputs in order to maximize their profits. Consumers demand goods to maximize their utilities. A linear logarithmic function

---

[22] It is important to note that the computational gains are also obtained by using differential models for analyzing transportation cost savings. These allow for unaffected variables to be eliminated from the data-gathering process; see Liew and Liew (1991, Section 3).

[23] Liew and Liew (1985) refer to the other approach as a 'producer equilibrium model' (p. 242).

[24] The choice between modes is based on fixed transportation costs per mode, and a spatial price equilibrium in which for each mode the producer price plus transportation price (i.e., price upon delivery) equals the purchase price.

represents the production technology. It describes the production of a primary and secondary output (multiproduct technology) and the use of intermediate and primary inputs. Intermediate deliveries are allowed between firms in one region as well as in different regions.

Constant returns to scale are assumed in these models. Produced products are purchased by industries (intermediate supply), households and final users. The latter is exogenous (usual in input-output models) and may include government, investment and fixed household spending. Producer equilibrium conditions, giving relationships between factor demands, supply and prices, are derived from a planning optimization problem of maximizing the sum of all industrial profits subject to the technological constraints and a market clearing condition. Next, price frontiers are derived, including $n*m$ linear relationships among an equal number of endogenous-price variables. The exogenous parameters in these are transportation cost factors, primary factor prices, and the parameters of the production frontier. Therefore, the prices may be solved independently from the quantities supplied and demanded. Given exogenous expenditures, and based on the equilibrium price solution, the primary production output vector can be calculated. Next, secondary output, intermediate deliveries, primary inputs, and finally consumption spending are calculated. Based upon this straightforward solution procedure differential models can be derived that provide relationships between the changes in endogenous variables resulting from changes in exogenous variables or parameters. Of course, differential models could also be calculated for more structural changes reflected by changed input-output tables over time. Liew and Liew (1991, p 169) suggest to project an IO table for a future target year, by using a RAS method, and to calibrate the model parameters on this projected table.

In the remainder of this section we consider one other approach to interregional trade that does not have a direct link to one of the approaches discussed up till now. Our presentation is based on Snickars (1982, p 61) who distinguishes between two different ways: comparative cost adaptation of interregional trade flows and network models of market equilibrium. Treyz et al (1980; referred to by Snickars) uses the following endogenous-price structure in the MEPA model (see also Treyz 1993).

Trade flows of product i from region o to region d ($t_{o,i}^{d}$) are determined according to a weighted average of comparative advantage terms $CA_{o,i}^{d}$:

$$t_{o,i}^{d} = \frac{CA_{o,i}^{d}}{\sum_{region} CA_{region,i}^{d}} \tag{2.13}$$

These terms may be interpreted as a measure of the comparative advantage of region o products in region d, taking both production and transportation costs into account. They may be set equal to $((q_{i,o}+c_{i,o}^{d})/p_{i}^{d})^{wi}$ where q is the producer price of product i in region k, c is the transportation cost of product i from region o (origin) to region d (destination), and p is the composite good price of commodity i in region d (w is a product-specific parameter). p is determined as the sum of the flows of product i from all sources times their respective per unit production and transportation cost:

$$p_i^d = \sum_{region} t_{region,i}^d * (q_{region,i} + c_{region,i}^d) . \tag{2.14}$$

The producer prices are determined by using a Cobb-Douglas function for sector o (F_o) of input prices (of labour $w_{i,o}$, capital $r_{i,o}$, energy $e_{i,o}$ and intermediate material inputs p_i^o):

$$q_{i,o} = F_o\left(r_{i,o}, w_{i,o}, e_{i,o}, p_i^o\right) \tag{2.15}$$

The system of equations given by (2.13) to (2.15) can be solved simultaneously for all endogenous prices. It is thereby assumed that all transportation cost and trade flows are known, and all primary factor prices are given.

This approach makes a step towards GE modelling with transport flows, as the Liew-Liew models. However, also here transportation costs are assumed given, while transport is not being modelled explicitly in the sense of decision-making actors and transport service supply-demand interactions.

2.8 Multi-regional Computable General Equilibrium Models

2.8.1 Introduction

A Computable or Applied GE (CGE/AGE) model is an operational or empirical general equilibrium model that can be numerically solved. As Dervis et al. (1982, p 3) state these models "... are structural models designed for policy analysis and cannot be used to make unconditional projections or forecasts ... [and] ... transparency is a key characteristic that an empirical general equilibrium model must have if it is to provide a framework for policy analysis". Many of the applied models do not take the individual actors as the decision units - as would be according to the Arrow-Debreu theory - but instead adopt a multisectoral approach, which was originated by Johansen (1960).[25]

Essential to the GE approach is the notion of a general equilibrium. The applied models are usually associated with neoclassical assumptions of smooth monotonous convex functions and competitive market assumptions which allow for a single equilibrium. Instead, one may however use model structures and functional specifications for which existence and uniqueness proof are not available. One may still find solutions then (see Dervis et al. 1982, p 170), but the interpretations are uncertain. One may also extend the GE approach to include rigid prices (i.e. quantity constraints, markets do not clear), and imperfect markets (i.e. especially oligopolistic or monopolistic supply functions; markets clear). Such models are still exceptions (see Burniaux and Waelbroeck 1992).

---

[25] Bergman (1990) gives a typology and overview of CGE model applications.

The GE may be approached from several of the aforementioned modelling types: from input-output variable price models or from generalized SPE models along the lines of the extensions proposed by Florian and Los (1982). We motivate our choice for a GE model as most flexible and helpful in estimating the economy-wide effects of transportation policies. In addition, on a more fundamental level, one may regard a general equilibrium framework incomplete when it does not address the important spatial features of interactions between economic agents. The latter, which is the result of spatial dispersion of actors and markets, requires transportation. In fact one might trace a development of more and more spatial realism introduced in GE models, by distinguishing between a-spatial GE models, open (multi)country GE models, (multi)regional GE models, network-GE models and combinations of these model types with transportation formulations (see also the Chapters by Elbers and Roson in this book; Roson and Westin 1990).

Single region macro-economic oriented type of models using endogenous-price multi-sectoral models include both pure GE models (see e.g. Despotakis 1988; Liew 1984; Westin 1988) and mixed GE-Macro models (see Harrigan and McGregor 1989). Early studies in this context were done by Moses (1960) and Hansen and Tiebout (1963). An important element of this approach, and its related or derived multi-regional approach dealt with, is that the level of analysis is close to that in international trade theory. Sometimes even elements are used from such theories, depending of course on the interpretation of 'region' (see, e.g. Taylor and Black 1974; Srinivasan and Whalley 1986). A common assumption in GE models of open economies (regions or nations) is that identical products of different origins are treated as imperfect substitutes. This is the Armington assumption, which will be discussed now in more detail.[26]

2.8.2 Trade, Transport and the Armington Assumption

Many GE models are concerned with small open economies. Small means that the economy is a price-taker on international markets[27]. Open means that trade is relatively important, i.e. it requires to be dealt with in the GE formulation. Especially thc pricc-taking aspect may present some inconsistency in a GE approach where prices are typically determined endogenously[28]. The perfect

---

[26] Snickars (1982) distinguishes between 5 types of treating interregional trade in multiregional economic models: (i) CGE approaches based on comparative advantage (Heckscher-Ohlin theories) dealing only with surplus trade (no cross-hauling); (ii) linear programming to find a transport equilibrium or cost-efficient shipment schedule; (iii) interregional input-output models (economic-base, Chenery-Mose, Isard, balanced Leontief) that have rigid interdependencies; (iv) gravity and entropy models, relating trade patterns to transportation costs and including network considerations; and (v) econometric models based on accessibility and potential concepts.

[27] A common distinction is into small open, large open and closed economies. The second type is not only influenced by world prices but has a significant impact on them as well.

[28] This inconsistency follows from the theoretical result established by Samuelson (1953) that with constant returns to scale in production and with domestic and foreign foreign identical goods being perfect substitutes the domestic economy will produce only as many different kinds of tradable goods as it has primary inputs (see Despotakis and Fisher, 1988).

substitutability assumption usually employed in trade theory exaggerates the power of trade policy over the domestic price system and economic structure[29]. Such an approximation is not a workable one in a GE context. An escape from both extreme specialization (with endogenous prices) and exogenous prices (i.e. smallness) is to assume that domestic and foreign products are imperfect substitutes. This is the basic element of the Armington assumptions that will be explained in more detail below. The advantage is that the domestic economy is still a pricetaker with regard to imports[30].

Generally, for every open economy, one has to deal with substitution between inputs of the same kind supplied by different origins, both in utility and in production functions. Usually this is solved by some assumption of imperfect substition. This is the so-called Armington assumption or specification which is by now almost standard in GE models to deal with interregional or international trade, both imports and exports. The basic idea, formulated by Armington (1968 and 1969), is that commodities of the same kind coming from different regions or countries are treated as imperfect substitutes. Together with a cost minimizing objective this leads to a mix of similar inputs from different sources. For exports in a multiregional system the Armington specification involves the linking of identical goods from different (domestic) regions. These are regarded imperfect substitutes for one another. Foreign buyers are assumed to choose a mix of such regional goods that minimizes their import expenses subject to a constraint that specifies the substitution possibilities.

One way of looking at the constraint (usually a CES function is chosen to specify it) is to regard it as the 'production' of a composite good (for the domestic/foreign market in the case of imports/exports), in a sort of 'demand plant'. This allows one to keep the aggregative commodity categories across countries, but to specify product differentiation by country of origin into the structure of demand. The price of this composite good is endogenous, since it is a function of the exogenous world market price, the endogenous cost of production, and the degree of substitution between the domestic and imported/exported commodity.

With regard to exports, the world as a whole can be assumed to behave according to the Armington assumption: minimizing costs of purchase subject to a CES composite world commodity. The world demand for an exported good can then be derived to be (see Dervis et al. 1982, p 226) (no commodity indices included):

[29] An exception are a few homogeneous products where quality and brand are irrelevant.

[30] It is noteworthy that Bergman's discussion of 'M-sectors' (1991, p. 44) indicates that the Armington assumption is related to some influence of domestic producers ('as a group') on export prices.

$$E = C * \left[\frac{P_{WA}}{P_{WE}} \right]^{\gamma} \tag{2.16}$$

with P_{WA} is the aggregate world price of the specific commodity type or class, C is a constant, γ is price elasticity of export demand, and P_{WE} is the dollar price of exported products, which is equal to the production costs times the export subsidy (one plus the ad valorem rate) times the exchange rate.

The export supply is unclear. It may be derived as the difference between domestic supply and domestic demand. Alternatively, one may assume a dependence of the export to domestic production ratio on the ratio of export and domestic prices. A continuous relationship may have the shape of a logistic curve, since the ratio falls in between zero and one.

The advantage of the Armington assumption is that it avoids the extreme patterns of specialization which may arise from, e.g. linear programming formulations of trade. One may overcome such an unrealistic feature by adding constraints on minimum production and/or import/export levels. However, the Armington assumption is more general, more flexible and consistent with substitution assumptions and cost minimization objectives.

2.9 Computable General Equilibrium and Transportation Studies

Rare examples of - rather different - GE studies with regional dimensions and transportation are Buckley (1992), Elbers (1992) and Horridge and Johnson (1990). In addition to the above studies other interesting ideas on how one can develop multiregional equilibrium-transport (MET) type of models were developed in a number of papers: Roson and Vianelli (1993), Mayeres and Proost (1993), Lundqvist (1993) and Van den Bergh (1993) provide analyses in the context of the MET approach (scc Roson 1993) (see also the Chapters by Elbers and Roson in this book). Elbers uses a network approach arising from international trade models, applied to a simple infrastructure network of Nepal. Other similar studies are performed for New Zealand.

The Horridge and Johnson study is based on the well-known ORANI model for Australia (see Dixon et al. 1982). This is a Johanson type of multiregional GE model that uses a linearized form model to calculate changes in variables. The ORANI model includes 112 industries, 4 categories of final demand, and 114 commodities, so that 116*114=13,224 flows result. Each flow has its 'margin' or transaction cost. Margin commodities/services include wholesale and retail trade, and four types of transport (road, rail, sea and air). The margin costs resulting from these are indirect and related to commodity flows only. In addition, there are 116 direct (non-margin) uses, related to both freight flows (e.g. private furniture movement) and for the most part to passenger flows. In total this gives 13,224+116=13,340 users of transport.

Especially the disaggregation choices in the ORANI study with regard to transportation are of interest for our survey. The road transportation industry is disaggregated into seven types: passenger by bus above and below 100 km distances; passenger by taxi or hire car; interstate freight transport; intrastate freight transport above and below 100 km distance; and freight transport brokerage (shipping). Six users of transportation services are distinguished: far and near intermediate users; and final users capital, housholds, export and other. And twelve commodity groups include: bulky goods (minerals), local services (by their nature not transported), far primary goods (produced in remote areas), near primary goods, state-specific goods (produced and consumed in similar amounts in each state), far secondary goods (produced in one or more states and consumed throughout the country), and near secondary goods (produced near locations of consumption). For each type of commodity different transportation sectors are included. It is clear that distance is an important criterion in classifying goods and consequently transportation services and activities. The users and commodities together give a matrix of possible commodity flows. Substitution is not possible between interregional (far) and intraregional (near) transport. Elasticities of substitution are assumed equal to one between road and rail modes (assuming implicitly that a constant share of the transport budget is spent on each mode).

2.10 Comparison and Evaluation

The foregoing sections have been devoted to a discussion of various model types which were able to address the relationship between economy, space and transport: (multi)regional input-output models; (multi)regional econometric models, 'hybrid-sequential' models (i.e. combining modules of economy, land use, transport, etc.), spatial price equilibrium models, linear endogenous price models, and (multi)regional (computable) general equilibrium models. The spatial equilibrium notion was an important concept that links most of these models. From the viewpoint of transport analysis it is interesting to make a comparison between these model types, in terms of weak and strong elements. Instead of going through many details, Table 2.1 is constructed to indicate main differences, using six criteria: whether product prices in the model are endogenous; whether transport prices or costs in the model are endogenous; whetehr the the transport activity has an impact on the other activities or is only considered a derived demand; whether interactions between different sectors are included (which may provide a link to transport); whether production of transport is made explicit (providing links to inputs and costs of transport); and finally, whether solutions to the model are easily obtained.

Notice in Table 2.1 that where under hybrid-sequential models we have listed 'possible' often we have indicated an unequivocal 'yes' under computable general equilibrium models. The reason for this is that when transport is included in a multiregional GE model it should necessarily be produced, have a variable endogenous price, and the transportation sector sells its services on the market, while it buys (and competes with other sectors) on the input markets. With regard to the intersectoral linkages' criterion, for the econometric model we have listed a 'no' since it commonly it does not estimate interlinkages as such. The SPE model receives a 'no' as a result of the fixed supply functions - the model is not including production using inputs. The adjoint 'possible' for GE model indicates that it is not required to have intermediate deliveries, although a treatment of freight transport depends very much on the inclusion of such features.

Table 2.1. A comparison of different multi-regional models for estimating the economy-wide effects of transportation costs and transportation policy.

TYPE	endogenous prices	endogenous transport price/cost	effect of transport activity on economic variables	inter-sectoral linkages	production of transport included	computa-bility easy and certainly solvable
1. input-output	no	no	no	yes	no	yes
2. econometric	no	no	no	no	no	yes
3. hybrid sequential	possible	possible	no	possible	possible	no/yes
4. spatial price equilibrium	yes	no	no	no	no	no
5. linear endogenous price	yes	no	no	yes	no	yes
6. computable general equilibrium	yes	yes	yes	possible	yes	no

The category 5 models are more general than the category 1 models. The category 3, 5 and 6 models are very flexible, although 3 and 6 somewhat more than 5. Category 2 models are the most restrictive with respect to multiregional or network features related to transportation. A provisional conclusion is that the GE and 'hybrid-sequential' models are most flexible in their treatment of transport, although this is at the cost of computability features such as possibility and cost of calculating equilibrium solutions. Although one must be careful in being too general, the hybrid-sequential models are usually somewhat more easy to solve, although it may involve heuristic methods and data transferring from one to another submodel. On the other hand, the GE model is of course theoretically consistent and powerful in its interpretative content. From the perspective of the simplicity in calculating numerical solutions the linear endogenous-price models of category 5 are attractive. So it seems that on the basis of the above six criteria a choice between model types for economic-transport analysis would come down to one of these three model types. Depending on the specific topic that one wishes to study in the context of economy-space-transport a specific choice for a model type will result. Possible topics are mentioned in the concluding section.

2.11 Concluding Remarks

The main modelling approaches that one may consider for studies of economic effects of transportation policy were discussed here. Up till now integrating transportation analysis with economy-wide analysis has received relatively little attention. It is, however, important since large changes are to be expected in both quantitative and qualitative characteristics of transport. Furthermore, this implies an important role for public policies in the fields of environment, other externalities (noise, health, congestion), growth and infrastructural conditions, and public transport provision. It is clear on forehand that the general, economy-wide analysis of transport policies and strategies leaves one with an ambitious task. Various topics to be studied in combination with policy analysis, will be mentioned here to complete the discussion:

- sectoral disaggregation;
- commodity classification;
- vertical and horizontal transportation sector composition;
- spatial disaggregation: multi-regional and/or network structure;
- market organization of transportation services;
- environmental-transportation policy representation;
- data, specification, estimation, and guesstimation issues;
- numerical solution methods.

There are several transport policy issues that spatially disaggregated (multiregional or network) equilibrium models can better deal with than most other transport, regional, or macroeconomic models, notably related to:

- interactions between transport and commodity markets via the price mechanism;
- evaluation of financial instruments for transport policy;
- social welfare trade-off between instruments or consequences;
- commodity flow reactions to transport parameter changes;
- household and multisectoral effects of transport policies;
- the factors determining transport demand and supply;
- integration of externalities in both the transport sector (congestion) and economy-wide (environmental) or mixed (accidents).

Finally, it is also interesting to note that these models can be used in studies of related policy issues, such as economy-wide effects of transport market interventions, land use policies, infrastructure investment, and general environmental policy.

The main extensions to existing spatial and economic equilibrium approaches that were indicated in the foregoing can be summarized as follows:

- modelling of transport markets as part of a wider system of markets for commodities and services;
- spatial disaggregation of activities and households in a general equilibrium context;
- intersectoral freight flows in a general equilibrium context;
- transportation as an activity with specific objectives and constraints: processes with inputs and outputs; and therefore costs and profits;

- transportation decision-making at two levels in a general equilbrium context, represented by shippers and carriers, i.e. producers and organizers of transport services;
- networks and congestion (of an internal and external character) in a general equilibrium context.

Some of the areas brought together in this survey are developing rapidly. In order that the above issues can be dealt with, it is also useful that much effort is put into the integration of these areas. Hopefully, some of the viewpoints presented here may be helpful to realize that. Actual integration may then proceed on both theoretical and case study levels. Although we have indicated that some models are more suitable than others in this respect, a large degree of freedom remains in making actual choices, depending on one's interest, expertise, case study characteristics, policy questions, available data and budget.

References

Adams FG, Glickman NJ (1980) Modeling the Multiregional Economic System. D.C. Heath, Lexington, MA.

Alonso W (1964) Location and Land Use. Harvard University Press, Cambridge, MA.

Amano K, Fujita M (1970). A Long-run Economic Effect Analysis of Alternative Transportation Facility Plan: Regional and National. Journal of Regional Science 10 No 3

Anas A (1982) Residential Location Markets and Urban Transportation. Academic Press, New York.

Anas A, Duann LS (1986) Dynamic Forecasting of Travel Demand, Residential Location, and Land Development: Policy simulations with the Chicago Area Transportation - Land Use Analysis System. In: Hutchinson and M. Batty (Eds.).

Armington PS (1968) A Theory of Demand for Products Distinguished by Place of Production. IMF Staff Papers 14: 159-178

Armington PS (1969) The Geographic Pattern of Trade and the Effects of Price Changes. IMF Staff Papers 14: 176-199

Batten DF, Westin L (1990). Modelling Commodity Flows on Trade Networks. Retrospect and Prospect. In: Chatterji M, Kuenne R. New Frontiers in Regional Science. Macmillan, London, pp 135-156

Hutchinson B, Batty M (1986) Advances in Urban Systems Modelling. North-Holland, Amsterdam.

Beckmann ML, Mcguire CB, Winsten CB (1956) Studies in the Economics of Transportation. Yale University Press, New Haven, CT

Berechman J, Gordon P (1986) Linked Models of Land Use Transport Interactions. A Review. In: Hutchinson B, Batty M (Eds)

Berechman J, Small K (1988) Modeling Land Use and Transportation. An Interpretive Review for Growth Areas. Environment and Planning 20: 1285-1310

Bergh JCJM van den (1993) Formalization of Transportation in a Teneral Equilibrium Model. Disaggregation, Region, Network and Externalities. Presentation in Workshop on Transportation and Spatial CGE Models, University of Venice, Venice, Italy. In: Roson (Ed) (1993) Chapter 3. Proceedings of the Venice Workshop

Bergman L (1990) The Development of Computable General Equilibrium modeling. In: Bergman L, Jorgenson DW, Zalai E. General Equilibrium Modeling and Economic Policy Analysis. Basil Blackwell, Oxford

Bergman L (1991) General Equilibrium Effects of Environmental Policy. A CGE-Modeling Approach. Environmental and Resource Economics 1:43-61

Bolton R (1985) Regional Econometric Models. Journal of Regional Science 25: 255-283

Boyce DE, LeBlanc LJ, Chon KS (1988) Network Equilibrium Models of Urban Location and Travel Choices. A Retrospective Survey. Journal of Regional Science 28: 159-182

Buckley PH (1992) A Transportation Oriented Interregional Computable General Equilibrium Model of the United States. Annals of Regional Science 26:331-348

Burniaux, Waelbroeck (1992). CGE and Imperfect Competition Model for EC. Journal of Policy Modeling 14:65-92

Dervis K, De Melo J, Robinson S (1982) General Equilibrium Models for Development Policy. Cambridge University Press, Cambridge

Despotakis KA (1988) Energy in a Regional Economy. A Computable General Equilibrium Model for California. Journal of Environmental Economics and Management 15:313-330

Dixon PB, Parmenter BR, Sutton J, Vincent DP (1982) ORANI. A Multisectoral Model of the Australian Economy. North-Holland, Amsterdam

Dixon PB, Parmenter BR, Powell AA, Wilcoxen PJ, Pearson KR (1991) Notes and Problems in Applied General Equilibrium Economics. North-Holland, Amsterdam

Elbers C, (1992) Spatial Disaggregation in General Equilibrium Models - With an Application to the Nepalese Economy. Vrije University Press, Amsterdam

Florian M, Los M (1982) A New Look at Static Spatial Price Equilibrium Models. Regional Science and Urban Economics 12:579-597

Friesz TL, Harker PT (1985) Freight Network Equilibrium. A Review of the State of the Art. In: AF Daughety (Ed) Analytical Studies in Transportation Economics. Cambridge University Press

Friesz TL, Tobin RL, Harker PT (1983) Predictive Intercity Freight Network Models. Transportation Research 17A:409-417

Ginsburgh VA, Waelbroeck JL (1981) Activity Analysis and General Equilibrium. North-Holland, Amsterdam.

Hansen W, Tiebout C (1963) An Intersectoral Flow Analysis of the California Economy. Review of Economics and Statistics 45:409-418

Harker PT, (1987) Predicting Intercity Freight Flows. VNU Science Press. Utrecht, The Netherlands

Harrigan F, McGregor PG (1989) Neoclassical and Keynesian Perspectives on the Regional Macro-Economy. A Computable General Equilibrium Approach. Journal of Regional Science 29: 555-573

Henderson JV (1988) Urban Development. Oxford University Press, New York

Horridge JM, Johnson D (1990) The Economy-Wide Effects of Fuel-Saving Technological Improvements in Road-Freight Transport. IAESR Working Paper 2-1990, The Institute of Applied Economic and Social Research, University of Melbourne, Parkville, Australia

Isard W (1951a) Interregional Input-Output Analysis. A Model of a Space Economy. Review of Economics and Statistics 33:318-328

Isard W (1951b) Location and Space Economy. MIT Press, Cambridge, MA

Issaev B, Nijkamp P, Rietveld P, Snickars F (1982) Multiregional Economic Modeling. Practice and Prospect. North-Holland, Amsterdam

Johansen L (1960) A Multisectoral Study of Economic Growth. North-Holland, Amsterdam

Kim TJ (1979) Alternative Transportation Modes in an Urban Land Use Model. A General Equilibrium Approach. Journal of Urban Economics 6:197-215

Krutilla K (1989) Tariff Burdens and Optimal Tariffs Under Alternative Transport Cost and Market Structures. Economic Letters 31:381-386

Labys WC, Takayama T, Uri ND (Eds) (1989) Quantitative Methods for Market Oriented Economic Analysis over Space and Time. Gower Publishing Company, Brookfield, VT

Leontief W, Strout A (1963) Multiregional Input-Output Analysis. In: Barna T (Ed) Structural Interdependence and Economic Development. St. Martin's Press, N.Y.

Liew LH (1984) A Johansen Model for Regional Analysis. Regional Science and Urban Economics 14:129-146

Liew CK, Liew CJ (1984) Multi-Modal, Multi-Output, Multi-Regional Variable Input-Output Model. Regional Science and Urban Economics 14:265-281

Liew CK, Liew CJ (1985) Measuring the Development Impact of a Transportation System. A Simplified Approach. Journal of Regional Science 25: 241-256

Liew CK, Liew CJ (1991) A Multiregional, Multiproduct, Household Interactive, Variable Input-Output Model. The Annals of Regional Science 25:159-177

Lundqvist L (1981) A Dynamic Multiregional Input-Output Model for Analyzing Regional Development, Employment and Energy Use. Dept. of Mathematics, Royal Institute of Technology, Stockholm

Lundqvist L (1993) Some Experiments from Modelling Price Sensitive Substitution Processes at Urban and Multiregional Level. Presentation in Workshop on Transportation and Spatial CGE Models. University of Venice, Venice, Italy

MacKinnon JG (1975) An Algorithm for the Generalized Transportation Problem. Regional Science and Urban Economics 5:445-464

Mayeres I, Proost S (1993) The Taxation of Uniform Externalities of the Congestion Type. Presentation in Workshop on Transportation and Spatial CGE Models. University of Venice, Venice, Italy

Mills ES (1967) An Aggregative Model of Resource Allocation in a Metropolitan Area. American Economic Review 197-210

Moses LN (1955) The Stability of Interregional Trading Patterns and Input-Output Analysis. American Economic Review 45:803-832

Moses LN (1960) A General Equilibrium Model of Production. Interregional Trade, and Location of Industry. Review of Economics and Statistics 373-397

Nagurney A (1987) Computation Comparisons of Spatial Price Equilibrium Methods. Journal of Regional Science 27:55-76

Nagurney A (1993) Network Economics. A Variational Inequality Approach. Kluwer Academic Publishers, Boston

Nijkamp P, Paelinck J (1982) Operational Theory and Application of Regional Economics. 2nd edition Gower, Aldershot

Polenske K (1980) The US Multiregional Input-Output Accounts and Model. DC Heath, Lexington, MA

Rietveld P (1982) Infrastructure and Regional Development. A Survey of Multiregional Economic Models. In: Issaev B et al. (Eds)

Rietveld P (1989) Infrastructure and Regional Development. A Survey of Multiregional Economic Models. Annals of Regional Science 23:255-274

Rietveld P (1994) Transport Infrastructure and the Economy. A Survey of Approaches at the Regional Level. Paper Presented at the OECD Conference on Capital Formation and Employment. 27, 28 January, 1994, Amsterdam. Forthcoming in OECD Report

Roson R, Westin L (1990) Production and Trade in Interacting Economies. A Survey of Applied Equilibrium Models. Research Paper 90:20. Department of Economics, University of Venice, Italy

Roson R (1991) The Adjustment of Interregional Input-Output Coefficients under Heterogeneous Price Sensitivity. A Linearized Model. Annals of Regional Science 25:101-114

Roson R (Ed) (1993) Transportation and General Equilibrium Models. Proceedings of the Venice Workshop

Roson R, Vianelli P (1993) Developing a Multiregional Network CGE Model for Freight Transport Analysis. Presentation in Workshop on Transportation and Spatial CGE Models. University of Venice, Venice, Italy

Rowse J (1981) Solving the Generalized Transportation Problem. Regional Science and Urban Economics 11:57-68

Samuelson PA (1952) Spatial Price Equilibrium and Linear Programming. American Economic Review 42:283-303

Scarf H, Shoven J(Eds) (1984) Applied General Equilibrium Analysis. Cambridge University Press, Cambridge

Shoven JB, Whalley J (1984) Applied General-Equilibrium Models of Taxation and International Trade. An Introduction and Survey. Journal of Economic Literature 22:1007-51

Smith C (1989) Integrated Multiregion Models for Policy Analysis. An Australian Perspective. North-Holland, Amsterdam

Snickars F (1982) Interregional Linkages in Multiregional Economic Models. In: Issaev B, et al. (Eds)

Srinivasan TN, Whalley J (1986) General Equilibrium Trade Policy Modeling. MIT Press, Cambridge, MA

Takayama T, Judge G (1964) Equilibrium Among Spatially Separated Markets. A Reformulation. Econometrica 32 Oct

Takayama T, Judge G (1971) Spatial and Temporal Price and Allocation Models. North-Holland, Amsterdam

Takayama T, Hashimoto H, Nguyen DH, Whitacre RC (1989) Application of the Spatial and Temporal Price and Allocation Model to the World Food Economy. In: Labys et al. (Eds)

Taylor L, Black SL (1974) Practical General Equilibrium Estimation of Resource Pulls under Trade Liberalization. Journal of International Economics 4:35-38

Thompson RL (1989) Spatial and Temporal Price Equilibrium Agricultural Models. In: Labys et al. (Eds)

Treyz G (1993) Regional Economic Modeling. A Systematic Approach to Economic Forecasting and Policy Analysis. Kluwer Academic Publishers

Treyz G, Friedlaender A, Stevens B (1980) The Employment Sector of a Regional Economic Policy Simulation Model. Review of Economics and Statistics 62:63-73

Wardrop JG (1952) Some Theoretical Aspects of Road Traffic Research. Proceedings of the Institution of Civil Engineers 1:325-378

Webster FV, Bly PH, Paulley NJ (1988) Urban Land-Use and Transport Interaction ISGLUTI. Avebury, Aldershot

Westin L (1988) A Short Run Inter-Regional Equilibrium Vintage Model. Australian Journal of Regional Studies 3:3-11

II

Regions, Trade and Transport

3 Re-exports in Spatial and Temporal Price and Allocation Models

Hideo Hashimoto[1]
University of Osaka

3.1 Introduction

This paper deals with what P. A. Samuelson omitted in the text of his article on spatial equilibrium although he mentioned it in its footnote. The footnote reads as: "... cargo going from London to San Francisco is not to be treated as both an import and export of Panama."[2] However, if goods are landed at Colón and shipped again from Panama City, they are considered to be Panama's re-exports.

Re-exports are not a trivial matter. In Jordan, for example, a large portion of goods landed at Aqabah is transported to the Iraqi territory by truck. Before the Gulf crisis, re-exports had made up 16 percent of Jordan's total exports. This is because transportation costs from Iraq's major trading countries through Aqabah to Baghdad are lower than those of transporting goods to Basra on the Persian Gulf, landing there and transporting them to Baghdad by land. After the United Nations imposed an embargo on Iraq, the Iraqis' potential demand for Jordan's re-exports has increased.

The purpose of this paper is to propose the correct formulation that can incorporate re-exports into the spatial and temporal price and allocation (STPA, hereinafter) models. For this purpose, we will first examine the difficulties related to re-exports in the transportation cost minimization model, and propose its reformulation to incorporate re-exports. Then we will examine re-exports in the spatial version of the STPA models. It will also be shown that the same arguments are relevant to the temporal models. The reformulation of the spatial models to incorporate re-exports has a wider range of applications than those suggested by the term of re-exports.

---

[1]This paper is based on the author's recent article, Programming Model ni okeru Saiyushutsu ni tsuite (On Re-exports in Programming Models). In: Gendai Keizai Shakai ni okeru Shomondai (in Japanese). The main new feature of this paper is on determination of re-exports.

[2]Samuelson (1952) p.293.

3.2 Transportation Cost Minimization Model

3.2.1 Formulation of the Model

The transportation cost minimization model, initiated by Koopmans-Hitchcock,[3] is a linear programming (LP) problem. In this model, no consumption is assumed at production points, and no production is assumed at consumption points. In addition, a maximum production capability is given at each production point, while a minimum consumption requirement is given at each consumption point. Then, given unit transportation costs between all the production and consumption points, and given capacity and requirement constraints, the problem is to minimize total transportation cost. In this setting, there is no room to incorporate re-exports.

In this paper, we make two modifications on the original transportation cost minimization problem. First, we allow all the points of activities to produce and consume. This modification is needed because we often deal with large areas of production and consumption such as those in world trade models of certain commodities. (In this paper, locations of activities are termed as countries, instead of points.) Second, we allow producers to trade with consumers in their own country (such trade is here described as x_{ii}.); and define each country's demand and supply balances separately, each of which contains x_{ii}. The reason for the latter treatment is, first, to make models encompass marketing margins and second, to facilitate comparison with the STPA models.

With these assumptions, let us formulate our transportation cost minimization model as an LP problem. The notations of the model are listed below:

x_{ij} quantity transported from country i to country j.

ρ_i demand price in country i.

ρ^i supply price in country i.

$\bar{y}_i$ demand quantity in country i.

$\bar{y}^i$ supply quantity in country i.

d_{ij} unit transportation cost from country i to country j (including marketing margins).

L the Lagrangean of the optimization problem.

---

[3]Koopmans TC (1951) and Hitchcock FL (1941).

The model with N countries and one commodity can be written as follows:

$$\text{minimize} \quad \sum_{i=1}^{N}\sum_{j=1}^{N} d_{ij}x_{ij}. \tag{3.1}$$

subject to;

$$-\bar{y}_i + \sum_{j=1}^{N} x_{ji} \geq 0 \text{ for all } i. \tag{3.2}$$

$$\bar{y}^i - \sum_{j=1}^{N} x_{ij} \geq 0 \quad \text{for all } i. \tag{3.3}$$

$$x_{ij} \geq 0 \quad \text{for all } i \text{ and } j. \tag{3.4}$$

The Kuhn-Tucker conditions of this optimization problem can be written as follows:

$$\frac{\partial L}{\partial x_{ij}} = \rho_j - \rho^i - d_{ij} \leq 0 \text{ and } \frac{\partial L}{\partial x_{ij}} x_{ij} = 0 \text{ for all } i \text{ and } j. \tag{3.5}$$

$$\frac{\partial L}{\partial \rho_i} = -\bar{y}_i + \sum_{j=1}^{N} x_{ji} \geq 0 \text{ and } \frac{\partial L}{\partial \rho_i}\rho_i = 0 \text{ for all } i. \tag{3.6}$$

$$\frac{\partial L}{\partial \rho^i} = \bar{y}^i - \sum_{j=1}^{N} x_{ij} \geq 0 \text{ and } \frac{\partial L}{\partial \rho^i}\rho^i = 0 \text{ for all } i. \tag{3.7}$$

It is assumed that readers are familiar with the interpretation of the equilibrium price-quantity relations described above. Only two points of explanation would be worthwhile here. First, in this LP model, equilibrium prices are solved only in relative terms unlike the STPA models that can solve absolute prices. Second, when $i=j$, x_{ij} stands for the quantity domestically produced for domestic consumption.

3.2.2. Difficulties with Re-exports

The following numerical example is presented to exemplify a case where the problem defined above does not work properly when re-exports take place.

Numerical example 1

Quantity of Capacity and Requirement

	Production	Consumption
Country 1	1,350	200
Country 2	600	450
Country 3	400	1,700

Unit Transportation Cost

from \to	Country 1	Country 2	Country 3
Country 1	3	15	45
Country 2	15	2	22
Country 3	45	22	1

Note that: $t_{13} > t_{12} + t_{23} - t_{22}$

that is, the 'Pythagorean' relation does not hold; thus, it is expected that country 2 imports and exports at the same time.

The solutions of this numerical example are as follows:

Prices

	Supply Price	Demand Price
Country 1	0	3
Country 2	23	15
Country 3	44	45

Quantity Transported

from \to	Country 1	Country 2	Country 3
Country 1	200	450	700
Country 2			600
Country 3			400

Note: total transportation cost = 52,450

Consider country 2. There are two prominent features in the solutions. First, there is no trade between producers and consumers (i.e., $x_{22}=0$); in other words, all the imports (450 units) are consumed and all the goods produced domestically (600 units) are exported. There are no re-exports. Second, supply price exceeds demand price. If importers (or consumers) in country 2 import and resell to producers (or exporters) in the same country, they would earn profits. This suggests that the solutions are not in equilibrium.

3.2.3 Reformulation of the Model to Incorporate Re-exports

In numerical example 1, why was the equilibrium solution not obtained? Let us examine the quantity balance of the demand side, i.e., equation (3.2). Equation (3.2) means that country i's consumption should not exceed the sum of that country's production and its imports from all other countries. If there is no re-export, this balance equation does not cause any problem; however, if re-exports are taken into account, the correct quantity balance is that country *i's* consumption should not exceed the sum defined above *minus re-exports*. Denote country *i's* re-exports as REX_i. Then, the quantity balance of the demand side should be:

$$-\bar{y}_i + \sum_{j=1, j\neq i}^{N} x_{ji} + (x_{ii} - REX_i) \geq 0 \quad \text{for all } i.$$

Similarly, the correct quantity balance of the supply side should be:

$$\bar{y}^i - \sum_{j=1, j\neq i}^{N} x_{ij} - (x_{ii} - REX_i) \geq 0 \quad \text{for all } i.$$

Note that both balance equations include x_{ii} - REX_i; therefore, denote x'_{ii} as follows:

$$x_{ii} - REX_i = x'_{ii} \quad \text{for all } i.$$

Verbally, x'_{ii} is equal to domestic production for domestic consumption minus re-exports. Furthermore, note that the sign of x'_{ii} is free, while x_{ii} is constrained to be non-negative.

Coming back to numerical example 1, we can see that, without REX_i, re-exports must be included in x_{ii}. In this example, the quantity domestically produced for domestic consumption is at most 450 units. Then, when re-exports take place, x_{ii} starts to decline from 450 units, but it is constrained to be non-negative; therefore, re-exports cannot exceed 450 units.[4] When re-exports reach that amount, x_{ii} becomes zero,[5] and the 'proper' relationship between demand and supply prices

---

[4]If production and consumption in country 2 are 1,301 units and 1,151 units respectively, re-exports are 1,150 units and $x_{22}=1(>0)$. The problems regarding re-exports do not occur.

[5]Now we know that x_{ii} is the quantity domestically produced for domestic consumption minus reexports. The solution: $x_{22} = 0$ in numerical example 1 means that (a) both domestic production for domestic consumption and reexports are 450 units, or (b) both of them are zero, or (c) somewhere between (a) and (b).

is lost. As shown in numerical example 2 below, the correct equilibrium re-exports of this problem should be at least 700 units.

In consideration of the argument above, we can revise the model in order to incorporate re-exports in the following manner:

Replace x_{ii} in (3.2) and (3.3) by x'_{ii} for all i.
Sign of x'_{ii} is free for all i.

Accordingly, (3.5) of the Kuhn-Tucker conditions should be replaced by:

$$\frac{\partial L}{\partial x_{ij}} = \rho_j - \rho^i - d_{ij} \leq 0 \ \text{ and } \ \frac{\partial L}{\partial x_{ij}} x_{ij} = 0 \ \text{ for all } i \text{ and } j \ (i \neq j).$$

$$\frac{\partial L}{\partial x'_{ii}} = \rho_i - \rho^i - d_{ii} = 0 \ \text{ for all } i.$$

Numerical example 2

A numerical example is made by using all the parameters in numerical example 1 on the new formulation of the problem.

The solutions are as follows:

Prices

	Supply Price	Demand Prices
Country 1	0	3
Country 2	13	15
Country 3	34	35

Quantity Transported

from \to	Country 1	Country 2	Country 3
Country 1	200	1,150	
Country 2		-700	1,300
Country 3			400

Note: total transportation cost = 45,450

Consider country 2 again. It now imports 1,150 units and consumes 450 units. This means that re-exports range from 700 units to 1,150 units. If 700 units are re-exported, all consumption of country 2 is supplied from imports, and all goods it produces are exported; in other words, there is no trade between producers and consumers within that country.[6] If 1,150 units are re-exported, 150 units produced

[6] In this case, $x_{22} = 0$ *and* $REX_2 = 700$.

in country 2 are exported, and the remaining 450 units are consumed domestically.[7] In either case, demand price is equal to the sum of supply price and marketing margin. Total transportation cost is smaller than in numerical example 1. The model works well with re-exports.

However, this model still has a problem, that is, it does not tell you how many units are re-exported, although the range of re-exports is solved. This problem will be discussed in detail in section 3.2.5.

3.2.4 Remarks on the Revised Model

Four remarks on the revised model are presented here. The first is on the interpretation of $d_{ii}\,x'_{ii}$ in the objective function. In the original model, x_{ii} is restricted to be non-negative; so $d_{ii}\,x_{ii}$ is always non-negative. The meaning of $d_{ii}\,x_{ii}$ is obvious. By contrast, in the revised model, $d_{ii}\,x'_{ii}$ becomes negative when $x'_{ii} < 0$. How can we interpret a negative $d_{ii}\,x'_{ii}$ in the objective function?

In consideration that $x'_{ii} = x_{ii} - REX_i$, let us write the objective function (i.e., total transportation cost; denote it as TC, hereinafter) as follows:

$$TC = \sum_{i=1}^{N} \sum_{j=1, j \neq i}^{N} d_{ij} x_{ij} + \sum_{i=1}^{N} d_{ii}(x_{ii} - REX_i)$$

The unit cost d_{ij} represents the sum of transportation cost from exporters in country i to importers in country j plus the marketing margin between importers and consumers (or exporters) in country j. The marketing margin regarding re-exports, which are the flows directly from importers to exporters, has been already included in d_{ij}. Therefore, if only $d_{ii}\,x_{ii}$ is included in TC, there would be double-counting of the marketing margin by $d_{ii}\,REX_i$. This is the reason why $d_{ii}\,REX_i$ is subtracted in TC.[8] In the TC equation, REX_i can be greater than x'_{ii}; then, $d_{ii}\,x'_{ii}$ becomes negative.

Second, as a result of a variable with a free sign in the objective function, the problem may become unbounded. Particularly, when

$$d_{ii} > d_{ij} \quad \text{for some } i \text{ and } j.$$

the problem may end up with 'unboundedness'. (For example, replace $d_{11} = 3$ by $d_{11} = 70$ in numerical example 2; then, the problem becomes unbounded.) This 'unboundedness' problem can be solved by a model presented in the following section.

Third, we have discussed the cases that the 'Pythagorean' relation with respect to transportation cost does not hold, and have shown that re-exports may occur in such cases. In addition, if there is an upper limit on a trade flow, such as:

$$x_{ij} \leq \bar{x}_{ij} \quad (\bar{x}_{ij} \text{ is given}).$$

re-exports may occur.

---

[7]In this case, $x_{22} = 450$ and $REX_2 = 1{,}150$.

[8]Note that all the marketing margins are assumed to be equal.

To understand this, let us denote the Lagrangean multiplier of the constraint above as θ_{ij}. Then, in the 'Pythagorean' relation, by using $d_{ij}+\theta_{ij}$ instead of d_{ij}. Even when the 'Pythagorean' relation with respect to transportation cost holds as:

$$t_{13} < t_{12} + t_{23} - t_{22}$$

but if the flow from country 1 to country 3 is restricted, and

$$t_{13} + \theta_{13} > t_{12} + t_{23} - t_{22}$$

holds, country 2 may re-export to country 3.

Fourth, the revised model has solved only a range of re-exports. The quantities of re-exports are not determined uniquely. In the following section, we will discuss a model that can determine the quantities of re-exports.

3.2.5. Determination of Re-exports

The objective of this section is to develop a model that can determine re-exports. Before moving to such a model, let us look at the trade routes within a country. They can be shown schematically below:

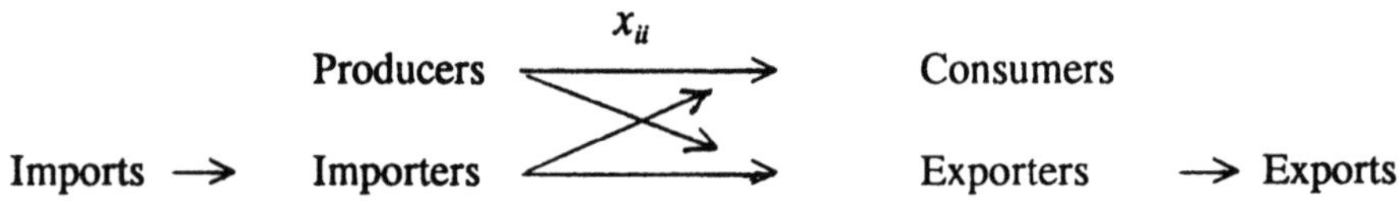

Within a country, there are four distinct trade routes. Re-exports are considered to be flows from importers to exporters. It is natural to assign different marketing margins to different routes. In the models discussed in the previous section, however, the marketing margins are assumed to be equal in all the routes. As a result, consumers as well as exporters are indifferent to sources of their goods: whether imported or domestically-produced goods. This results in indeterminacy of re-export quantities.

To solve for re-export quantities, it is necessary to distinguish the marketing margins of various trade routes within a country. The notations of the model that can determine re-exports are as follows:

x_{ii} — domestic production for domestic consumption in country i, i.e., flow from producers to consumers in country i.

$x_{i(m,c)}$ — flow from importers to consumers in country i.

$x_{i(p,x)}$ — flow from producers to exporters in country i.

$x_{i(m,x)}$ — flow from importers to exporters in country i.

t_{ij} — unit transportation cost (excluding any marketing margin) from country i to country j.

$d_{i(p,c)}$ — marketing margin between producers and consumers in country i.

$d_{i(m,c)}$ marketing margin between importers and consumers in country i.

$d_{i(p,x)}$ marketing margin between producers and exporters in country i.

$d_{i(m,x)}$ marketing margin between importers and exporters in country i.

The model can be rewritten as follows:

$$\text{minimize} \sum_{i=1}^{N} \sum_{j=1, j \neq i}^{N} t_{ij} x_{ij}$$

$$+\sum_{i=1}^{N} [d_{i(p,c)} x_{ii} + d_{i(m,c)} x_{i(m,c)} + d_{i(p,x)} x_{i(p,x)} + d_{i(m,x)} x_{i(m,x)}]$$

subject to;

$$-y_i + x_{ii} + x_{i(m,c)} \geq 0 \quad \text{for all } i.$$

$$y^i - x_{ii} - x_{i(p,x)} \geq 0 \quad \text{for all } i.$$

$$(\sum_{j=1}^{N} x_{ji} - x_{ii}) - x_{i(m,c)} - x_{i(m,x)} \geq 0 \quad \text{for all } i.$$

$$-(\sum_{j=1}^{N} x_{ij} - x_{ii}) + x_{i(p,x)} + x_{i(m,x)} \geq 0 \quad \text{for all } i.$$

$$x_{ij(i \neq j)} \geq 0 \quad \text{for all } i \text{ and } j.$$

$$x_{ii} \geq 0,\ x_{i(m,c)} \geq 0,\ x_{i(p,x)} \geq 0,\ x_{i(m,x)} \geq 0 \ for \text{ all } i.$$

Numerical example 3

To illustrate how this model works, we developed two cases of the model described above by using the following unit transportation costs (excluding marketing margins) and marketing margins, while keeping the quantity constraints of numerical example 1 intact.

Unit Transportation Cost[9]

from \to	Country 1	Country 2	Country 3
Country 1	0	10	41
Country 2	10	0	19
Country 3	41	19	0

[9]The numbers of this matrix are exactly equal to those in the unit transportation cost matrix of numerical example 1. Apparent differences are due to the fact that the former excludes marketing margins, while the latter includes them.

Unit Marketing Margin in Country 2[10]

	Case A	Case B
$d_{2(p,c)}$	2.0	1.9
$d_{2(m,c)}$	1.9	2.0
$d_{2(p,x)}$	1.9	2.0
$d_{2(m,x)}$	2.0	2.0

A part of the solutions are shown below;

Flows in Country 2

	Case A	Case B
x_{22}	0	450
$x_{2(m,c)}$	450	0
$x_{2(p,x)}$	600	150
$x_{2(m,x)}$	700	1,150

In Case A, where the marketing margins between importers and consumers as well as those between producers and exporters are less than those between producers and consumers, domestic consumption is filled with imports, and all the goods produced domestically are exported. There is no trade between producers and consumers. In Case B, where the marketing margin between producers and consumers is less than those of other routes, part of domestically produced goods is used for domestic consumption, and the remainder is exported. In both cases, the re-exported quantities are determined uniquely.

As for the 'unboundedness' problem, from the model structure it is clear that the problem does not occur in this model.

The cost of expansion of the model is that the number of variables has increased sharply. This may add unmanageable complexity to model builders, particularly when the model is applied to the STPA models.

[10]All the marketing margins in countries 1 and 3 are assumed to be 3.0 and 1.0 respectively.

3.3 The STPA Models

3.3.1 Difficulties with Re-exports

It is assumed that readers are familiar with the STPA models; therefore, the formulation of the STPA models is not presented here. The readers should refer to the Takayama-Judge book.[11] Here, we simply present a case where the STPA models do not work properly when re-exports take place.

Numerical example 4

The parameters of this model are presented in a tableau form in Table 3.1. Note that:

$$d_{13} > d_{12} + d_{23} - d_{22}$$

thus, it is expected that country 2 imports and exports at the same time.

The solutions are as follows:

	Quantity		Price	
	Supply	Demand	Supply	Demand
Country 1	20.00	11.00	6.00	6.50
Country 2	5.38	9.00	12.77	7.00
Country 3	27.54	32.92	15.27	15.77

Quantity Transported

from \to	Country 1	Country 2	Country 3
Country 1	11.00	9.00	
Country 2			5.38
Country 3			27.54

Note: net social quasi-welfare = 474.83

Consider country 2 again. This country imports 9.00 units from country 1 and exports 5.38 units to country 3. All imports are consumed and all the goods produced domestically are exported. There is no trade between producers and consumers. Supply price exceeds demand price. This example has the same problem as numerical example 1.

[11] Takayama T, Judge GG (1971); particularly, Chapter 7.

Table 3.1. Parameters of Numerical Example 5

	y_1	y_2	y_3	y^1	y^2	y^3	x_{11}	x_{12}	x_{13}	x_{21}	x_{22}	x_{23}	x_{31}	x_{32}	x_{33}	ρ_1	ρ_2	ρ_3	ρ^1	ρ^2	ρ^3	
y_1	.5															1						−12
y_2		1															1					−16
y_3			.25															1				−24
y^1				.25															−1			1
y^2					2															−1		2
y^3						.5															−1	1.5
x_{11}																−1			1			.5
x_{12}																	−1		1			1
x_{13}																		−1	1			10
x_{21}																−1				1		1
x_{22}																	−1			1		.5
x_{23}																		−1		1		3
x_{31}																−1					1	10
x_{32}																	−1				1	3
x_{33}																		−1			1	.5
ρ_1	−1						1			1			1									
ρ_2		−1						1			1			1								
ρ_3			−1						1			1			1							
ρ^1				1			−1	−1	−1													
ρ^2					1					−1	−1	−1										
ρ^3						1							−1	−1	−1							

Note: six parameters of the upper-left cells are the slope coefficients of the inverse demand and supply functions. Six numbers of the furthest-right column represent the intercepts of the inverse demand and supply functions. The following nine numbers of the same column are unit transportation costs.

3.3.2 Re-formulation of the Model to Incorporate Re-exports

As in the case of the transportation cost minimization model, the reformulation can be made by replacing x_{ii} with x'_{ii} and making the sign of x'_{ii} free.

Numerical example 5

A numerical example was made by using all the parameters of numerical example 4 on the new formulation. The solutions are as follows:

	Quantity		Price	
	Supply	Demand	Supply	Demand
Country 1	32.07	4.96	9.02	9.52
Country 2	3.76	5.98	9.52	10.02
Country 3	21.04	45.93	12.02	12.52

Quantity Transported

from \to	Country 1	Country 2	Country 3
Country 1	4.96	27.11	
Country 2		-21.13	24.89
Country 3			21.04

Note: net social quasi-welfare = 541.06

Consider country 2 again. The solutions indicate that country 2 imports 27.11 units from country 1 and exports 24.89 units to country 3. Re-exports in country 2 range from 21.13 units to 24.89 units, and domestic production for domestic consumption ranges from 0 units to 3.76 units. Demand price is equal to the sum of supply price and the marketing margin. The net social quasi-welfare is larger than in numerical example 4. The model works well.

3.3.3 'Irregular' Cases

One reason that, in the STPA models, the quantity balances are specified on both demand and supply sides separately is to encompass the 'irregular' cases.[12] The irregular cases mean those in which demand (or supply) quantity in a certain country is zero, and the market equilibrium price is not located on the demand (or supply) function. Therefore, we must check whether the revised formulation can work on the 'irregular' cases.

Numerical example 6

For that purpose, a new example is made by reducing the constant term in country 2's inverse demand function by 10 and increasing the constant term in country 3's inverse supply function by 15. The solutions are as follows:

	Quantity		Price	
	Supply	Demand	Supply	Demand
Country 1	37.81	2.10	10.45	10.95
Country 2	4.48	0.00	10.95	11.45
Country 3	0.00	40.19	13.45	13.95

[12]See Takayama T, Judge GG (1971), pp 135-137.

Quantity Transported

from \to	Country 1	Country 2	Country 3
Country 1	2.10	35.71	
Country 2		-35.71	40.19
Country 3			

Note: net social quasi-welfare = 401.74

Country 2's consumption and country 3's production are zero. Country 2 imports 35.71 units from country 1. Because there is no consumption in country 2, all of its imports are re-exported to country 3. In addition, all of its production (4.48 units) is also exported to country 3. Country 3's consumption is totally supplied from its imports. The relationship between supply and demand prices are correct. The revised model formulation works properly in the 'irregular' case.

3.3.4 Application to Temporal Equilibrium Models

The STPA temporal equilibrium models can be considered to be those with 're-exports.' To explain this, the simplest version of the temporal model is written, and its notations are listed below:

$z_{(t-1,t)}$ carry-over quantity from the (t-1)th period to the t-th period.

y_t demand quantity in the t-th period.

y^t supply quantity in the t-th period.

x_t quantity consumed in the t-th period from the quantity produced in the t-th period.

The quantity balances are as follows:

$$-y_t + z_{(t-1,t)} + x_t \geq 0 \quad \text{for all } t.$$

$$y^t - z_{(t,t+1)} - x_t \geq 0 \quad \text{for all } t.$$

In this formulation, goods carried over from the (t-1)th to the t-th period, i.e., $z_{(t-1,t)}$, and those from the t-th to the (t+1)th period, i.e., $z_{(t,t+1)}$, can be considered as the t-th period's 'import' from the (t-1)th period and its 're-export' to the (t+1) period. The quantity consumed in the t-th period from the quantity produced in the same period, i.e., x_t plays the same role as x_{ii} in the spatial models, which generates the same difficulty as in the spatial models.[13] The difficulty can be solved by freeing the sign of x_{ii} as in the spatial models.

The same treatment can also apply to the spatial and temporal version of the STPA models.

3.4 Conclusions

This paper explains that, in a broader framework of programming models, the STPA models (both spatial and temporal) do not work properly when re-exports and inventories are taken into account, and proposes the reformulation in which the non-negativity condition on x_{ii} is relaxed (in section 3.2.3). With this formulation, the model works properly. However, it still has some problems. One of them is that the revised model cannot determine re-export quantities uniquely. To cope with these difficulties, we present an alternative model (in section 3.2.5).

In practical uses of any model, we face a familiar trade-off between simplicity and completeness. In the STPA models, simplicity is particularly important because of the dimensionality problem. That is, when we develop an N country model, we must deal with N x N trade flows, and the model dimension increases sharply. It is imperative to reduce the number of variables. In actual applications, therefore, the relaxation of the non-negativity condition of x_{ii} would be a practical way to incorporate re-exports, without being bogged down in the awesome dimensionality problem.

Acknowledgements

The author is grateful to Professor Takashi Takayama of the University of Western Australia, Professor Ronald Duncan of the Australian National University and Dr. Jerome Kreuser of the World Bank for their comments on an earlier draft of this paper. Any remaining errors and omission are the sole responsibility of the author.

---

[13]Guise JB (1979) first presented a case where the STPA models did not work properly, by pointing out the problems generated by the quantity balances regarding inventories. He was rejoined by Takayama T, Hashimoto H (1979). Hashimoto (1985) generalized the difficulties of both the spatial and temporal versions of the STPA models, by attributing the cause of the difficulties to the treatment of 're-exports". He (1994) further examined the difficulties regarding re-exports in a general framework of programming models.

References

Guise JWB (1979) An Expository Critique of the Takayama-Judge Models of Interregional and Intertemporal Market Equilibrium. Regional Science and Urban Economics 9:83-95

Hashimoto H (1985) Transshipment and Inventories in the Takayama-Judge Model. Regional Science and Urban Economics 15:365-381

Hashimoto H (1994) 'Programming Model ni okeru Saiyushutsu ni tsuite (On Re-exports in Programming Models). In: Oishi Yasuhiko Kyoju Koki Kinen Ronbunshu. Gendai Keizai Shakai ni okeru Shomondai, (Various Issues in the Contemporary Economies and Societies. A Collection of Treatises in Honor of Professor Yashuhiko Oishi) Toyo Keizai Shimposha. Tokyo 1:217-245

Hitchcock FL (1941) Journal of Mathematics and Physics. Massachusetts Institute of Technology 20:224-230

Koopmans TC (1951) Activity Analysis of Production and Allocation, John Wiley, New York

Kuhn H, Tucker A (1951) Nonlinear Programming. In: Newman J (Ed) Proceedings of the Second Berkeley Symposium. University of California Press pp 481-492

Samuelson PA (1952) Spatial Price Equilibrium and Linear Programming. American Economic Review 42:283-303

Takayama T, Hashimoto H (1979) A Rejoinder. Regional Science and Urban Economics 9:97-104

Takayama T, Judge GG (1971) Spatial and Temporal Price and Allocation Models. North-Holland, Amsterdam

4 Le Châtelier Principle and the Flow Sensitivity of Spatial Commodity Models

Walter C. Labys
West Virginia University

C. W. Yang
Clarion University of Pennsylvania

4.1 Introduction

One of the major tools of spatial interaction analyses is, needless to say, the spatial equilibrium model developed from (1) the linear programming transportation problem by Hitchcock (1941), Kantorovich (1942) and Koopmans (1949), and (2) an analogue of Kirchhoff's law of electric circuits by Enke (1951). Samuelson (1952) showed that the Enke problem could be converted into a mathematical programming model. In 1964 Takayama and Judge reformulated it into an operationally efficient concave quadratic programming model. Since then, there has been a proliferation of theoretical advances that improve and extend the basic linear programming transportation and spatial equilibrium models. These include their transformation into a new algorithm by Liew and Shim (1978) and Nagurney (1986); Thore's formulation with income (1982); sensitivity analyses of model performance by Yang and Labys (1981, 1982), Irwin and Yang (1982), Chao and Friesz (1984), Daffermos and Nagurney (1984), and Tobin (1984, 1987); computational evaluations by Meister, Chen and Heady (1978), and Nagurney (1987b); iterative solution methods by Pang and Chan (1982), and Irwin and Yang (1982); a linear complementarity formulation by Uri (1976) and Takayama and Uri (1983); sensitivity analysis of the complementarity problem by Tobin (1984) and Yang and Labys (1985); a sufficient condition for the complementarity problem by Smith (1984); a flow dependent spatial equilibrium problem by Smith and Friesz (1985); nonlinear complementarity models by Irwin and Yang (1983) and Friesz, et al. (1983); variational inequality solutions by Pang and Chang (1982), Daffermos (1983), Harker (1984), Tobin (1986) and Nagurney (1987a); a path dependent spatial equilibrium model by Harker (1986); transhipment and location selection problem by Tobin and Friesz (1983, 1984); and evaluations of the spatial equilibrium models and Maxwell-Boltzmann entropy models by Yang (1990). For the detailed description of the advances in the spatial equilibrium models, readers are referred to Labys, Takayama and Uri (1989) and Labys and Yang (1991).
Numerous empirical applications of spatial equilibrium models have been witnessed in different industries: livestock feed by Fox (1951), eggs by Judge (1956), apples by Fuchs, et al. (1974), oil by Kennedy (1974), pork by Martin,

etc. (1975), electricity by Uri (1976), policy evaluations in the corn belt by Taylor and Frohberg (1977), Appalachian steam coal by Labys and Yang (1980), citrus produce by Chao and Friesz (1984), world aluminum market by Newcomb, et al. (1989), aluminum and bauxite by Takayama and Hashimoto (1989), European community feed grain model by Peeters (1990), and the impact analysis of the spatial ad valorem coal taxes by Yang and Page (1993). The successful applications of the spatial models are of unequalled importance, especially in energy and agricultural industries (see Labys (1989) and Thompson (1989)). A comparative sensitivity analysis, i.e., the impact of a change in policy parameters on decision variables, of the relative performance of these models, nonetheless, is lacking. For instance, one of the earlier applications of the linear programming transportation problem can be found in Henderson's US coal model (1954), while many more recent applications are found in the market-oriented Takayama-Judge formulation (e.g., natural gas model by Sohl (1984)). In addition, two of the earlier applications by Fox (1951) and Judge (1956) are fixed-supply spatial equilibrium models. Hence, an investigation into the performance of different spatial equilibrium methodologies is warranted, especially in the wake of the growing deregulation of various industries. For instance, various energy resource industries captured in the spatial equilibrium models by Uri (1976 and 1989) and natural gas models by Sohl (1984), and Yang and Labys (1985) been subject to deregulation. Given the various forms of regulations found in the socialist economies of Eastern Europe, the transition to free markets in terms of more privatization in these countries extends the application of spatial models in a deregulation context. In the era characterized by an explosion of free-trade arrangements (e.g., the European Common Market, the North American Free Trade Agreement, and the Pacific Rim Economic Region) theoretical investigations of the flow sensitivities and welfare changes play an important role; and they can be examined in the context of the famed Le Châtelier principle.

Our modest objective of this paper is to compare a family of spatial equilibrium models via a well-known theorem in thermodynamics. In what follows, we present the linear programming, fixed-demand spatial equilibrium, and the Takayama-Judge spatial equilibrium models; we then analyze the sensitivity properties of the model and invoke the Le Châtelier principle. Policy implications are drawn in the conclusions section, based on comparisons of previous steam coal model analysis by Labys and Yang (1980) and Yang and Labys (1982).

4.2 A Sensitivity Analysis of Flow Variables

In order to facilitate comparisons of the family of spatial models, we first present three different spatial models: (1) the Hitchcock-Kantorovich-Koopman linear programming transportation model in which all consumption (demand) levels and production (supply) capacities are fixed points, (2) the fixed-demand spatial equilibrium model in which only consumption levels are predetermined, and (3) the Takayama-Judge spatial and temporal price allocation (STPA) model in which both demand and supply relations are endogenous to the market system (see Takayama-Judge 1971). The fixed-demand spatial equilibrium model is employed since (1) the demand relations are more difficult to estimate, especially in the energy market, and (2) deregulations may well occur via the demand price (hence, quantity demanded).[1]

Given m supply regions and n consumption regions, the Hitchcock-Kantorovich-Koopmans linear programming transportation model can be represented as a minimization problem:

$$\underset{x_{ij} \epsilon R_+}{\text{Minimize}} \quad TC = \sum_{i \epsilon I} \sum_{j \epsilon J} t_{ij} x_{ij} \tag{4.1}$$

$$\text{subject to} \quad \sum_{i \epsilon I} x_{ij} \geq D_j \ \forall \ j \epsilon J \tag{4.2}$$

$$\sum_{j \epsilon I} x_{ij} \geq S_i \ \forall \ i \epsilon I \tag{4.3}$$

$$x_{ij} \geq 0 \ \forall \ ij \epsilon \ (I \times J) \tag{4.4}$$

where x_{ij} = volume of shipment from supply region i to demand region j

t_{ij} = unit transportation cost from supply region i to demand region j

D_j = fixed level of consumption in demand region j

S_i = fixed level of production capacity in supply region i

I, J are positive integer sets of $(1 \ldots m)$ and $(1, \ldots n)$

$I \times J$ is the Cartesian product of I and J

R_+ represents a set of all the nonnegative real numbers

On the other hand, the objective of the Takayama-Judge STPA model is to maximize the sum of additive economic surpluses:

$$\underset{y_j, x_i, x_{ij}}{\text{maximize}} \ NSP = \sum_{j \epsilon J} a_{jy_j} - \tfrac{1}{2} \sum_{j \epsilon J} b_j y_j^2 - \sum_{i \epsilon I} c_{ix_i} - \tfrac{1}{2} \sum_{i \epsilon I} d_i x_i^2 - \sum_{i \epsilon I} \sum_{j \epsilon J} t_{ij} x_{ij} \tag{4.5}$$

[1]Coal supply equations can be estimated using geological and engineering data (see Newcomb and Fan, 1980). However, the effect of deregulation can also be demonstrated by dealing with the supply price variables e.g. well price of the natural gas.

$$\text{subject to } \sum_{i\epsilon I} x_{ij} \geq y_i \ \forall \ j\epsilon J \tag{4.6}$$

$$\sum_{j\epsilon J x_{ij}} \leq x_i \ \forall \ i\epsilon I \tag{4.7}$$

$$x_{ij} \geq 0 \ \forall \ ij\epsilon \ (I \times J) \tag{4.8}$$

where a_j = price intercept of the inverse demand function in region j

b_j = the slope of the inverse demand function in region j

c_i = price intercept of the inverse supply function in region i

d_i = the slope of the inverse supply function in region i

y_j = quantity demanded in region j

x_i = quantity supplied in region i

Clearly, demand and supply functions are linear and separable or f: $R_+ \rightarrow R_+$ such that if y_j is a nonnegative real number in R (or $y_j \epsilon R_+$), then there must be a corresponding nonnegative demand price in region j or $p_j \epsilon R_+$. Similarly g: $R_+ \rightarrow R_+$ such that for $x_i \epsilon R_+$, there exists a nonnegative supply price in region i (or $p_i \epsilon R_+$). The welfare evaluated along the demand and supply equations ($p_j = a_j - b_j z_j$ and $p_i = c_i + d_i x_i$) can therefore be measured and used in the objective function. Similarly, the fixed-demand spatial equilibrium model can be represented as:

$$\underset{x_i, s_{ij}}{\text{Maximize}} \ NSPP = -\sum_{i\epsilon I} c_{ix_i} - \tfrac{1}{2}\sum_{i\epsilon I} d_{ix_i^2} - \sum_{i\epsilon I} \sum_{j\epsilon J} t_{ij} x_{ij}$$

which is equivalent to the minimization of the sum of production and transportation costs or

$$\underset{x_i, X_{ij}}{\text{Minimize}} \ TPC = \sum_{i\epsilon I} c_i x_i + \tfrac{1}{2}\sum_{i\epsilon I} d_i x_i^2 + \sum_{i\epsilon I} \sum_{j\epsilon J} t_{ij} x_{ij} \tag{4.9}$$

$$\text{subject to} \quad \sum_{i\epsilon I} x_{ij} \geq D_j \ \forall \ j\epsilon J \tag{4.10}$$

$$\sum_{j\epsilon J} x_{ij} \leq x_i \ \forall \ i\epsilon I \tag{4.11}$$

$$x_{ij} \geq 0 \ \forall \ ij\epsilon \ (I \times J) \tag{4.12}$$

Note that D_j is predetermined and hence y_j is dropped out of equation (4.5). Obviously, if the actually observed consumption level in the *jth* region (D_j) equals the optimum consumption level $\bar{y}_j$ and actually observed supply level in the *ith* region (S_i) equals the optimum production level $\bar{x}_i$, the spatial equilibrium and linear programming transportation (LPT) models are equivalent. Similarly, if $D_j = \bar{y}_j$ for all $j \epsilon J$, the spatial equilibrium and fixed-demand spatial equilibrium models become equivalent. Within this framework, we examine two characteristics in the LPT model. First, there cannot exist more than $m+n-1$ numbers of positive commodity flows in the optimum solution set. A typical linear programming theorem states that the number of positive-valued decision variables cannot exceed the number of independent constraints. In the LPT model, one of the constraints is redundant since

$$\sum_{i \epsilon I} \sum_{j \epsilon J} x_{ij} = \sum_{j \epsilon J} \sum_{i \epsilon I} x_{ij}$$

and hence the maximum number of positive-valued x_{ij}'s is $m+n-1$. If the number of $x_{ij} > 0$ falls short of $m+n-1$, it is commonly referred to as a case of degeneracy which leads to isolated trade patterns (Samuelson 1952). Second, a small or moderate change in the unit transportation cost in general will not alter the optimum volumes of commodity flows *within a given base*, i.e. trade pattern. An excellent description of these properties can be found in Gass (1985).

Likewise, for a set of well-behaved linear demand and supply functions the number of optimum positive commodity flows in the STPA model cannot exceed $m + n - 1$. This is due to the fact that for a given set of optimum y_j and x_i, the SE model can be converted to an LPT model (Silberberg 1970). This property also holds for the fixed-demand spatial equilibrium model for the same reason. However, unlike that in the LPT model, the impact of a change in the unit transportation cost t_{ij} on $m+n-1$ positive flows is in general significant (see Silberberg 1970; Irwin and Yang 1982; Daffermos and Nagurney 1984; and Tobin 1987). In the case of the fixed-demand spatial equilibrium model, the sensitivity result is lacking in the literature. As shown in Tables 4.1 and 4.2, we employ the data from Labys and Yang (1980). The simulation results of the Appalachian Steam Coal Market are reported in Tables 4.3 through 4.5. In what follows we propose the following preposition for flow sensitivity.

Proposition #1: In the fixed-demand spatial equilibrium model with $m \leq n$ at least one optimum commodity flow is not responsive to a change in unit transportation cost(s).

Proof: Since the number of positive flows cannot exceed $m+n-1$, there must be at least one positive $x_{ij} - D_j$. For instance for the case of $m = n = 7$ shown in Table 4.4, the maximum number of positive flows is 13. Obviously, at least one positive flow must equal a demand requirement.[2] A perusal of Table 4.4 indicates that there are two such flows, i.e., $x_{51} = D_1$ and $x_{64} = D_4$. As a consequence, $\partial x_{64}/\partial t_{ij} = \partial D_4/\partial t_{ij} = 0$ since D_1 and D_4 are constant.

---

[2]The inputs for the perturbation analysis are reported in Tables 4.1 and 4.2. A software package by Cutler and Pass (1971) was employed for computation.

Many applications of the spatial equilibrium models are formulated with $m = n$ (e.g. Campbell et al 1980; Labys and Yang 1980; and Newcomb et al 1989). The sensitivity result in Proposition 1 applies to all fixed-demand STPA models with $m \leq n$. Exactly the same line of reasoning applies to all fixed-supply STPA models with $m \geq n$. The sensitivity result indicates that in the case of the fixed-demand or fixed-supply spatial equilibrium models, part of the optimum commodity flows possesses the property of the LPT model, i.e., $\partial x_{ij}/\partial t_{lk} = 0$ for a given trade pattern; and the rest of the flows are responsive to the changes in some unit transportation cost(s) or $\partial x_{ij}/\partial t_{lk} \neq 0$. In the next section, we explore the Le Châtelier's principle in the three models.

Table 4.1 Unit Transportation Cost (cents per 10^6 BTU)

FROM TO	1	2	3	4	5	6	7
1	19.1	22.2	20.2	20.5	21.2	24.3	26.8
2	14.5	18.3	13.5	16.9	17.6	20.4	23.3
3	13.2	13.8	11.8	16.6	20.3	16.7	19.8
4	18.8	19.8	19.6	22.5	21.8	17.1	18.1
5	15.1	13.3	13.7	12.6	17.2	13.4	15.7
6	21.1	22.5	21.8	19.2	14.1	16.9	7.0
7	21.0	24.0	20.9	16.5	14.7	13.6	14.3

Source See Mutchler, etc and Labys and Yang (1980)

Table 4.2 Estimates on regional demand and supply

Demand Intercepts	=	(69.8, 86.4, 57.5, 49.7, 61.3, 47.4, 61.6)
Demand Slopes	=	(-462.73, -33.03, -20.5, -40.08, -12.71, -17.79, -22.13)
Supply Intercepts	=	(27, 26.3, 25, 30.4, 23, 27, 28.2)
Supply Slope	=	(4.46, 4.221, 14.259, 25.07, 24.35, 2.6028, 59.62)

*Quantities are measured in 10^{15} BTU, see Labys and Yang (1980)

Table 4.3 The spatial equilibrium optimum solution (in 10^{15} BTU) of the Appalachian Steam Coal Market

From / To	PA MD	OH	Northern WV	Southern WV	VA	East KY TN	AL	Total Demand
New England					(0.042)			(0.042)
	[0.0416]							[0.0416]
					((0.0415))			((0.0415))
Mid-Atlantic	(1.025)		(0.195)					(1.22)
	[0.960]		[0.264]					[1.224]
	((1.034))		((0.185))					((1.22))
IN-MI		(0.304)	(0.336)					(0.64)
	[0.386]	[0.26]						[0.645]
	((0.29))	((0.348))						((0.638))
North Central						(0.053)		(0.053)
						[0.051]		[0.051]
						((0.052))		((0.052))
Ohio Valley		(0.708)		(0.035)		(0.628)		(1.371)
		[0.602]		[0.039]		[0.723]		[1.364]
		((0.732))		((0.036))		((0.61))		((1.368))
South Central					(0.081)		(0.139)	(0.22)
					[0.075]		[0.140]	[0.215]
					((0.129))		((0.0891))	((0.219))
South Atlantic					(0.138)	(0.653)		(0.792)
					[0.191]	[0.597]		[0.788]
					((0.092))	((0.698))		((0.79))
Total Supply	(1.025)	(1.012)	(0.531)	(0.035)	(0.262)	(1.334)	(0.139)	
	[1.002]	[0.987]	[0.524]	[0.039]	[0.266]	[1.371]	[0.14]	
	((1.034))	((1.022))	((0.534))	((0.036))	((0.263))	((1.35))	((0.089))	

( ) = Spatial Equilibrium Solution. Value of objective function = 58.857
[ ] = Spatial Equilibrium Solution with Δt_{25} = 0.2. Value of objective function = 58.726
(( )) = Spatial Equilibrium Solution with Δt_{76} = 3. Value of objective function = 58.515

Table 4.4 The fixed-demand spatial equilibrium optimum solution (in 10^{15} BTU) of the Appalachian Steam Coal Market

From To	PA MD	OH	Northern WV	Southern WV	VA	East KY TN	AL	Total Demand
New England	[0.042]				(0.042) ((0.042))			(0.042)
Mid-Atlantic	(1.025) [0.963] ((1.037))		(0.195) [0.258] ((0.184))					(1.22)
IN-MI		(0.304) [0.374] ((0.29))	(0.336) [0.266] ((0.35))					(0.64)
North Central						(0.053) [0.053] ((0.053))		(0.053)
Ohio Valley		(0.708) [0.616] ((0.734))		(0.035) [0.039] ((0.037))		(0.628) [0.716] ((0.6))		(1.371)
South Central					(0.081) [0.080] ((0.131))		(0.139) [0.141] ((0.0893))	(0.22)
South Atlantic					(0.138) [0.186] ((0.091))	(0.653) [0.606] ((0.701))		(0.792)
Total Supply	(1.025) [1.004] ((1.037))	(1.012) [0.99] ((1.024))	(0.531) [0.525] ((0.535))	(0.035) [0.039] ((0.037))	(0.262) [0.266] ((0.264))	(1.334) [1.375] ((1.354))	(0.139) [0.141] ((0.083))	

( ) = Fixed-Demand Spatial Equilibrium Solution. Value of objective function = 183.677.
[ ] = Fixed-Demand Spatial Equilibrium Solution with $\Delta t_{25} = 0.2$. Value of objective function = 183.81.
(( )) = Fixed-Demand Spatial Equilibrium Solution with $\Delta t_{76} = 3$. Value of objective function = 184.019.

Table 4.5 The linear programming optimum solution (in 10^{15} BTU) of the Appalachian Steam Coal Market

From To	PA MD	OH	Northern WV	Southern WV	VA	East KY TN	AL	Total Demand
New England	 [0.0415]				(0.042) ((0.042))			(0.042)
Mid-Atlantic	(1.025) [0.984] ((1.025))		(0.195) [0.237] ((0.195))					(1.22)
IN-MI		(0.304) [0.346] ((0.304))	(0.336) [0.294] ((0.336))					(0.64)
North Central						(0.053) [0.053] ((0.053))		(0.053)
Ohio Valley	(0.708) [0.666] ((0.703))		(0.035) [0.035] ((0.035))		(0.628) [0.67] ((0.628))		(1.371)	
South Central				(0.082) [0.082] ((0.082))		(0.139) [0.139] ((0.139))		(0.22)
South Atlantic				(0.138) [0.18] ((0.138))	(0.653) [0.612] ((0.653))		(0.792)	
Total Supply	(1.025)	(1.012)	(0.531)	(0.035)	(0.262)	(1.334)	(0.139)	

( ) = Linear Programming Solution. Value of objective function = 58.782.
[ ] = Linear Programming Solution with Δt_{25} = 0.2. Value of objective function = 58.92.
(( )) = Linear Programming Solution with Δt_{76} = 3. Value of objective function = 58.515.

4.3 The Le Châtelier's Principle

According to the physical law of thermodynamics, the decision variables respond less sensitively to parametric perturbations (e.g., temperature) as the system is gradually constrained. The system tends to adjust itself to a new equilibrium by counteracting the effect of the change as far as possible. This law, known as Le Châtelier's principle, has been applied to economics by Samuelson (1949, 1960, and 1972) and Silberberg (1971, 1974, and 1978). However, this principle has not been employed thus far in the study of the class of STPA models. We state the following proposition:

Proposition #2: The sensitivity of the positive flows (those not equal to demand requirements) in response to the change of its corresponding unit transportation cost in the STPA model is greater than that of the fixed-demand spatial equilibrium model, whose corresponding flow sensitivity is in turn greater than that of the linear programming transportation model. For the flows that equal the demand requirements, the flow sensitivity in the spatial equilibrium model is greater than that of both the fixed-demand spatial equilibrium and the linear programming transportation models.

Proof: Let indices rs and uv denote the trade patterns whose flows do not equal and equal demand requirements respectively such that $rs \cup uv = i'j' \in (I \times J)$, where $i'j'$ is the flow pattern with $m+n-1$ positive optimum flows. For a given optimum set of y_j, x_i, x_{ij}, as more y_j's are constrained in the spatial system, the objective functions or the net social payoff (NSP) changes less rapidly since more y_j's are fixed and cannot adapt to the change in a unit transportation cost. As a result,

$$abs\left[\frac{\partial^2 NSP}{\partial t_{rs}^2}\right] > abs\left[\frac{\partial^2(-NSPP)}{\partial t_{rs}^2}\right] > abs\left[\frac{\partial^2(-TC)}{\partial t_{rs}^2}\right] \tag{4.13}$$

where $NSP = f(y_j, x_i, x_{ij}, t_{ij})$ such that $f\colon R_{++}^{m+n+(m+n-1)} X\ R_{++} \rightarrow R_{++}$

$NSPP = g(x_i, x_{ij}, t_{ij})$ such that $g\colon R_{++}^{m+(m+n-1)} X\ R_{++} \rightarrow R_{++}$

$TC = h(x_{ij}, t_{ij})$ such that $h\colon R_{++}^{m+n-1} X\ R_{++} \rightarrow R_{++}$

R_{++}^{m+n-1} the positive orthant of $m+n$-1 dimensional Euclidean space and *abs* denotes absolute value.

Via the envelope theorem (Silberberg 1978; Dixit 1990), it can be shown that[3]

$$\frac{\partial NSP}{\partial t_{rs}} = x_{rs}^{*} < 0 \tag{4.14}$$

which is, in turn, determined by model parameters. It follows immediately from equation (4.13) that

$$abs\left[\frac{\partial x_{rs}^{*}}{\partial t_{rs}}\right] > abs\left[\frac{\partial x_{rs}^{**}}{\partial t_{rs}}\right] > abs\left[\frac{\partial x_{rs}^{***}}{\partial t_{rs}}\right] \tag{4.15}$$

where x_{rs}^{*}, x_{rs}^{**}, and x_{rs}^{***} are optimum positive commodity flows in the spatial and temporal price allocation, fixed-demand and linear programming transportation models.

The Le Châtelier principle of equation (4.15) states that in the STPA model, the system will react to a change in the unit transportation cost in such a way that the impact of the shock is minimized. In this case, the flow of the Takayama-Judge STPA model will change its volume more than that of its corresponding fixed-demand spatial equilibrium model which will, in turn, change its volume more than that of the linear programming transportation model.

The impact of a change in the unit transportation cost t_{ij} on the flow x_{uv} can be shown as

$$abs\left[\frac{\partial x_{uv}^{*}}{\partial t_{ij}}\right] > abs\left[\frac{\partial x_{uv}^{**}}{\partial t_{ij}}\right] = abs\left[\frac{\partial x_{uv}^{***}}{\partial t_{ij}}\right] = 0 \tag{4.16}$$

since x_{uv} equals fixed demand requirement D_v for some v.

To verify the result implied by the Le Châtelier principle in the three spatial models, we employ the data and structure of models previously constructed for the Appalachian coal industry by the authors, e.g., see Labys and Yang (1980), Yang and Labys (1985), and Yang (1990). As shown in Table 4.1, this model features coal trade flows from seven Appalachian coal supply regions to seven major demand regions, the latter representing electrical utilities or the final consumers.

The adaptation necessary for comparing the models is that we purposely make (1) $D_j = y_j$ for all $j \epsilon J$ in the case of the fixed-demand spatial equilibrium model; and (2) $D_j = y_j$ for all $j \epsilon J$, and $S_i = x_i$ for all $i \epsilon I$ in the case of the linear programming transportation model. Within this framework, the three unperturbed spatial models generate an identical set of $m+n-1 = 13$ positive optimum flows (see Tables 4.3, 4.4, and 4.5).

We first increase transportation costs t_{25} by 0.2 cents per million BTU in all three models. The consequent reductions in x_{25} are 0.708 - 0.602 = 0.106, 0.708 - 0.616 = 0.092, and 0.708 - 0.666 = 0.042 (unit in 10^{15} BTU). These changes occur in the Takayama-Judge spatial and temporal price allocation, fixed-demand spatial equilibrium and linear programming transportation models respectively.

---

[3]The proof via the indirect objective functions appears in standard microeconomic texts, i.e. Silberberg (1978), Dixit (1990) or Varian (1992). Hence, it is not repeated here. However, we must rule out the beginning and ending points in each basic solution set since the indirect objective function of the LPT model is piece-wise linear (stepfunction) with respect to the change in the unit transportation cost.

Note that in this case, the trade pattern has undergone changes, i.e., x_{51} disappears while x_{11} emerges. Hence, $\Delta x_{25}/\Delta t_{25} < 0$ even for the linear programming transportation model. Further, the decreases in the volume of x_{64} for the three spatial models are 0.053 - 0.051 = 0.002, 0, and 0 respectively. This is because $x_{64} = y_4 = D_4$ in the linear programming and fixed-demand spatial equilibrium models is exogenous to the system. Consequently, $\Delta x_{64}/\Delta t_{25} = \Delta D_4/\Delta t_{25} = 0$ since both D_4 and t_{25} are constant. Hence, both results of the Le Châtelier principle or equations (4.15) and (4.16) are verified even with the altered trade patterns.

In the case of an unchanged trade pattern, we increase t_{76} by 3 cents per million BTU; the consequent reductions in x_{76} are thus 0.139 - 0.0891 = 0.0499, 0.139 - 0.0893 = 0.0497, and 0 respectively. The reductions of x_{51} and x_{64} for the spatial equilibrium model are 0.042 - 0.0415 = 0.0005 and 0.053 - 0.052 = 0.001 respectively. Hence, the values of x_{51} and x_{64} under the fixed-demand spatial equilibrium and linear programming transportation models remain unchanged, since $x_{51} = y_1$ *and* $x_{64} = y_4$, which are consistent with equation (4.16). The same principle can be derived in the case of a degeneracy solution, i.e., there exists less than m+n-1 positive flows. Consequently, there are more than one independent trade patterns. Within each trade pattern, exactly the same results of the Le Châtelier principle equations (4.15) and (4.16) apply.

4.4 Concluding Remarks

Past sensitivity analyses involving, for example, reciprocity conditions or perturbation designs of the STPA model have been extensively investigated. However, comparative sensitivity analyses employing the Le Châtelier principle are lacking. In this paper, we investigate this principle for spatial models. Our results show that the flow sensitivity of the STPA model is greater than that of the fixed-demand spatial equilibrium model, which in turn is greater than that of the linear programming transportation model. For an additional set of flow variables where demand requirements are met, the flow sensitivity equals zero within a given trade pattern under the fixed-demand spatial equilibrium and linear programming transportation models. The same results are also valid for a degenerate spatial system that consists of two or more independent subsystems. By exactly the same reasoning, the results of the Le Châtelier principle can be extended to the fixed-supply STPA model.

The Le Châtelier principle can also be applied to welfare levels. For a decreasing unit transportation cost, the Le Châtelier principle implies that the welfare level will increase most rapidly in the Takayama-Judge STPA model, followed by the fixed-demand spatial equilibrium model, and by the linear programming transportation model. It would appear from Proposition 2 that via the envelope theorem, the flow sensitivity also follows that order. The results of this paper also imply that the welfare level will be higher in the case of price deregulation (hence the quantity demanded) coupled with a decreased transport or transaction costs in large scale STPA models.

It comes as no surprise that the results from applying the Le Châtelier principle yield prospects for modeling the gradual deregulation in the utility, airline, energy, banking and other industries which have recently occurred globally. With technological advances (i.e. substantial potential reductions in the costs perhaps due to advances in the electronic transmission of information or the high speed of transport modes) economic welfare will improve most significantly in a spatial system with the least constraints. This is because, according to the Le Châtelier principle, the least constrained system can absorb exogenous shocks most effectively in terms of the welfare measure via flexible adjustment processes. The growing trend of industry deregulation, demise of the socialistic economies, and advances in technology suggest that future modeling efforts be concentrated on a spatial system with the least constraints.

References

Campbell T C, Hwang MJ, Shahrokh F (October 1980) Spatial Equilibrium in the US Coal Industry. Energy Economics 230-236

Chao GS, Friesz TL (1984) Spatial Price Sensitivity Analysis. Transportation Research 18:423-440

Cutler L, Pass DS (1971) A Computer Program for Quadratic Mathematical Models to be Used for Aircraft Design and Other Applications Involving Linear Constraints. Rand Report R-516-PR. Santa Monica

Daffermos S (1983) An Iterative Scheme for Variational Inequalities. Mathematical Programming 26:40-47

Daffermos S, Nagurney A (1984) Sensitivity Analysis For the General Spatial Economic Equilibrium Problem. Operations Research 32:1069-1086

Dixit AK (1990) Optimization in Economic Theory. 2nd edition. Oxford University Press, Oxford

Enke S (1951) Equilibrium Among Spatially Separated Markets. Solution by Electric Analogue. Econometrica 19:40-47

Fox KA (1951) A Spatial Equilibrium Model of the Livestock Feed Economy. Econometrica 19:547-566

Friesz TL, Tobin RL, Smith TE and Harker PT (1983) A Nonlinear Complementarity Formulation and Solution Procedure for the General Derived Demand Network Equilibrium Problem. Journal of Regional Science 23:337-359

Fuchs, HW, Farrish ROP, Bohall W (1974) A Model of the U.S. Apple Industry. A Quadratic Interregional Intertemporal Activity Analysis Formulation. American Journal of Agricultural Economics 56:739-750

Gass SI (1985) Linear Programming Methods and Applications. 5th edition. McGraw-Hill Inc. New York

Harker PT (1984) A Variational Inequality Approach for the Determination of Oligopolistic Market Equilibrium. Mathematical Programming 105-111

Harker PT (1986) Solving the Path-Dependent Spatial Price Equilibrium Problem. Socio-Economic Planning Sciences 20:299-310

Harker PT (1988) Dispersed Spatial Price Equilibrium. Environment and Planning A 20:353-368

Henderson JM (1958) The Efficiency of the Coal Industry. An Application of Linear Programming Harvard University Press, Cambridge

Hitchcock FL (1941) Distribution of a Product from Several Sources to Numerous Localities. Journal of Mathematics and Physics 21:224-230

Irwin CL, Yang CW (1982) Iteration and Sensitivity for a Spatial Equilibrium Problem with Linear Supply and Demand Functions. Operations Research 30: 319-335

Irwin CL, Yang CW (1983) Iteration and Sensitivity for a Nonlinear Spatial Equilibrium Problem. In: Lecture Notes in Pure and Applied Mathematics. Vol 85 Fiacco A (Ed) Marcel Dekker Inc. New York 91-107

Judge GG (1956) A Spatial Equilibrium Model for Eggs. Connecticut Agricultural Experiment Station. Storrs, CT

Kantorovich LV (1942) On the Translocation of Masses. Doklady Adad. Nauk SSR, Vol. 37. Translated in Management Science 5(1) (1958)

Kennedy M (1974) An Economic Model of the World Oil Market. The Bell Journal of Economics and Management Science 5:540-577

Koopmans TC (1949) Optimum Utilization of the Transportation System Econometrica 71:136-146

Labys WC (1982) Measuring the Validity and Performance of Energy Models. Energy Economics July:159-168

Labys WC (1989) Spatial and Temporal Price and Allocation Models of Mineral and Energy Markets. In: Quantitative Methods for Market Oriented Economic Analysis Over Space and Time. Labys WC, Takayama T, Uri ND (Eds) Gower Publishing Company Limited, Aldereshot, UK

Labys WC, Takayama T, Uri ND (1989) Quantitative Methods for Market Oriented Economic Analysis Over Space and Time. Gower Publishing Company Limited, Aldershot, UK

Labys WC, Yang CW (1980) A Quadratic Programming Model of the Appalachian Steam Coal Market. Energy Economics 2:86-95

Labys WC, Yang CW (1991) Advances in the Spatial Equilibrium Modeling of Mineral and Energy Issues. International Regional Science Review 14(1):61-94

Lewis HR, Papadimitriou CH (1981) Elements of the Theory of Computation. Prentice-Hall Inc, Englewood Cliffs NJ

Liew CK, Shim JK (1978) A Spatial Equilibrium Model. Another View. Journal of Urban Economics 5:526-534

Martin LJ, Zwart AC (1975) A Spatial and Temporal Model of the North American Pork Sector for the Evaluation of Policy Alternatives. American Journal of Agricultural Economics 57:55-66

Meister AD, Chen CC, Heady EO (1978) Quadratic Programming. Models Applied to Agricultural Policies, Ames. Iowa State University Press

Nagurney A (1987a) Competitive Equilibrium Problems, Variational Inequalities and Regional Sciences. Journal of Regional Science 503-517

Nagurney A (1987b) Computational Comparison of Spatial Price Equilibrium Methods. Journal of Regional Science 55-76

Nagurney A (1986) An Algorithm for the Single Commodity Spatial Price Equilibrium Problem. Regional Science and Urban Economics 16:573-588
Newcomb R, Fan J (1980) Coal Market Analysis Issues. EPRI Report EA-1575. Electric Power Research Institute, Palo Alto, CA
Newcomb R, Reynolds SS, Masbruch TA (1989) Changing Patterns of Investment Decision-Making in World Aluminum. Resource and Energy 1:261-297
Peng JS, Chan D (1982) Iterative Methods for Variational and Complementarity Problems. Mathematical Programming 24:284-313
Peeters L (1990) A Spatial Equilibrium Model of the EC Feed Grain Sector European Review of Agricultural Economics 17(4):365-386
Samuelson PA (1949) The Le Châtelier Principle in Linear Programming. Rand Corporation Monograph, Santa Monica
Samuelson PA (1952) Spatial Price Equilibrium and Linear Programming. American Economic Review 42:283-303
Samuelson PA (1960) An Extension of the Le Châtelier Principle. Econometrica April:368-379
Samuelson PA (1972) Maximum Principles in Analytical Economics. American Economic Review 62(3):249-262
Silberberg E (1970) A Theory of Spatially Separated Markets. International Economic Review, 11(2):334-348
Silberberg E (1971) The Le Châtelier Principle as a Corollary to a Generalized Envelope Theorem. Journal of Economic Theory 3:146-155
Silberberg E (1974) A Revision of Comparative Static Methodology in Economics or How to do Economics on the Back of an Envelope. Journal of Economic Theory 7:159-172
Silberberg E (1978) The Structure of Economics. A Mathematical Analysis, McGraw-Hill Book Company, New York
Smith TE (1984) A Solution Condition for Complementarity Problems. With an Application to Spatial Price Equilibrium. Computer and Operations Research 61-69
Smith TE, Friesz TL (1985) Spatial Market Equilibrium with Flow Dependent Supply and Demand. The Single Commodity Case. Regional Science and Urban Economics 15:181-218
Sohl JE (1984) An Application of Quadratic Programming to the Deregulation of Natural Gas. In: Spatial Price Equilibrium. Advances in Theory. Computation, and Application. Harker P (Ed) Springer-Verlag, New York 196-207
Takayama T, Hashimoto H (1989) A Comparative Study of Linear Complementarity Programming Models and Linear Programming Models in Multi-Region Investment Analysis. Aluminum and Bauxite. In: Quantitative Methods for Market-Oriented Economic Analysis Over Space and Time. Labys WC, Takayama T, Uri NB (Eds) Gower Publishing Company, Aldershot, UK 129-163
Takayama T, Judge G (1964) Equilibrium Among Spatially Separated Markets. A Reformulation. Econometrica 32:510-524
Takayama T, Judge G (1971) Spatial and Temporal Price and Allocation Models, Amsterdam. North-Holland Publishing Company
Takayama T, Uri ND (1983) A Note on Spatial and Temporal Price and Allocation Modeling. Regional Science and Urban Economics 455-470

Taylor CR, Frohberg KK (1977) The Welfare Effects of Erosion Controls Banning Pesticides, and Limiting Fertilizer Application in the Corn Belt. American Journal of Agricultural Economics 59:25-35

Thompson RL, Spatial and Temporal Price Equilibrium Agricultural Models. In: Quantitative Methods for Market Oriented Economic Analysis Over Space and Time. Labys WC, Takayama T, Uri ND (Eds) Gower Publishing Company Limited, Aldershot, UK 49-65

Thore S (1982) The Takayama-Judge Spatial Equilibrium Model with Endogenous Income. Regional Science and Urban Economics 12:351-364

Tobin RL, Friesz TL (1982) Formulating and Solving the Spatial Price Equilibrium Problem with Transshipment in Terms of Arc Variables. Journal of Regional Science 2(2):187-198

Tobin RL, Friesz TL (1984) A New Look at Spatially Competitive Facility Location Models. In: Spatial Price Equilibrium. Advances in Theory, Computation and Application, Harker P (Ed) Springer-Verlag, New York 1-19

Tobin RL (1984) General Spatial Price Equilibria. Sensitivity Analysis for Variational Inequiality and Nonlinear Complementarity Formulations. In: Spatial Price Equilibrium. Advances in Theory, Computation and Application. Harker P (Ed) Springer-Verlag, New York 158-195

Tobin RL (1986) Sensitivity Analysis for Variational Inequalities. Journal of Optimization Theory and Applications. January:191-204

Tobin RL (1987) Sensitivity Analysis for General Spatial Price Equilibrium. Journal of Regional Science 77-102

Uri ND (1989) Linear Complementarity Programming. Electric Energy as a Case Study. In: Quantitative Methods for Market-Oriented Economic Analysis Over Space and Time. Labys WC, Takayama T, Uri ND (Eds) Gower Publishing Company, Aldershot, UK 165-188

Uri ND (1976) A Spatial Equilibrium Analysis of Electrical Energy Pricing and Allocation. American Journal of Agricultural Economics 58:653-662

Yang CW, Labys WC (1985) Stability of Appalachian Coal Shipments Under Policy Variations. The Energy Journal 2(3):111-128

Yang CW, Labys WC (1982) A Sensitive Analysis of the Stability Property of the QP Commodity Model. Journal of Empirical Economics 7:93-107

Yang CW, Labys WC (1985) A Sensitivity Analysis of the Linear Complementarity Programming Model. Appalachian Steam Coal and Natural Gas Markets. Energy Economics 7(3):145-152

Yang CW (1990) An Evaluation of the Maxwell-Boltzmann Entropy Model of the Appalachian Steam Coal Market. Review of Regional Studies 20(1):21-29

Yang CW, Page WP (1993) Sensitivity Analysis of Tax Incidence in a Spatial Equilibrium Model. Annals of Regional Science August:1-17

5 Spatial Equilibrium in Labour Markets

Martin J. Beckmann
Brown University

5.1 Introduction

Farmers live on their land or in villages adjacent to their acreage. All over the world small merchants reside in or above their working premises, both for security and for convenience. In the early days of industrialization workers were settled next to the factories, particularly when steady working habits had yet to be developed. It is modern urbanized mankind that has escaped from this trap of living where one works. Cheap transportation and above all the mobility of the automobile have enabled modern man to choose where to live from a typically wide array of residential areas. Even when housing markets are tight or non-functioning, say because of rent control, one is not compelled to work right next door. Commuting between residence and workplace is thus a normal pattern of modern life.

This paper explores some of the economic questions arising from this common fact. How is labour from dispersed sources to be allocated efficiently among workplaces in different locations? How are residential locations to be compared and valued with regard to their availability of attractive working opportunities? What incentives have firms or labour to locate so as to reduce or even minimize expensive commuting costs to the economy, to pose just a few economic questions? They require a special type of spatial equilibrium analysis (3.3).

5.2 Model

We consider first an efficient spatial allocation of labour. The objective to be maximized is the total product of labour minus commuting cost. The production technologies considered are of the fixed coefficient type: firm k (the firm in location k) may employ labour of a given type at constant marginal productivity v_k up to a (employment) capacity limit c_k. Our activity variable is x_{ik}, the number of commuters (of this type of labour) from residences i to firms k. Let r_{ik} be the commuting cost, a measure of distance, from i to k (and return).

$$\max_{x_{ik}\geq 0} \sum_k v_k \sum_i x_{ik} - \sum_{ik} r_{ik} x_{ik} \tag{5.1}$$

The constraints are as follows:

Availability of workers in residential location i

$$\sum_k x_{ik} \leq a_i \tag{5.2}$$

employment capacity of firm k

$$\sum_i x_{ik} \leq c_k. \tag{5.3}$$

The system (5.1), (5.2), (5.3) represents a linear program. It is always feasible, eg, with zero commuting $x_{ik} \equiv 0$.

5.3 Analysis

We now study the Koopmans efficiency conditions (Koopmans, 1951). Consider the Lagrange function

$$\begin{aligned} L = \sum_{ik} (v_k - r_{ik}) x_{ik} + \sum_k \lambda_k (a_i - \sum_k x_{ik}) \\ + \sum_k \mu_k (c_k - \sum_i x_{ik}) \end{aligned}$$

It involves dual variables or efficiency prices

$\lambda_i \quad i = 1,\ldots m$ and

$\mu_k \quad k = 1,\ldots n$

associated with residential and firm locations respectively. The Koopmans efficiency conditions are obtained by maximizing L with respect to $x_{ik} \geq 0$ and minimizing with respect to $\lambda_i \geq 0$, $\mu_i \geq 0$.

$$x_{ik} \left\{ \begin{matrix} = \\ \geq \end{matrix} \right\} 0 \Leftrightarrow v_k - r_{ik} - \lambda_i - \mu_k \left\{ \begin{matrix} < \\ = \end{matrix} \right\} 0 \tag{5.4}$$

$$\lambda_i \begin{Bmatrix} = \\ \geq \end{Bmatrix} 0 \Leftrightarrow \sum_k x_{ik} \begin{Bmatrix} < \\ = \end{Bmatrix} a_i \tag{5.5}$$

$$\mu_i \begin{Bmatrix} = \\ \geq \end{Bmatrix} 0 \Leftrightarrow \sum_i x_{ik} \begin{Bmatrix} < \\ = \end{Bmatrix} c_k \tag{5.6}$$

The right part of (5.4) may also be written

$$\lambda_i = \max_k (v_k - \mu_k - r_{ik}) \tag{5.4a}$$

or

$$\mu_k = \max_i (v_k - r_{ik} - \lambda_i) \tag{5.4b}$$

The two equations (5.4a) and (5.4b) represent two possible institutional arrangements for implementing competition in spatial labour markets. In (5.4a) the market sets wage rates $w_k = v_k - \mu_k$ in the various work places. Workers then choose employers by comparing net wages after commuting costs and taking the best offer

$$\lambda_i = \max_k (w_k - r_{ik}) \tag{5.4c}$$

where

$$w_k = v_k - \mu_k$$

is the market wage. This is in effect a sellers' market, most likely to emerge in the event of a labour shortage

$$\sum_k c_k > \sum a_i.$$

Alternatively, competition or government may fix workers' net earnings λ_i and employers then seek out labour at lowest marginal cost.

$$v_k - \mu_k = \min_i (\lambda_i + r_{ik}) \tag{5.4d}$$

and thus maximize rent

$$\mu_k = \max_i (v_k - \lambda_i - r_{ik}) \\ = v_k - \min_i (\lambda_i + r_{ik}) \tag{5.4b}$$

This is the appropriate scenario for a buyer's market with unavoidable unemployment since

$$\sum_k c_k < \sum a_i.$$

When unemployment is widespread, ie occurring in all residential locations

$$\sum_k x_{ik} < a_i$$

then

$$\lambda_i \equiv 0$$

which must be interpreted as net wages λ_i being at the subsistence or legal minimum level.

With λ_i uniform the employers objective becomes that of hiring so as to minimize total commuting cost, which means hiring only from the nearest residential locations.

To implement this, employers may provide the transportation of workers to their plants in company buses, say, a practice common in India and other countries of the Third World.

In any event competition determines both λ_i and μ_k and they are consistent provided (5.4) holds implying both (5.4a) and (5.4b). The set of firm locations k in which labour's net wage is maximized is the labour market area of residential location i. Conversely, the set of residential locations i where labour will maximize employers rent μ_k is the labour supply area of firm k. Observe, however, that when aggregate labour demand and supply are matched

$$\sum_i a_i = \sum_k c_k$$

the levels of employers rent μ_k and labour net wage λ_i are no longer uniquely determined since a (positive or negative) constant C may be added to all λ_i provided it is subtracted from all μ_k. (This leaves a range of indeterminacy for C limited by the requirement that all $\lambda_i \geq 0$, $\mu_{ik} \geq 0$.)

$$0 \leq \min_i \lambda_i - C$$
$$0 \leq \min_k \mu_k + C$$

or

$$\max_{k} - \mu_k \leq C \leq \min \lambda_i .$$

Notice also that in the balanced case just considered the v_k are irrelevant since the allocation problem is equivalent to the Koopmans transportation problem.

$$\min_{x_{ik} \leq 0} \sum_{ik} r_{ik} x_{ik} \tag{5.8}$$

such that

$$\sum_{k} x_{ik} = a_i$$
$$\sum_{i} x_{ik} = c_k$$

5.4 Residential Choice

In the medium run workers of every type j may choose among the available residential locations i and then reconsider their places of employment k. Assume demand for housing by worker type j to be h_j (square meters of housing space) and let a_i now denote the available housing space in residential location i. Let x_{ji} workers of type j choose to live in i under a housing availability constraint

$$\sum_{j} h_j x_{ji} \leq a_i \tag{5.9}$$

When several types of labour are considered, the production levels of the firms k become important, since they determine the (joint) demand for the various types j of labour. Let x_{jik} be the flow of labour j to firm k and let x_k be its output and b_k value added (excluding labour cost) per unit output. Then the object of an efficient spatial allocation of labour is

$$\max_{x_k > 0, x_{jik} > 0} \sum_{k} b_k x_k - \sum_{j,ik} r_{ik} x_{j,ik} \tag{5.10}$$

subject to firms' capacity limits

$$0 \leq x_k \leq c_k \tag{5.11}$$

and

$$\sum_{i} x_{jik} \geq a_{jk} x_k \tag{5.12}$$

$$\sum_{j,k} h_j x_{jik} \leq a_i \tag{5.9}$$

This is once more a feasible linear program, one feasible solution being $x_k \equiv 0$, $x_{jik} \equiv 0$. Its Lagrange function is

$$\begin{aligned} L = \sum_k b_k x_k - \sum_{j,ik} r_{ik} x_{jik} + \sum_{jk} \omega_{jk}\Big(\sum_i x_{jik} - a_{jk} x_k\Big) \\ + \sum_k v_k(c_k - x_k + \sum_i \rho_i(a_i - \sum_{jk} h_j x_{jik}) \end{aligned} \tag{15.13}$$

generating the efficiency conditions

$$x_k \begin{Bmatrix} = \\ \geq \end{Bmatrix} 0 \Leftrightarrow b_k - v_k - \sum_j \omega_{jk} a_{jk} \begin{Bmatrix} < \\ = \end{Bmatrix} 0 \tag{5.14}$$

$$x_{jik} \begin{Bmatrix} = \\ \geq \end{Bmatrix} 0 \Leftrightarrow - r_{ik} + \omega_{jk} - \rho_i h_j \begin{Bmatrix} < \\ = \end{Bmatrix} 0 \tag{5.15}$$

Here the efficiency prices ω_{jk} are wage rates for labour j in (firm) location k; ρ_i is the housing rent per square meter bid in residential location i; and v_k is profit or rent of the firm in location k. The right part of (5.14) may be rewritten

$$\rho_i = \max_{j,k} \frac{\omega_{jk} - r_{ik}}{h_j}$$

to show how the bid rent ρ_i is determined. Residential space in location i will be occupied by that type of worker j employed in k who can afford the highest bid in rent. In a monocentric city with all employment in the centre so that commuting cost is the same for all worker types j low-wage earners can outbid others for proximate locations only if their housing demand h_j is small enough, a well-known observation in urban economics.

We will not consider the long-run problem of firm relocation since it essentially involves an integer programming problem. We record, however, the implication of (5.4b) that firms are sensitive to commuting cost and should thus attempt to move closer to residential locations with available labour, *ceteris paribus*, when considering relocation. If one wants to visualize spatial competition in terms of iso-price-lines, or iso-wage-lines, then a continuous flow model is appropriate (Beckmann, 1993).

References

Beckmann M (1993) Potentials and Bottlenecks in Spatial Development. A Reconsideration of Urban Rents and Commuting.. Nijkamp P, Kohno H (Eds) Springer-Verlag, Berlin pp 23-30

Koopmans TC (1951) Analysis of Production as an Efficient Combination of Activities. Activity Analysis of Allocation and Production pp 33-97

Takayama T, Labys WC (1985) Spatial Equilibrium Analysis. Nedlands, Western Australia

6 Modelling the Economic Effects of Environmental Policy Measures Applied to Transport

Jeroen C.J.M. van den Bergh and Piet Rietveld
Vrije Universiteit

6.1 Introduction

Transport is an economic activity characterized by intensive energy use and direct emissions of various environmentally and health damaging substances. Environmental policy may therefore include specific attention to measures aimed at ameliorating such negative and external effects of transport, as well as from related physical infrastructure. From a policy point of view it is relevant then not only to know the direct effect of applying policy measures to transport but also economy-wide effects. The latter may include both macro-aggregate indicators as well as more or less structural aspects of alterations in micro-economic compositions of an economy. This can take the form of processes on individual markets, changes in relative prices, and substitutions between inputs or outputs. The importance of the micro-observation level derives from the fact that transport is entangled with most economic activities, and can even be regarded as a physical dimension of interaction between such activities, both spatially and sectorally. The two levels of economic observation are not separate of course, and, as we will see later, the economic methodology of general equilibrium analysis can serve as a basis for accomplishing an adequate linkage.

In order to deal with problems and questions of transport and environment one may choose from a variety of approaches. Important criteria for making a choice out of these are the spatial content and the range of economic relationships and effects that is endeavoured. For short periods a number of spatial phenomena are not considered, i.e. the assumption is made that various spatial data are fixed over time, for instance locations, infrastructure, available modes of transport, and transport costs. Lengthening of the period of analysis compels one to make some of these spatial elements variable over time, of course at the cost of increasing complexity of the analysis. The second criterion is not entirely independent from the first. The economic scope of an analysis of transport and environment is linked to its time horizon. But apart from such a link one may enforce certain restrictions on it such as those related to the completeness of economic contents, the level of sectoral aggregation, and the endogeneity of economic processes modelled. For this reason we will devote separate sections later on to approaches towards transport, economy and environment ranging from a partial to complete reach and from a planning to a market orientation.

Our intention with this paper is to provide a background for studies of transport and environment - oriented towards both conceptual and methodological aspects. The studies that we have in mind differ from existing ones in the sense that they devote explicit attention to economy-wide effects of specific environmental policy goals and measures applied to transport. In other words, we do not take the economic structure as given, since there is a clear need for approaching the problem of environment and transport in a wider context. In Section 6.2 we will introduce the relationship between transport and environmental effects. The remainder of the paper deals with models for analysis of economic implications of environmental policy directed towards alleviating the negative environmental effects of transport activity. Throughout the text we will strive for a clear account of the issues that have to be contemplated in order to make an adequate selection of assumptions and elements making up a final model for analysis of transport and environment. When different roads are open we will as far as possible take notice of them and concisely discuss the significance and implications of each. With regard to the relationship between environment and transport our interest is primarily focused on combining micro-level decisions affecting transport flows with micro and macro conditions concerned with environmental objectives. Such a combination can be accomplished either from the micro-level or the macro-level, i.e. taking a market or a planning approach, respectively. A concise overview of spatial interaction, transport planning and market-transport models is given in Section 6.3. Models are discussed here with increasing levels of integration of transport and economic features, and extended towards environmental policy analysis. Section 6.3.1 deals with spatial interaction models. Section 6.3.2 gives an account of planning oriented transport modelling approaches. Section 6.3.3 explores the potential for an operational analysis of transport with applied general equilibrium (AGE) models. Concluding remarks are given in a closing section.

6.2 Transport and Environment

Environmental and health impacts of transport are numerous. In the first place different modes of transport cause different types of pollution and disturbance to environmental mediums, in natural areas as well as in cities. A general type of classification of environmental effects is into use of space, causation of barriers (e.g. roads cutting through natural areas), disturbance (e.g. noise, smell, unsafety), direct energy use, air pollution from emission of chemicals (notably CO, CO_2, C_xH_2, NO_x, SO_2, and aerosols), and production of noise (see Van Gent and Rietveld 1990). More difficult to take into account are indirect or delayed effects caused by materials and energy use in the construction of transport infrastructure and fabrication of vehicles, the generation of material waste (discarded tires, wrecks of cars, ships, planes, etc.) and by such scattered issues as oil spillage at gas stations leading to soil and water pollution. Lastly, there is a clear visual and aesthetic negative impact of transport and infrastructure which has to be taken especially serious since it goes along with irrevocable changes in the landscape.

In the second place there is a continuous increase in transport flows. This is the result of a myriad of trends such as related economic growth (production and income growth), international demand for domestic transport services, labour and housing market developments, demographic trends (individuality and ageing), spatial policy (locations, parking, infrastructure), public transport provision, and patterns of preferences with respect to, e.g. recreation. Technological development is also rapid in this field (e.g. introduction of catalytic converters, other types of fuel, cleaner motors, less noisier cars and roads, and also telecommunication). For certain types of pollutants such as acid deposition the net effect is favourable, but for others such as noise the situation seems to worsen.

In OECD countries the relative contribution in environmental stress caused by transport is as follows (see Barde and Button 1990): 90% of carbon monoxide and 40 % of carbon dioxide emissions, 50% of nitrogen oxide emissions, 40-50% of hydrocarbon emissions, 50% of lead atmospheric emission (100% in urban areas). This is of course strongly related to energy consumption which is 30 % of total and of which 82% is consumed by road transport. This is of course not independent from the fact that 130 million cars occupy the roads of Europe. 16 % of the OECD population is exposed to unacceptable noise levels (above 65 db (A)) arising from road, rail and air traffic. Barde and Button provide also estimated social costs (private plus external costs) of transport as a percentage of GDP for OECD countries and arrive at a figure of 18.3 %. This includes also accidents, time spent and user expenditure.

Important for empirical research of transport and environment is to establish the importance of various relationships between economic activities, transport and environment, by taking notice of certain possible classifications, in the following order:

- i the role of transport in generating environmental effects in comparison with other economic sectors; this information provides a basis for deciding upon the environmental effect or type of pollution to concentrate on in one's analysis;
- ii the role of transport for domestic activities and consumption in generating environmental effects in comparison with the throughflow transport (from-to abroad);
- iii the role of transport modes in generating environmental effects in comparison with the total transport sector;
- iv the role of freight versus passenger transport in generating environmental effects;
- v the link between transport between different sectors of the economy and the amplitude of environmental effects: natural resource materials deliveries, intermediate deliveries, final goods deliveries, and waste transport; i.e. the relationship between the part of chain of economic production and the transport related environmental stress; more generally, this is about the classifications of sectors and goods, such that links between sectors (causing goods' flows) coincide with significant environmental effects from associated transport flows;

and

vi the link between single goods transport and environmental effects, considering alternative routes, substitution possibilities, etc.

Insight into these relationships allows one to make choices with respect to important and solvable problems of transport and environment, along with the magnitude of economic impacts. Following that one may use data on the chosen classification level to perform certain analyses, based on suitably aggregated models. The most significant impact of passenger transport for instance in the Netherlands arises from short-distance trips. So from an environmental policy point of view one may concentrate on these trips and leave long-distance trips aside. Since we are interested here in economic sectoral allocative impacts of environmental policy of transport, there is a strong argument to concentrate on freight transport since it affects production decisions on a sectoral level through its impact on cost structures.

In order to solve environmental problems related to transport activity one can take various routes. Policy scenarios may, for instance, include the following elements:

- stimulating technologies which are cleaner and more quiet than existing or used ones;
- efficient use of transport networks to reduce congestion and negative environmental and health effects;
- encouraging the use of alternative, environmentally more friendly modes of transport;
- price policies to reflect external costs of transport;
- spatial policies, such as route guidance and parking regulation;
- priority rules for different types of transport to minimize the ratio of environmental impact to economic productivity;
- additional provision and promotion of use of public transport;
- logistic planning for environmental and economic conditions;
- locational policy with respect to industries, markets, resources and housing;
- economic policy to find an optimal or acceptable mix of economic activities, transport and environmental quality.

Substitution processes between transport and economic inputs of or conditions for production underlie these policy directions. Isard (1956, p. 94) mentions 3 types of such substitution possibilities, based on a spatial perception, namely between transport and inputs in production, between sources (origins) of a product, and between places of delivery (destinations). These type of substitutions are much dependent on average cost curves of production and of transport. In the presence of economies of scale one can expect average production cost to show for relatively small activity size a sharper decline than the increase in transport costs, which can cause concentration tendencies of economic activities (see Figure 6.1).

When environmental cost are accounted for in transport the optimal scale falls, given by the minimum of the total cost curve in Figure 6.1, i.e. where marginal transport cost equal minus marginal production cost. Additional possible types of substitution are related to a change of transport modes. In a broader economic perspective one may add the whole range of usual economic-technological substitutions open to alleviate the need for certain transportation. Barde and Button (1990, pp. 17-18) express the opinion that "... environmental protection measures alone will not suffice because prevailing intervention failures lead to an oversupply of transport, exemplified by the fact that the rapid and continuous growth of especially road traffic outstrips efforts to reduce the environmental effects of individual vehicles". The conclusion is thus that an appropriate transport-related environmental policy, i.e. one that meets its environmental objectives, requires a coordination between transport and environmental policies. In fact, what is required is a balanced development of transport and economy, such that environmental objectives are met, i.e. a sort of sustainable mobility (see Ruijgrok 1992) encapsulated in a multisectoral sustainable development (see van den Bergh 1991).

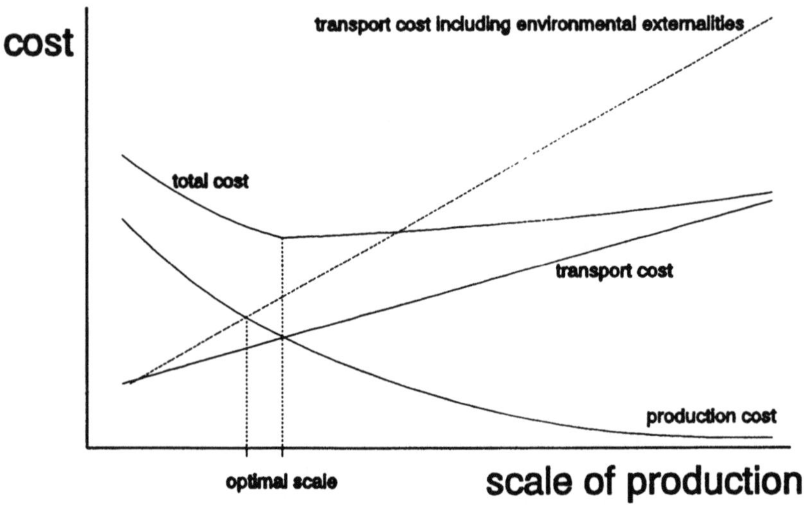

Fig 6.1. Optimal production scale: production and transport cost

Substitution mechanisms have an intricate impact in terms of environmental consequences. It is important therefore on the one hand to distinguish clearly between transport as measured in distance units, weights or number of passengers moved, costs and environmental effects. On the other hand relationships will have to be established between such variables, so that in the end economic and environmental consequences of transport policy can be clearly exposed.

6.3 Spatial and Economic Models of Transport with Environmental Components

6.3.1 Gravity Type Models

The spatial interaction which is the reason or the result of transport or interregional flows, can be taken as the starting point for analyses. Instead of taking cost-benefit considerations as the origin of transport flows, as will be pursued in Section 6.3.3, one may then use entirely different concepts - based on theories in physics - such as gravity, potential and entropy formulations. More extensive overviews than here are provided by e.g. Nijkamp (1979), and by Batten and Boyce (1986). These models explain or predict patterns of social and economic interaction (including passenger traffic, telephone calls, migration flows, commuting flows, trade flows, etc.) over geographic space.

The gravity model allows for a representation of the number of interactions between two elements of space. Gravitation is then taken to be a function increasing in mass and decreasing in distance. A general specification of a spatial analogy of gravitational attraction (or the size of a flow, or number of interactions) n_{ij} is given in (6.1), where O_i and D_j denote indicators (e.g. flows) at origins and destination, and where f is a distance deterrence function ($f'<0$). The resulting specification is a special type of weighted geometric average.

$$n_{ij} = kO_i^{\alpha}D_j^{\beta}f(d_{ij}) \quad k,\alpha,\beta > 0. \tag{6.1}$$

The distance parameter d_{ij} is measured as economic costs of transporting from i to j (c_{ij}). This may be generalized by including the impact of environmental taxes levied when the trajectory i-j is used (e_{ij}). In that case $d_{ij} = c_{ij} + e_{ij}$ when environmental costs are independent from the length of a link, and $d_{ij} = c_{ij}(1+e_{ij})$ when this is not the case.

Potential models, as shown in (6.2), represent the influence exercised by a set of spatial points on a specific spatial point, so that they may give an indication of the direction in which an element will move. The potential V_i at point i is a weighted sum of indicator variables h_j (masses in the physical analogy) for each influencing spatial point, where the weights are reciprocals of the distance to the point i. Again these weights can be based on normal transport costs and environmental taxes. This specification is a special type of a weighted arithmetic average.

$$P_i = k\sum_j \frac{h_j}{d_{ij}}. \tag{6.2}$$

In order to have all flows sum up to the total flows in the spatial system where the interactions take place one has to attach additivity conditions to the model in (6.1) of the form given in (6.3) and (6.4).

$$\sum_{j} n_{ij} = O_i, \ \forall i \tag{6.3}$$

$$\sum_{j} n_{ij} = D_j, \ \forall j \tag{6.4}$$

In order to ensure that these equalities hold true, the model has to be extended with balancing factors A_i and B_j so that the formulation in (6.5) results.

$$n_{ij} = kA_iO_i^{\alpha}B_jD_j^{\beta}f(d_{ij}) \tag{6.5}$$

Models for spatial interaction have also been approached by using the thermodynamic concept of entropy, in the probabilistic sense of being the most likely macro state of a system, i.e. that macrostate which corresponds to the largest number of possible microstates (classified by Batten and Boyce as the statistical approach). Finding the most probable macrostate of spatial interactions subject to the flow additivity conditions and a transport budget restriction will give rise to an optimization problem the solution of which is essentially identical to model (6.5). The conclusion reads that the gravity model can be based on a maximization principle, operating at the system level. There is still another way to link the gravity model to a maximization principle, i.e. by making use of stochastic utility maximization at the level of individual actors being located in origin i and making a choice among the alternatives j taking into account transport costs or distances. Anas (1983) shows that when a logit model is used, the resulting aggregate flows are again identical to the ones described in (6.5). This provides a really behavioural basis to the gravity model.

As opposed to the behaviourial approach of spatial interaction modelling, which requires estimation from disaggregate data, the statistical approach is generally based on an estimation from aggregate data. However, both approaches can be transformed in a shape that allows for dealing with data of the other level of aggregation. From this we may infer that both approaches are consistent with one another, and since they give both rise to analogous formulations to the gravity model, one may choose the most convenient approach in view of the model at hand. The micro-behaviourial approach may be used in the market model approach discussed in Section 6.3.3. However, one may leave explicit formulation of behaviourial elements out and turn immediately to the gravity model as a description of spatial interactions in a market model.

6.3.2 Transport Planning Models

In Section 6.3.1 we focused on descriptions or explanations of spatial interactions. In this section we will see how, given a set of possible interactions or routes between a given set of points[1], one can set up planning problems for transport flows for given transport costs. The planning objective can take various forms, e.g. minimizing total system transport costs. Other conditions, on required inflows or outflows (production input, demands, produced output or supply) can be added. An overview of such models can be obtained from Takayama and Judge (1971) and Batten and Boyce (1986). In the following some characteristic examples from this literature will be presented in a format that we have expanded to deal with environmental policies or costs. As in the foregoing section, also here the statement applies that many approaches which take a different starting point can be reconciled within a single framework. This is in fact the emphasis in the presentation by Batten and Boyce (1986). An important difference between the gravity type model and transport planning models is that cross-hauling does not occur in the latter. Thus when $x_{ij}>0$, then $x_{ji}=0$.

First we consider the situation with fixed transportation costs, where a distinction can be made between transportation and activity analysis models. The first class consists of (discrete) spatial allocation or distribution model for a single commodity shown in (6.6), that is based on sectoral revenue, total production costs and transport costs (a variation on a net revenue maximization model in Takayama and Judge 1971, Chapter 3). Here the problem is to find x_{ij} which are commodity shipments from point or region i to j. The economic environment is fixed, i.e. the quantities available or produced at i (supply), x_i, and the quantities demanded at j, y_j, are all given. Furthermore, all prices are fixed: the sales price of a product in region j coming from region i, p_{ji}; the cost of producing goods in i for j, c_{ij}; the cost of transporting one unit of the commodity from i to j, t_{ij}. Variables i and j assume integer values in the range 1 to the number of regions, so that each region can be a demander and supplier. Various formulations of this model can be given to include environmental objectives. Presently, the model in (6.6) includes an environmental tax per unit of commodity transport, e_{ij}, reflecting the environmental cost of transporting one unit of commodity from i to j. This can be interpreted as a fixed markup cost, since both transport and environmental costs include already the impact of the distance and other relevant characteristics of the connection between i and j.

$$\begin{aligned} & \underset{x_{ij}\geq 0}{\text{maximize}} \quad \sum_i \sum_j (p_{ji}-c_{ij}-t_{ij}-e_{ij})x_{ij} \\ & s.t. \quad \sum_i x_{ij} \geq y_j \ , \ \forall j, \\ & \qquad \sum_j x_{ij} \leq x_i \ , \ \forall i. \end{aligned} \tag{6.6}$$

It is interesting to look at the dual problem associated with (6.6) the solution of which gives the optimal price system at demand and supply regions, so that

[1]In a pure production approach as discussed in section 3 it is difficult to include the network aspects of infrastructure underlying transport (see Rietveld, 1989). Routes can not be distinguished, since one cannot go beyond the specification of input-output relationships that reflect costs of transport (per unit of distance and weight).

comparative location and price (dis)advantages are made explicit. One condition of the dual is that $p_{ji}-q_j+q^d \leq t_{ij}+c_{ij}+e_{ij}$, where q_j can be interpreted as the zero-profit price at the production unit, and q^d as the subsidy to satisfy the demand quota. This relationship is interesting because it states the exclusion of reallocation of the product by arbitragers. It shows that environmental taxation will lead to higher optimal prices, or positive subsidies, or it will in the presence of demand quota stimulate lower production costs. Of course, endogenous demand prices would give rise to lower demand levels (see next section).

$$
\begin{aligned}
&\underset{x_{ij}\geq 0}{\textit{maximize}} \quad \sum_i \sum_j (p_{ji}-c_{ij}-t_{ij})x_{ij} \\
&\text{s.t.} \ \sum_i x_{ij} \geq y_j \ , \ \forall j, \\
&\qquad \sum_j x_{ij} \leq x_i \ , \ \forall i, \\
&\qquad (a) \ \sum_i \sum_j e_{ij}x_{ij} \leq E, \\
&\qquad (b) \ \sum_i e_{ij}x_{ij} \leq E_j \ , \forall j, \ \ \sum_j e_{ij}x_{ij} \leq E_i \ , \forall i, \\
&\qquad (c) \ e_{ij}x_{ij} \leq E_{ij}, \ \forall i,j.
\end{aligned}
\tag{6.7}
$$

Alternatively, the model in (6.6) may be replaced by the one in (6.7) that represents physical regulation rather than environmental taxing. Here one or more constraints are added to a general optimal transportation problem - subject to the conditions in (6.6) - to reflect an upper bound to environmental impacts. These additional constraints may be formulated for the total system level as shown in (6.7a), but also for less aggregated environmental indicators linked to origins or destinations of spatial flows, as in (6.7b), or to each single spatial connection, as in (6.7c). The righthand sides E, E_i, E_j and E_{ij} in (6.7a-c) denote environmental regulations on various levels of aggregation. The variables e_{ij} in (6.7) are not necessarily similar to the ones in (6.6). This is only true when E, E_i, E_j, E_{ij} are measured in monetary units.

$$
\begin{aligned}
&\underset{x_{ij}\geq 0}{\textit{minimize}} \quad \sum_i \sum_j e_{ij}x_{ij} \\
&\text{s.t.} \ \sum_i x_{ij} \geq y_j \ , \ \forall j, \\
&\qquad \sum_j x_{ij} \leq x_i \ , \ \forall i, \\
&\qquad \sum_i \sum_j (p_{ji}-c_{ij}-t_{ij})x_{ij} \geq r \ .
\end{aligned}
\tag{6.8}
$$

Finally, the model in (6.8) shows an even more drastic alteration than from (6.6) to (6.7). The objective is to minimize environmental effects (costs). The constraints added to this objective are the ones in (6.6) including a constraint stating that net benefits should exceed some mimimum return r (i.e. moving the function in the present objective in (6.6) to the constraints part of the optimization model). This exchange of objectives and constraints can be interpreted as an alteration in priorities from economic to environmental goals. Of course, the objective function can also be based on the less aggregate conditions (b) and (c) in (6.7), in which case a multi-objective framework results.

A second type of model, shown in (6.9) and (6.10), includes both production and allocation. In this case, technology and resources are to be included as well. Demands are also here assumed to be given. An important addition to the foregoing problem are spatial flows of primary products (or intermediate products, since we are dealing with a unisectoral approach) serving as inputs of production units (one type of input is considered here). Given locations in this context means thus that production capital or other inputs are non-mobile. The model gives two types of information: (i) spatial flows of inputs and final goods; and (ii) levels of production of final goods and use of inputs at each location. The dual again gives the optimal prices.

$$\begin{aligned} \underset{x_{ij},x_i,z_{ij}\geq 0}{maximize} \quad & \sum_i \sum_j (p^x_{ji} - t^x_{ij}) x_{ij} \\ & - \sum_i \sum_j (p^z_{ji} + t^z_{ij}) z_{ij} - \sum_i c_i x_i \\ & - \sum_i \sum_j (x_{ij} e^x_{ij} + z_{ij} e^z_{ij}) \end{aligned} \tag{6.9}$$

$$s.t. \quad (6.10), \ \forall i.$$

The first line of the objective function in (6.9) represents revenue from selling final goods net of transport costs from i to j. The second line represents cost of production, composed of purchase and transport costs of inputs and operations costs at the production unit. The final line in (6.9) is the environmental cost, which includes the revenue of taxes levied on transport of inputs to production units and final goods to consumers.

$$\begin{aligned} & x_i \geq \sum_j x_{ij}, \\ & s_i + \sum_j z_{ji} \geq a^z_i x_i + \sum_j z_{ij}, \\ & k_i \geq a^k_i x_i, \\ & \sum_j x_{ji} \geq y_i. \end{aligned} \tag{6.10}$$

The four requirements in (6.10) are to be interpreted as follows: the first condition states that for each region the amount of final goods produced is sufficient to allow for the transport flows of final goods to all regions; the second condition assures that the regional availability of inputs s_i and the transport inflow of inputs from other regions together are sufficient to satisfy regional production requirements and regional transport of inputs to all regions (for the production of one unit of output in i one needs a_i^z units of input); the third condition states that the immobile production capital and other resources are sufficient for regional production; the last constraint reflects the satisfaction of given regional demands for the final good. The foregoing statements on the environmental policy regulations in relation to model constraints apply here as well. No matter how environmental policy is included, the present spatial activity analysis allows for comparison of environmental policies directed at different transport flows: input or intermediate products versus final goods; or passenger versus freight, i.e. labour (as a special type of resource input) versus final goods. Finally, we have only considered here the case of constant transport costs. It may however be the case that transportation costs are flow dependent. But in essence, this can be considered a special type of network problem. As opposed to the foregoing models, network equilibrium planning problems are associated with different routes between two given spatial centers. In such cases it is from an environmental policy point of view interesting to stimulate route choices that give rise to a minimum of environmental impact. At this point we are however more interested in the relationship between transport and economic considerations, so that we will abstain from these network models.

In the long run spatial economic problems of optimum locations are relevant, which in a sense can be regarded as an expansion to the transport problems of cost minimization or revenue maximization. While in the transport problems all locations are fixed, including markets, resource deposits and firms, the location problem assumes that only markets and resource deposits are fixed. This makes both problems very similar. Finally, relaxation of all fixed factors gives rise to a complex integrated choice problem of locations, production levels and interactions, for given markets and resource deposits (see Batten and Boyce 1986). This approach is however still very much producer oriented since consumer demand is included as fixed data. Only a general equilibrium approach can relax both fixed producer and consumer data.

Theoretical underpinning of spatially extended general equilibrium approach can be found in Takayama and Judge (1971, Chapter 16). The general planning formulation of such a model, represented in (6.11), showing a one commodity multi-regional system, generates consumption and production levels for each region and all interregional transport flows, such that a competitive and spatial equilibrium is reached.

The equivalence of this programming formulation with the general spatial equilibrium model formulation was first shown by Samuelson (1952). This model has a maximand which equals the consumer satisfaction less the sum of production cost and transport costs. The consumer satisfaction is determined for goods in every region and per good as the area under the demand function D. The production cost for the total economy is calculated as the area under the supply

$$\underset{y_i,x_i,z_{ij}\geq 0}{maximize}\sum_i\left[\int_0^{y_i}D_i(p_i)dp_i - \int_0^{x_i}S_i(p_i)dp_i - \sum_j (t_{ji}+e_{ji})z_{ji}\right]$$
$$s.t.\quad y_j \leq \sum_i z_{ij},\ \forall j, \qquad (6.11)$$
$$x_i \geq \sum_j z_{ij},\ \forall i.$$

function S. The transport costs are calculated for all flows of final goods (z_{ij}), based on transport and environmental costs, t_{ij} and e_{ij}, respectively. The term within the larger brackets, which is summed over i, can thus be interpreted as the contribution to 'net social payoff' (Samuelson 1952) by the production, transport and consumption of region i.

6.3.3 Market Models

As indicated before, transport can best be considered as a production activity requiring inputs, that supplies services suitable for intermediate deliveries to other activities and consumers. Transport can in addition be regarded as an input next to the usual set of factors of production. The upshot of this is that one has to opt for an approach that links micro-economic processes to transport. Input-output modelling allows for such an approach but has the disadvantage of not allowing for price processes and substitution mechanisms. Applied or computable general equilibrium (AGE/CGE) modelling provides an extension that takes away these deficiencies (see, e.g. Bergman et al 1990; Dixon et al 1991). It is clear that this approach is based on strict assumptions that allow one to formally include: economic rationality of actors; and the contact between supply and demand. However, in order to have both endogenously determined sectoral variables and intersectoral linkages no other such method for analysis is available to which one can turn to. Consequently, in spite of its deficiencies, one can benefit from this powerful tool for economy-wide analysis. Although the pure general equilibrium approach is focused on clearing of markets by way of 'right' prices, applied models can also deal with price rigidities and inelastic demand or supply curves (rationed markets).

Of special importance in the context of AGE models is the explicit treatment of prices, namely by both including the interaction of demand and supply, and the impact of prices on producer and consumer decisions derived from constrained profit and utility maximization. Transport may be built in such a context by regarding it as an activity that renders services which have a price, and the production of which requires inputs. Subsidies, price-regulations (e.g. road-pricing or fuel-taxes) can be taken into account, and the model can serve to analyse the consequences of such regulations in terms of sectoral distribution, transport level and environmental effect, or even (welfare) distribution effects. The latter is certainly relevant since transport policy, in specific with regard to public transport, has to some extent served the purpose of influencing welfare or income distribution by using instruments such as subsidies or price-discrimination. Furthermore, transport is usually associated with price rigidities and quantity restrictions. This is partly due to delays in supply reactions. These can be dealt with in general equilibrium models, under the condition that equilibrium is maintained.

Especially the representation of goods flows between economic sectors requires that such micro-economic processes can be dealt with, so that an AGE model approach is a attractive choice. In order to pursue the AGE approach we have to consider micro-economic aspects of transport. Then special concern is required for the embedding of freight transport in a multi-sectoral context. This is a topic which has, curiously enough, received only little and ad hoc attention (an interesting exception and fundamental contribution is Elbers 1992). The basic problem is of course that it is difficult to find a pure neoclassical economic foundation for freight transport in relation to economic production and consumption decisions. The difficulties in such an exercise are probably caused by the complex influence of a myriad of factors, such as economies of scale, costs of inventories and imperfect competition. So, even if it is clear that micro-economic elements have to be supportive of transport modelling, there does not seem to be a straightforward way to it. For a nice study of AGE with transport see Horridge and Johnson (1990).

The micro-economic behavioural optimization modules integrated in a AGE framework can be extended by including either price repercussions of environmental taxes and so, or constraints arising from direct environmental regulation, that the individual decision-makers, be it producers or consumers, are confronted with. While in the planning formulation of the previous section one could choose from both macro and micro environmental conditions, the AGE framework allows only for micro conditions.

The advantages of using AGE models in formalizing relationships between economic sectors, transport and environmental policy can be grouped as follows:

- a separate, economically based treatment of transport of freight and passengers;
- deriving the demand for transport as the result of an interaction between availability of infrastructure, the cost of using transport modes, the benefits of transporting goods, and the alternative of not transporting goods;
- accounting for the impact of transport and infrastructure on economic production and consumption patterns;
- estimating the direct and indirect economic-sectoral effects of transport and environment-transport policies;
- dealing with the five different levels of aggregation/classification mentioned in Section 6.2, sometimes in one model or by using different models.

Some problems emerge when considering the inclusion of transport in AGE models. First, the lumpiness of transport infrastructure investments gives rise to relatively long periods of excess supply of or demand for transport services. This means that the equilibrium approach has to be adjusted by rationing mechanisms or price rigidities. Second, the positive externalities of infrastructure are not included explicitly, and are difficult to monetize, so that the choice is to leave them out or include them in a less satisfying manner - from a theoretical point of view. Third, the effects of (investment in) transport infrastructure are long term, and require a dynamic approach. Direct and short term effects such as construction giving rise to intermediate deliveries, crowding out effects of financing through higher interest rates, and operations plus maintenance costs can be included more easily than long run spin-offs of infrastructure (see Rietveld 1989). In a static AGE model one has to leave such effects out. In dynamic models the problem of specifying investment behaviour has to be solved, for instance by estimation of future benefits accruing to activities heavily dependent on infrastructure conditions. Finally, the public provision of infrastructure is not a market process. The decision of private transport actors is however conditional on public infrastructure. This means that one part of the transport capital can be linked in a straightforward manner in an AGE model, while the other part has to be dealt with in a specific way, based on what type of analyses are required (i.e. what kind of questions are to be answered).

Development of a general equilibrium type of model that includes transport can include the following steps:

- start by choosing an existing AGE model, with an adequate sectoral disaggregation;
- add one or more sectors (e.g. for different modes of transport) that produce transport services; in dynamic models one may also add capital producing sectors for allowing changes in infrastructure and transport vehicles;
- extend the production functions with transport services as an input;
- extend the utility functions with transport services as a special type of consumption good;
- concentrate on short-distance passenger transport (most frequent);
- separate international transport flows of fixprice traded and national intersectoral flows of flexprice non-traded goods; this is especially relevant to small open ecnonomies (see, Whalley 1984);
- reflect in the model such transport specific notions as: rigid prices, government subsidies, taxes/levies, and rationing;
- include the infrastructure capital and where necessary and possible features of networks; this may be done in a simple manner by using gravity model formulations, or alternatively by adding more extended planning model modules;
- include a module for the accounting of environmental effects; this provides information to judge outcomes, or may be used for feedback to environmental policy instruments such as regulation;
- allow in the model for use of different types of environmental policy instruments, such as regulation, fuel taxes, road pricing, etc.

Two critical assumptions may be taken initially in order to reduce model complexity. The first is that intrasectoral freight transport is not considered at all, or only included in a very simple manner as an exogenously determined level. This probably implies a choice for long-distance freight transport. The second is that the long-distance passenger transport has only a small contribution in the total environmental stress caused by transport. Furthermore, traffic can be modelled in a simple manner in a medium-term analysis, since then locations and infrastructure are fixed.

What can we eventually learn from analyses with AGE based models with transport and environmental policy variables? Equilibrium analysis is essentially welfare analysis, in the sense of comparing small changes in policies (e.g. taxes) in terms of changes in the welfare distribution for the new equilibrium in comparison with the original one. The distribution can apply to economic production units, consumer groups. Furthermore, one may investigate consequences of environment-transport policies for the the total level of welfare or production in the economy. In a spatially disaggregated AGE model it is also possible to examine changes in spatial interactions from one equilibrium to another. Finally, one can analyse the environmental effects under different equilibria.

For the relationship between planning and general competitive formulations see Takayama and Judge (1971, Chapter 2). Both approaches generate a quantity and price system, and a transport system. In the next section we have provided for a stylized two region two goods general equilibrium model with transport and environmental policy.

6.3.4 A Two Region Two Goods Model With Transport and Environmental Policy

In order to give an idea of the main structure of a general equilibrium model with transport and environmental policy we give a stylized model here. For an extensive treatment of a two-sector general equilibrium model one may look into Dinwiddy and Teal (1988). A more theoretical approach is presented in chapters 2 and 16 of Takayama and Judge (1971).

Let us assume a country with two regions, each region having one consumer. There are two production sectors located in different regions, and each producing a single unique consumption good. A third national production activity generates transport services. The consumer problem of maximizing utility from consuming the regionally produced and transported good, constrained by income from capital and labour, is given in (6.12). R denotes the region (1,2). Commodity prices p_i^R (i=1,2; R=1,2), capital prices r_i (i=1,2,z), and prices at location of production plant q_i (i=1,2) may be regionally differentiated. The wage rate and transport price t are the same for both regions.

$$\begin{aligned} &\textit{maximize} \quad U^R(C_1^R, C_2^R) \\ &s.t. \quad p_1^R C_1^R + p_2^R C_2^R = Y^R. \end{aligned} \tag{6.12}$$

The production decisions are separated in a choice of inputs by minimizing costs, and choice of output to maximize profits. This is shown in (6.13) and (6.14), respectively. The regional index R is not included because good one (two) is produced in region one (two).

$$\begin{aligned} &\textit{minimize} \quad TC_i = r_i K_i + wL_i \\ &s.t. \; X_i = F_i(K_i, L_i). \end{aligned} \tag{6.13}$$

$$\begin{aligned} &\textit{maximize} \quad \Pi_i = q_i X_i - TC_i \\ &s.t. \; X_i = F_i(K_i, L_i). \end{aligned} \tag{6.14}$$

The price variable q_i in (6.15) denotes the per unit sales price which can be derived by relating sales and prices of the two regions for each product. Thereby it is assumed that commodity units are chosen such that transport cost per unit are equal for both commodities.

$$\begin{aligned} p_i^i &= q_i, \; i=1,2, \\ p_1^2 &= q_1 + t \;, \\ p_2^1 &= q_2 + t \;. \end{aligned} \tag{6.15}$$

The transport activity generates services that are used to transport an amount Z_{12} of good 1 from region 1 to 2, and an amount Z_{21} of good 2 in the opposite direction. This may be done with two trucks at the same time or with one truck at different times, driving to and fro. Z is the total of transport services supplied. The optimization model for the transport activity is shown in (6.16) and (6.17).

$$\begin{aligned} &\textit{minimize} \quad TC_z = r_z K_z + wL_z \\ &s.t. \; Z = F_i(K_z, L_z). \end{aligned} \tag{6.16}$$

$$\begin{aligned} &\textit{maximize} \quad \Pi_z = tZ - TC_z - eZ \\ &s.t. \; Z = F_z(K_z, L_z). \end{aligned} \tag{6.17}$$

The profits are calculated by taking environmental costs into account. They may reflect for instance governmental use of taxes (e). We assume that these are exogenous to the model. This environmental policy on transport flows brings us immediately to the government sector given in (6.18). Revenues and outlays are balanced, while all outlays are used to pay wages at the national rate to labour, which is the only production factor of the government.

$$\begin{aligned} G &= eZ \\ wL_g &= G \end{aligned} \tag{6.18}$$

The market clearing conditions for all regional and national markets are formulated in (6.19). They state that supply through production and transport have to satisfy demands for each good in each region.

$$
\begin{aligned}
&C_1^1 + Z_{12} \leq X_1, \\
&C_2^1 \leq Z_{21}, \\
&C_1^2 \leq Z_{12}, \\
&C_2^2 + Z_{21} \leq X_2, \\
&Z_{12} + Z_{21} \leq Z.
\end{aligned}
\tag{6.19}
$$

The last condition in (6.19) represents the market clearing condition for transport services. It may be replaced by (6.20) when a transport service is regarded as a trip from one to another region and back. In that case the maximum amount of interregional trade determines the required transport service level (on the condition that time-planning allows for taking commodities in two directions by one truck).

$$MAX(Z_1, Z_2) \leq Z . \tag{6.20}$$

Initial resource endowments are given in (6.21), where it is assumed that capital is sector specific (or immobile for the regional activities), while labour is assumed to be completely mobile. Fixed initial stocks (denoted by a superindex *) indicate inelastic supply curves (i.e. short run model).

$$
\begin{aligned}
&K_i \leq K_i^* , \ i=1,2,z. \\
&L_1 + L_2 + L_z + L_g \leq L^* .
\end{aligned}
\tag{6.21}
$$

Finally, consumer income in each region is based on wage income, capital rents and profits, as is represented in equation (6.22). For this purpose it is assumed that L_i denotes the regional labour supply in and by region i (i.e. migration can take place in the short run), to avoid problems of commuting and income spending. Labour income, capital rents and profits of the national transport industry, as well as labour income earned in the government sector, are assumed to be divided equally over both regions.

$$Y^R = w(L_R + 0.5L_z + 0.5L_g) + r_R K_R + 0.5 r_z K_z + \Pi_R + 0.5\Pi_z . \tag{6.22}$$

Solutions of this model consist of a set of consumed goods in each region, produced goods in each region, nationally produced transport services, resource input levels in all production processes, transport flows between regions, and prices.

The model can be transformed into a system of 36 equations by taking first order conditions of the optimization problems, the profit, costs and production functions, the income formation relations, and the other constraints. The models in (6.12) give relationships for the commodity quantities demanded in each region. (6.13) give the demand functions for production factors of each sector in each region.

(6.14) give supply functions of each commodity (and in each region). (6.16) and (6.17) give the factor demand functions of the transport sector and the transport services supply function, respectively. The model contains 11 price variables, 19 other endogenous variables and 6 exogenous variables. This means that the model is solvable. If a price numeraire is added as a constraint (since the model will only solve for relative prices otherwise) one equation can be left out. Numerical results of computer simulations with a model of this style are contained in Van den Bergh (1994).

6.4 Concluding Remarks

Which approach should be taken for operational analysis of economic effects of environmental policy of transport? How can multiple micro level objectives be integrated in a single macro level objective? Which of these frameworks allows best for environmental policy inclusion? And is there a possibility for combining the two? The foregoing has made clear that planning and micro-economic models both have potential for application on policy questions. It is clear that the specific model used will depend on the following choices: (i) which environmental indicators; (ii) which transport modes; (iii) sectoral disaggregation; (iv) time horizon for policy analysis or planning; (v) markets, rigid prices and rationing; (vi) approach to and level of description of real-world transport system networks. Certain choices narrow down the choice spectrum of others. For instance, a choice of (i) has implications for (ii), e.g. when acidification is chosen as the emphasis of attention then road transport deserves emphasis as being the major contributor.

The review and extension with environmental policy variables of suitable planning and market models is useful both to see what tools are available, and what extensions are required and possible. It is clear that environmental policy, by way of economic instruments or physical regulation, causes a reduction in the number of transport flows. The models can assist one to obtain insight into the interregional and intersectoral relationships, allocation patterns and size of various activities. Finally, inclusion of both on-site environmental effects (pollution, noise, etc.) of activities, and transport related environmental effects (see Section 6.6.2), allows one to analyse the trade-off - under various policies - between locational choice and transport. Then both economically and environmentally relevant locational and spatial characteristics can be of importance for transport policy analysis. In this way the wider environmental objective of spatially sustainable development can be supported with adequate scientific information. This can be successful if spatial information on economic and transport activities is compared with spatial environmental sensitivity.

References

Anas A (1983) Discrete Choice Theory, Information Theory and the Multinomial Logit and Gravity Models. Transportation Research B 17:13-23

Barde J-P, Button K (1990) Transport, Policy and the Environment. Six Case Studies. Earthscan for the OECD, London and Paris

Batten DF, Boyce DE (1986) Spatial Interaction, Transportation, and Interregional Commodity Flow Models. In: Nijkamp P. Handbook of Regional and Urban Economics 1. North Holland, Amsterdam pp 357-406

Bergh JCJM van den (1991) Dynamic Models for Sustainable Development. Ph.D. Dissertation. Thesis Publishers/Tinbergen Institute, Amsterdam

Bergh JCJM van den (1994) Formalization of Transportation in a General Equilibrium Model. Disaggregation, Region, Network, and Externalities'. In: Roson R (Ed) (1994) Transportation and General Equilibrium Models. Proceedings of an International Workshop, University of Venice Press, Venice, pp 1-24

Bergman L, Jorgenson DW, Zalai E (1990) General Equilibrium Modeling and Economic Policy Analysis. Basil Blackwell, Oxford

Dinwiddy CL, Teal FJ (1988) The Two-Sector General Equilibrium Model. A New Approach. Philip Allan/St. Martin's Press

Dixon PB, Parmenter BR, Powell AA, Wilcoxen PJ, Pearson KR (1991) Notes and Problems in Applied General Equilibrium Economics. North-Holland, Amsterdam

Elbers C (1992) Spatial Disaggregation in General Equilibrium Models - With an Application to the Nepalese Economy, Vrije University Press, Amsterdam

Gent HA van, Rietveld P (1990) Road Transport and the Environment in Europe Milieu 6:242-247

Horridge JM, Johnson D (1990) The Economy-Wide Effects of Fuel-Saving Technological Improvements in Road Freight Transport. IAESR Working Paper No. 2/1990. The Institute of Applied Economic and Social Research, University of Melbourne

Isard W (1956) Location and the Space-Economy. A General Theory Relating to Industrial Location, Market Areas, Land Use, Trade, and Urban Structure. The MIT Press, Cambridge, MA

Nijkamp P (1979) Gravity and Entropy Models. The State of the Art. In: Jansen GRM, Bovy PHL, Est JPJM van, Le Clercq F. New Developments in Modelling Travel Demand and Urban Systems - Some Results of Recent Dutch Studies. Saxon House, Westmead, England, pp 201-224

Rietveld P (1989) Infrastructure and Regional Development. A Survey of Multiregional Economic Models. Annals of Reg. Science 23:255-274

Ruijgrok CJ (1992) Towards Sustainable Mobility in Europe. Paper Presented at the 4th European Transport Planning Colloquium, March. Brussels

Samuelson PA (1952) Spatial Price Equilibrium and Linear Programming. American Economic Review 42:283-303

Shoven JB, Whalley J (1984) Applied General-Equilibrium Models of Taxation and International Trade. An Introduction and Survey. Journal of Economic Literature 22:1007-51

Takayama T, Judge GG (1971) Spatial and Temporal Price and Allocation Models North-Holland, Amsterdam
Whalley J (1984) Comments to Chapter 12 by Dixon et al In: Scarf H, Shoven JB Applied General Equilibrium Analysis. Cambridge University Press, Cambridge

7 The Structure of Airline Equilibrium Networks

Joseph Berechman
Tel Aviv University

Oz Shy
The University of Michigan

7.1 Introduction

In this paper, we explore why airline firms restructure their route network as a hub-and-spoke instead of a fully connected network. The difference between the two networks is that under a fully connected network all passengers are serviced by direct flights between origins-destinations city pairs, whereas under a hub-and-spoke network all city pairs are linked via a hub. We show that when a monopoly airline firm finds it profitable to operate a hub-and-spoke, it will increase frequency of service, compared to the case of a fully-connected type network. Our analysis differs from existing studies in two major respects: (i) We employ 'frequency of flights' as the measure of airline's output instead of the commonly used measures like number of passengers or available-seat-km. (ii) Whereas previous studies have shown similar results based on airline's cost function analysis in this paper we focus on demand conditions which cause the airline firm to restructure its network as a hub-and-spoke.

The objective of this paper is to examine the economic foundations which underlie the behaviour of airline firms in designing their route network given demand, cost and market structure conditions. In particular, we are interested in the rationale for the hub-and-spoke network which has emerged as the dominant network type. The motivation for this analysis stems from our aim to explain, by means of microeconomic analysis, major phenomena in the evolution of the aviation industry during the last two decades by focusing our analysis on the endogenous determination of equilibrium aviation networks.

The passing of the comprehensive Airline Deregulation Act in the USA in 1978, marked a pivotal milestone in the history of world aviation. Not only did this reform revolutionize the structure of domestic aviation markets in the USA, it also carried profound implications for intercontinental and other aviation markets. The present legislation by the European Union which extensively liberalizes European aviation, to a large extent, stems from the numerous changes in traffic patterns, operating procedures, airfare structure and market organization, induced by the US deregulation.

Two of the most significant results from the US deregulation were the substantial decline in the number of airline firms and the emergence of a hub-and-spoke as the predominant type of network structure. Doganis (1991) estimates that between 1978 and 1984 the percentage of passengers changing their carrier or interlining on route to their final destinations has fallen from 23 to 8 percent. Using a Gini Index approach to measure changes in route concentration, Reynolds (1992) has shown a significant increase in concentration for major US carriers between 1978 and 1988. Bailey et al (1985) and McShan and Windle (1989) report that between 1977 and 1984, hub-and-spoke routing increased by over 48 percent. Thus, the airline industry seems to have moved towards becoming a highly oligopolistic one where a relatively small number of carriers control most of the market and operate a route network which is almost totally central-hub oriented.[1]

In this paper we explicitly assert that these two major outcomes of the USA deregulation are interrelated as, to a large extent they emanate from the strategic behaviour of airline firms in deregulated markets. That is, we attribute part of the increase in market concentration to successful entry deterrence by major airline firms through the choice of a hub-and-spoke type network and hub-airport dominance. Yet we also conjecture that the emergence of a hub-and-spoke as the leading network type would have occurred (and indeed did occur) even if deregulation would have not taken place provided, of course, that there were no external constraints on such economic behavior of airlines.[2] Thus, we intend to show that irrespective of market competition, under typical demand and cost conditions the profitability of an airline increases when it switches from a fully (or even partially) connected network structure to a hub-and-spoke one. Analysis of this phenomenon including the specification of the equilibrium conditions which sustain this result constitutes the main objectives of this paper.[3]

Other major results from the US deregulation and the deregulation of specific markets (e.g. London-Amsterdam, Dublin-London) were a general decline in fares (compared with the pre-deregulation fare level), and a significant increase in flight frequency (defined as the number of aircraft departures per time unit, e.g. a week) (see Barrett 1992). The US Department of Transportation has estimated about 20

---

[1]McShan and Windle (1989) define a hub as a centrally located airport serving as a focal point for the airline's route map. The CAB defines a large hub as an airport enplaning over one percent of the total number of revenue passengers departing US airports. It defines a medium size hub as a community enplaning between 0.25 and 0.99 percent of that total, and a small hub as enplaning 0.05 to 0.24 percent, and a non hub enplaning less than 0.05 percent.

[2]It is important to realize that in the USA the use of network structure to achieve economic gains was largely restricted prior to deregulation by the controlling agency (the Civil Aeronautic Board (CAB). Bailey et al (1985) discuss how the CAB policy on route structure has a limiting effect on the extent to which airlines could alter their route system. Existing airlines had to show how proposed new routes would benefit the public and that airlines already serving these routes, will not be adversely affected.

[3]In this regard it is interesting to point out that faced with imminent market liberalization conditions, European airlines are presently engaged in extensive mergers and collusion activities while increasingly attempting at the same time to consolidate their operations around major hub airports (Pucher et al 1993).

percent decline in the fare level (in 1980 prices) between 1980 and 1990 (US DOT 1992). However, several authors have pointed out that fares tend to be highest on routes where the hub is the origin or destination airport, in particular, when they are dominated by one airline (Borenstein 1989; Hurdle et al 1989; Standard and Poor 1992). Spiller (1989), argued that this is not necessarily the result of airlines' market dominance but, in the presence of capacity constraints, it represents an efficient price structure.[4]

In this paper we use frequency - number of flights per time unit - as the principal measure of output of airline firms. The reason is that this measure of output significantly affects consumers' utility and the airline's cost of operation, and it is amenable to empirical analysis. Morrison and Winston (1986), for example, have found that the largest source of welfare gains from the US deregulation occurred through the increase in (route weighted) departure frequencies. In this paper we argue that the increase in frequency is related to the phenomena of the increased dominance of the hub-and-spoke network pattern and of the differential pricing on hub routes. The rationale is that higher frequency can be used by an airline firm to attract passengers who are willing to pay higher fares for a more frequent service (given a network structure). Given the demand, the switch to a hub-and-spoke network enables the airline firm to provide services at higher frequencies at the same or at lower unit cost thereby increasing profits. Hence, the use of frequency as an output measure is extremely useful for demonstrating the conjectured association between air-travel demand, frequency, fare level, cost of operation and the equilibrium network structure. A number of studies attempted to explore the relationships between these variables. They can be grouped into two main classes: those which have empirically estimated results from deregulation; and those which have examined the rationale for the choice of network structure and the impact of this structure on airlines' costs and passengers' welfare (see Call and Keeler 1985, for a general review). Beginning with the first, Toh and Higgins (1985) and Bailey and Williams (1988) found positive relationships between the extent of hubbing and airline profitability. In a detailed study McShan and Windle (1989) analyzed the effect of a hub-and-spoke network on airline costs in the USA.[5] Using airlines panel data for the period 1970-1984 and a translogarithmic cost function model,[6] they found that a 1 percent increase in hub-and-spoke routing has resulted in reduced unit cost of 0.11 percent. They further concluded that the lack of space in hub airports confers significant cost advantage on airlines

---

[4]Berry (1992) claims that this is the result of product differentiation rather than of market power, so that a dominant firm with many connections from its hub airport to outlying destinations provides differentiated products for which it charges higher fares.

[5]They define the degree of hubbing as the proportion of an airline's total departures leaving the 3 percent most utilized airports (points served) in the airline's network.

[6]They used total cost as a function of the airline's output, input prices (labor, fuel, capital, and materials) and computed point estimates around the variables means, while taking into account airline's operational characteristics (average stage length, average load factor), network characteristics (number of points served, degree of hubbing), year shifts and airline shifts.

which dominate these hubs (see Bailey and Williams (1988) for a similar conclusion) and, as a result, also reduce the threat of entry.[7]

The main focus of the more theoretical class of models is on the assessment of the benefits to airline firms and potential gains or losses of passengers' welfare, from the adoption of a hub-and-spoke network. Morrison and Winston (1985) have shown that conditions of scope economies, stemming from aircraft-size economies, lead to the choice of this network pattern (given demand elasticities). Following a similar line of analysis, Bruckner and Spiller (1991) show that entry in a city-pair market within a hub-and-spoke network of a monopolist, can lead to welfare losses in other city-pair markets served by this network resulting from passengers switching to the entrant's services, thereby raising the incumbent's marginal costs of serving other routes in his network. Bittlingmayer (1990), Hendricks et al., (1992), and Starr and Stinchcombe (1992), explained the foundations of hub-and-spoke networks on the basis of network economies which emanate from economies of traffic density (given aircraft size) and from multi-output production conditions which lower the costs of joint production of services on city-pair routes within a network, below their stand-alone cost.

In the modeling efforts reviewed above, the major forces driving the hub-and-spoke network structure are cost conditions. Demand conditions such as passengers' response to greater flight frequency under a hub-and-spoke system or their propensity to pay a premium for direct, non-hub flights are, in general, not emphasized. In this paper we show that such demand characteristics can explain the choice of a hub-and-spoke network by an airline firm even when the airline's cost function is regarded as linear in output. Also, in this paper we do not consider the effect of potential entry and entry deterrence strategies on network structure since we show that even without deregulation airlines may restructure their network as a hub-and-spoke.[8]

The design of the paper is as follows. In Section 7.2 we present a simple network model of an airline and distinguish between two basic network types, a fully connected and a hub-and-spoke. Based on this model, in Section 7.3 we examine the economic behaviour and results of a regulated monopoly airline firm operating a fully-connected network. Subsequently, in Section 7.4 we carry out a similar analysis for the hub-and-spoke case. A comparison of these two alternative network structure relative to level of frequency, load factor and fares is performed in Section 7.5. Section 7.6 summarizes major results and suggests possible extensions of the model.

---

[7]Johnson (1985) tried to assess the effect of US deregulation on networking. Utilizing a sample of US city-pairs, his results show that a successful entry on a particular route is dependent on the airline's initial network structure.

[8]In Berechman and Shy (1991) we have explicitly assumed entry deterrence behaviour by the incumbent monopoly firm following deregulation and explored how network structure is used to that end.

7.2 The Model

Consider an economy composed of three cities, labeled *A*, *B*, and *C*. Each city pair is connected by an airline route denoted by i, $i = 1, 2, 3$. Thus, a route i is defined as a link connecting a specific pair of cities on which n_i passengers wish to travel. A passenger can be transported by an airline firm either directly from his city of origin to his city of destination, or indirectly though a third city. Such a city will be called a hub.

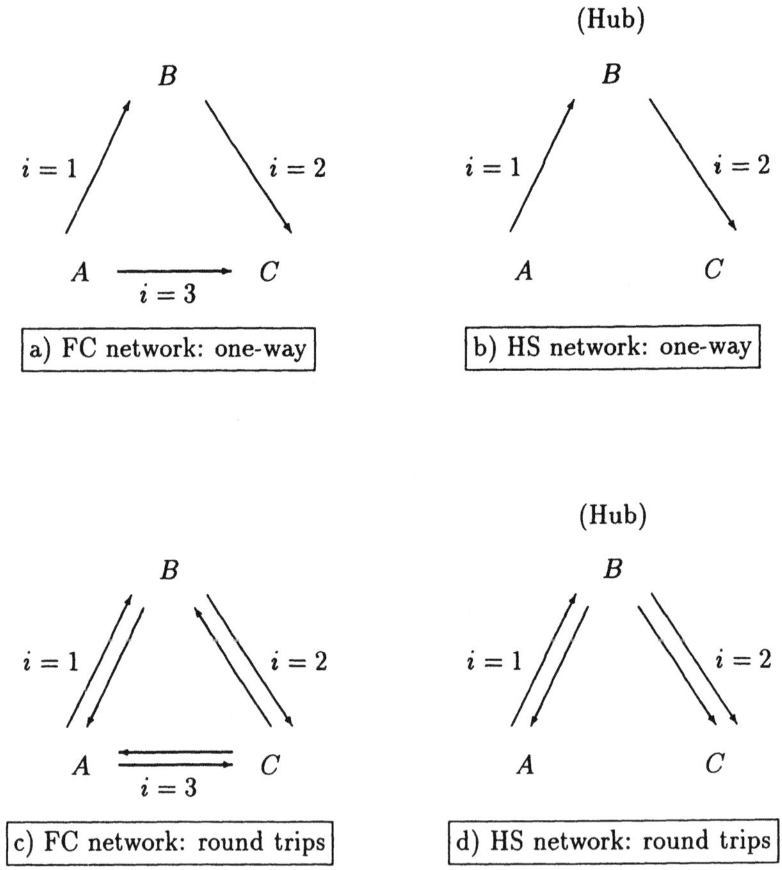

Fig. 7.1. Network types

The output of an airline firm is measured by flight frequency, i.e. the number of flight-departures per unit of time (e.g. a week). Let $f_i \geq 0$ denote the flight frequency on route i. We distinguish between two basic types of networks that the firm can operate: a Fully-Connected-network (FC) where all passengers are serviced via direct flights between origins-destinations city pairs; and a Hub-and-Spoke network (HS) where all city pairs are linked via a hub. Thus, under the HS network, there are no direct services between city pairs except for those passengers whose final destination or point of departure is the hub itself. By definition, a firm is operating an HS network if flight frequency on each route, not directly connecting origin cities and a hub, is zero. A FC network is, therefore, defined as a network where flight frequency on each route is strictly positive. Figures 7.1a and 7.1b illustrate a FC network and a HS network with a hub located at city B.[9] Thus, there are three routes: route 1 connecting A to B, route 2 connecting B to C, and route 3 connecting A to C. To simplify the notation and the analysis we disregard round-trips, and in what follows we analyze the economy described in Figures 7.1a and 7.1b only. In the Appendix section we show that the Round trips environment illustrated in Figures 7.1c and 7.1d is very similar to the one-way trip environment, and therefore a separate analysis is not need

The airfare per trip on route i is denoted by p_i. Let $d_i \in \{0, 1\}$, denote whether a flight is a direct one, $d^i = 1$, or not, $d_i = 0$, (e.g. from city A to C via B). On each route i, each of the n_i passengers is assumed to travel only once (per-time unit). The utility of a passenger on route i is affected by the fare (p_i), by the frequency of flights (f_i), and by whether the flight is direct or indirect. Formally, the utility function of a passenger on route i, U_i, is given by

$$U_i \equiv \begin{cases} d_i\delta + \dfrac{\lambda}{\gamma}(f_i)^\gamma - p_i & \text{if } p_i \le d_i\delta + \dfrac{\lambda}{\gamma}(f_i)^\gamma \text{ and } f_i \ge 0 \\ 0 & \text{otherwise} \end{cases} \tag{7.1}$$

where $\lambda > 0$, is the dollar value consumers attribute to frequency which corresponds to their value of waiting time; the function $(f_i)^\gamma$ represents consumer's utility gain from frequency. Assuming that $0 < \gamma < 1$, consumers' preferences exhibit diminishing marginal utility of frequency on route i. Although in this model each consumer travels only once during a given time period (e.g. one week), frequency is still important to consumers simply because it allows them greater flexibility in commuting and time saving, thereby increasing their productivity. The parameter δ is the extra dollar amount a consumer is willing to pay for a direct flight;

The airline firm's cost function is composed of two parts: airport charges and operating cost. Let ϕ_j be the per departure fee charged by the airport located at city j, $j = A, B, C$; F be fixed cost such as annual payments for leasing slots; c denotes the per flight-departure operating cost. Altogether, the airline's total cost TC is

[9]In this paper we do not discuss the reasons for choosing city B as the location for the hub. This choice can have a mix of economic, historical, and institutional reasons. See Hansen and Kanafani (1990).

$$TC\,(f_1, f_2, f_3 \equiv F + \emptyset_A(f_1 + f_3) + \emptyset_B f_2 + c\sum_{i=1}^{3}(f_i)^{\alpha}, \tag{7.2}$$

where α ($0 < \alpha < 1$) denotes the degree of output (frequency) scale economies on route i.[10]

In what follows, we assume that there are no airport fees (i.e., $\emptyset_j = 0$ for every j), that $F = 0$, and that $\alpha = 1$. For an analysis where $\alpha \neq 1$, see an earlier version (Berechman and Shy 1991).

Let $\tilde{k}$ denote the capacity of an aircraft (maximum number of passengers that can board an aircraft). In what follows, we assume that on each route there is an equal number of (potential) passengers. That is, $n_i = n$ for every i, $i = 1,2,3$. We further assume that the number of passengers on each route is considerably larger than the capacity of a single aircraft, i.e. $n > \tilde{k}$.

We now define a single airline equilibrium.

Definition 1

A single airline equilibrium is the vectors of route frequencies (f_1^e, f_2^e, f_3^e) and route airfares (p_1^e, p_2^e, p_3^e) that maximize

$$\max \pi = \sum_{i=1}^{3} p_i n_i - TC(f_1, f_2, f_3)$$

subject to

a. aircraft capacity constraint: $n_i/f_i \leq \tilde{k}$, and

b. threshold passenger utility: $U_i \geq 0$ for all $i = 1,2,3$,
   where U_i is given in (7.1).

The first constraint means that the number of passengers aboarding each flight, which equals the total number of passengers divided by the frequency of flights, should not exceed aircraft capacity.

7.3 A Single Firm Operating FC Network

In this section we examine the behaviour of an airline firm operating under a regulatory regime which permits only one firm to provide services to all cities. This type of regime is observed in many European countries where only one

[10] The asymmetry in (7.2) follows from our simplifying assumption of one-way trips.

airline firm is allowed to provide all domestic flights. We assume that this firm behaves as a monopoly, i.e. that it charges monopoly fares.[11]

Since all flights are direct, the monopoly firm extracts all consumer surplus by setting $p_i = \delta + \frac{\lambda}{\gamma}(f_i)^\gamma$. By symmetry, all routes will be served with equal frequency, and hence we set $f = f_i$ for every i. Thus, the single airline chooses f to

$$\max_f \pi^d = n(p_1+p_2+p_3) - TC(f_1, f_2, f_3) = 3n\left[\delta \ \frac{\lambda}{\gamma}(f)^\gamma\right] - 3cf \ s.t. \ \frac{n}{f} \leq \bar{k} \tag{7.3}$$

where π^d denotes the profit of a monopoly firm providing service on a FC (direct flights) network. We denote by $\hat{n}$ the 'threshold' number of passengers on each route which is defined by

$$\hat{n} \equiv \left[\frac{c}{\lambda(\bar{k})^{1-\gamma}}\right]^{\frac{1}{\gamma}}. \tag{7.4}$$

Hence, the monopoly's frequency, price, and profit levels on each route i, i = 1,2,3, are given by

$$f_i^d = \begin{cases} \left[\dfrac{\lambda n}{c}\right]^{\frac{1}{1-\gamma}} & \text{if } n \geq \hat{n} \\ \dfrac{n}{\bar{k}} & \text{if } n < \hat{n} \end{cases} \qquad p_i^d = \begin{cases} \delta + \dfrac{\lambda}{\gamma}\left[\dfrac{\lambda n}{c}\right]^{\frac{\gamma}{1-\gamma}} & \text{if } n \geq \hat{n} \\ \delta + \dfrac{\lambda}{\gamma}\left[\dfrac{n}{\bar{k}}\right]^{\gamma} & \text{if } n < \hat{n} \end{cases}$$

$$\pi_i^d = \begin{cases} 3n\delta + 3\left[\dfrac{\lambda n}{c^\gamma}\right]^{\frac{1}{1-\gamma}}\left[\dfrac{1-\gamma}{\gamma}\right] & \text{if } n \geq \hat{n} \\ 3n\delta + \dfrac{3\lambda(n)^{1+\gamma}}{\gamma(\bar{k})^\gamma} & \text{if } n < \hat{n} \end{cases} \tag{7.5}$$

Figure 7.2 illustrates the behaviour of the airline firm in the price-frequency space.

---

[11] One could argue that a single regulated firm is not allowed to charge monopolistic prices. Here we associate a single regulated firm with a monopoly since assuming otherwise complicates the analysis in the sense that there is no consensus of what are the objectives of a single regulated airline firm.

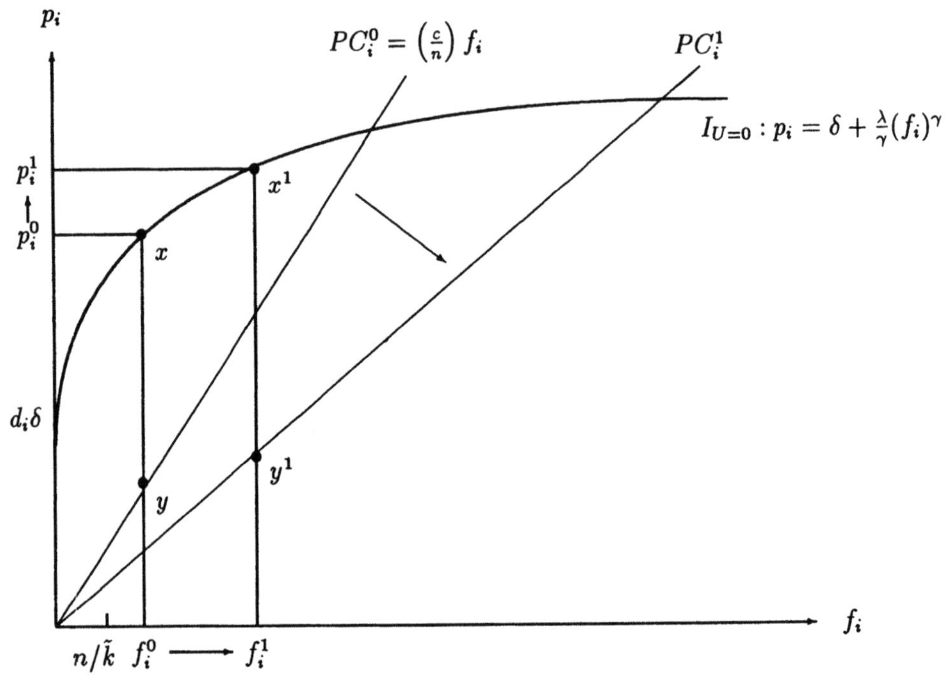

Fig. 7.2. Monopoly's profit maximizing frequency for $n \geq \tilde{n}$

In this figure, the ray from the origin is the flight operating cost per-passenger on route i, denoted by $PC_i(f_i) \equiv \left[\frac{c}{n}\right] f_i$. Since the monopoly airline sets prices which completely extract all consumer surplus, the curve $I_{U=0}$ is an indifference curve of a passenger on route i with a zero utility level. The curve $I_{U=0}$ also shows the maximum price a monopoly can charge for different levels of flight frequencies. This price increases with flight frequency since it is assumed that passengers are willing to pay more for higher frequency service.[12] In this space indifference curves representing higher utility levels ($U > 0$) would lie on the south-east side of the $I_{U=0}$ indifference curve, reflecting a combination of a lower price with a higher frequency.

The monopoly chooses a flight frequency which maximizes the difference between the airfare it charges and the cost of operation per passenger. In Figure 7.2, the profit maximizing frequency is f_i^o, where the segment connecting the points x and y is the per-passenger profit level. Disregarding for the moment aircraft capacity constraints, an increase in the number of passengers n (or a decrease in the cost per-flight, c; would tilt PC_i^0 to PC_i^1, showing an increase in

[12]Note that if a passenger on route i travels indirectly, $d_i = 0$, the zero utility indifference curve would start from the origin. In the *FC* case, $d_i = 1$.

the profit maximizing flight frequency, and the profit level per-passenger.

In sum, when the number of passengers increases, the monopoly will provide service at a higher frequency f_i^l, will charge a higher price p_i^l, and as a result, will realize higher per-passenger profits illustrated by the segment connecting x^l and y^l.

Given these conclusions it is interesting to ask what is the profit maximizing number of passengers per flight for the monopoly airline, given aircraft's capacity, $\bar{k}$. Denoting by k_i the actual number of passengers per flight, then,

$$k_i^d = \begin{cases} \dfrac{n}{f_i^d} = \left[\dfrac{c}{\lambda n^\gamma}\right]^{\frac{1}{1-\gamma}} & \text{if } n \geq \hat{n} \\ \bar{k} & \text{if } n < \hat{n} \end{cases} \qquad (7.6)$$

Equation (7.6) shows that when the aircraft capacity constraint is not binding, the profit maximizing number of passengers per flight on route i, k_i^d, decreases with the total number of people wishing to travel on this route n_i, (and increases with the cost per departure, c). The intuition for this conclusion is that since consumers prefer higher frequency, the monopoly airline can charge a higher price for providing a greater frequency. Hence, as the number of passengers increases, the monopoly airline finds it profitable to increase the frequency of flights proportionally more than the increase in the number of passengers. If the condition in the upper part of (7.6) holds ($n \geq \hat{n}$), then a monopoly airline operating a FC network will not operate at full aircraft capacity (i.e. operate at less than 100 percent load factor).

If $n < \hat{n}$, then (7.6) implies that capacity is binding. In this case, under the FC network, the airline's frequency on route i is $\tilde{f}^d = n/\bar{k}$.

Figure 7.3 illustrates the effects of number of passengers and aircraft capacity constraint on the actual number of passengers-flown per-flight k_i^d, and on the frequency f_i^d.

The horizontal axis in both parts of Figure 7.3 shows the number of passengers on route i, n_i. The threshold value $\hat{n}$ is associated with the case when the profit maximizing number of passengers per flight equals aircraft's capacity (i.e. $k_i^d = \bar{k}$). When n is greater than this threshold, the airline firm operates at less than 100 percent load factor ($k_i^d/\bar{k} < 1$). In this case, the optimal frequency increases at a decreasing rate with n. When the number of passengers is smaller than the threshold value, aircraft operate at a full capacity, so $f_i = n/\bar{k}$.

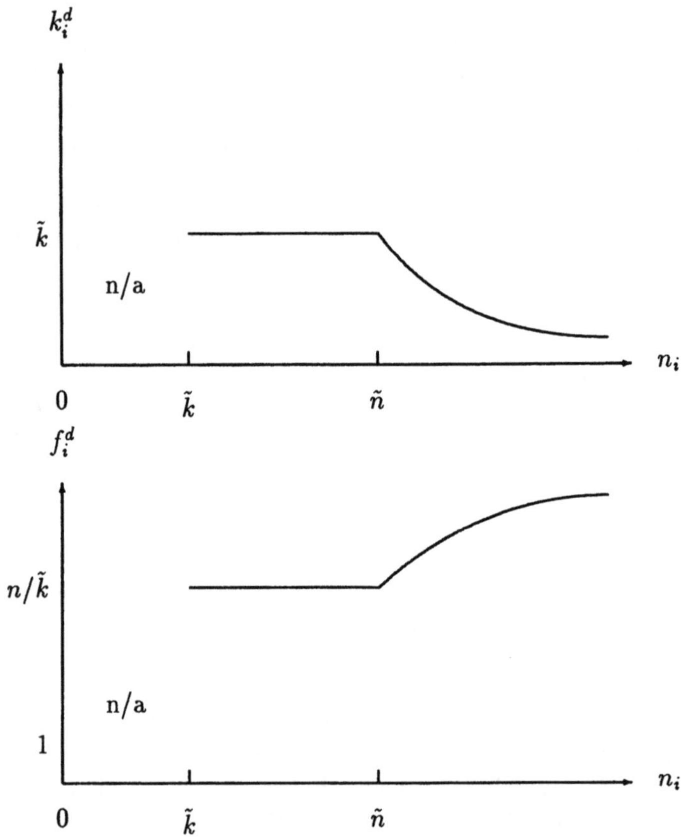

Fig. 7.3. Optimal number of passengers per departure and optimal frequency

7.4 A Single Firm Operating a Hub-and-Spoke Network

Suppose now that the monopoly airline does not provide a direct flight on route $i = 3$, but instead transports all passengers via a hub at city B (see Figure 7.1). Since now passengers on route 3 travel indirectly $(d_3 = 0)$, $p_3 = \dfrac{\lambda}{\gamma}(\hat{f})^{\lambda}$, where $\hat{f} \equiv \min\{f_1, f_2\}$; whereas $p_i = \delta + \dfrac{\lambda}{\gamma}(f_i)^{\gamma}$, $i = 1, 2$. Still assuming that $n_i = n$ for all i, the airline chooses a common frequency f to maximize total profit.[13]

[13]The superscripts h denotes values under hub-and-spoke network. Also, to reduce the amount of writing, we ignore the aircraft capacity constraint $n/f \leq \tilde{k}$. The only difference it makes is that the threshold value given by $\tilde{n}$ in (7.4) is replaced by $n \equiv \{2c/[3\lambda(\tilde{k})^{1-\gamma}]\}^{1/\gamma} = (2/3)^{1/\gamma}\tilde{n}$.

$$\max_{f} \pi^h = n(p_1 + p_2 + p_3) - 2cf = 2n\delta + 3n\frac{\lambda}{\gamma}(f)^\gamma - 2cf. \qquad (7.7)$$

Hence, under the HS network, the monopoly's frequency on each served route, prices, and profits are given by

(7.8)

$$f^h = \left[\frac{3\lambda n}{2c}\right]^{\frac{1}{1-\gamma}} \text{ and } p_i^h = \delta d_i + \frac{\lambda}{\gamma}\left[\frac{3\lambda n}{2c}\right]^{\frac{\gamma}{1-\gamma}}, \text{ for } d_1 = d_2 = 1,\ d_3 = 0$$

and

$$\pi^h = 2n\delta + \left[\frac{3\lambda n}{2^\gamma c^\gamma}\right]^{\frac{1}{1-\gamma}} \left[\frac{1-\gamma}{\gamma}\right].$$

Given a sufficiently large number of passengers $n_i = n$, the profit maximizing number of passengers per flight is,

$$k^h \equiv \frac{2n}{f^h} = 2\left[\frac{2c}{3\lambda n^\gamma}\right]^{\frac{1}{1-\gamma}}. \qquad (7.9)$$

Proposition 1 confirms Borenstein's (1989) empirical finding of the effect of deregulation on prices also for the case when markets are regulated. From equation (7.8) we can state:

Proposition 1

Under the hub-and-spoke network, the monopoly's airfare for flights originating or ending at a hub city exceeds the airfare paid by passengers whose destination is not a hub city. Formally, $p_3^h < p_1^h = p_2^h$.

This result emerges because consumers value a direct flight more than an indirect one. Consequently, there is a lower surplus that the monopoly airline firm can extract from those passengers who fly indirectly, thus, charging them a lower fare. Of course, one could argue that facing these prices, a passenger from *A* to the hub city *B* would be better off purchasing a ticket for travelling from *A* to *C* (via *B*), and get off at point *B*. However, such consumer behaviour is rarely observed, since it involves some inconveniences such as being unable to pick up luggage, and facing the risk of stopping at a different hub.

7.5 A Single Airline: A Comparison of FC with HS Networks

We now compare the monopoly's frequency, number of passengers per-flight, and prices on each route under the FC and the HS networks. A comparison of equations (7.5) with (7.8) yields the following proposition:

Proposition 2

1. A monopoly airline will operate a higher frequency under the HS network compared with the FC network. Formally, $f^h > f^d$.

2. If $\gamma < \gamma^*$ where $\gamma^* = 2 - \dfrac{\ln 3}{\ln 2} \approx 0.42$, the number of passengers per flight under the HS is greater than that under the FC network ($k^h > k^d$).[14]

3. The airfare set by the monopoly for passengers who start or end their trip at the hub city *B* is higher under the HS than under FC. Formally, for routes $i = 1,2$, $p_i^h > p_i^d$.

4. If passengers' valuation of direct flights (δ) is higher than a critical value, the airfare for passengers travelling from city *A* to a non-hub destination at city *C* is lower under the HS than under the FC network. Formally, there exists δ^* such that for every $\delta > \delta^*$, $p_3^h < p_3^d$.[15]

The intuition behind part 4 of Proposition 2 is that under the HS, route 3 passengers gain lower utility since they are transported indirectly via a hub. Therefore the airline must reduce the airfare for route 3 passengers. Note that the reduction in the airfare is lower than δ since under the HS the airline provides the passengers with a higher frequency service. For this reason, the reduction in p_3 occurs only for sufficiently high values of δ. Table 7.1 summarizes the results of Proposition 2.

Table 7.1 shows that when operating a HS network, the monopoly airline would provide a higher frequency, than under FC and would operate with a higher load factor. Under HS the airfare for passengers flying to and from the hub would be higher than under FC while it would be lower for passengers flying between non-hub cities. If aircraft capacity is binding, then frequency provided under the HS is exactly twice the frequency provided under the FC network.

---

[14]For the case of m cities with a single hub, $\gamma^* = (m - 1) - \dfrac{\ln 0.5m(m-1)}{\ln(m-1)}$.
Since we assume $0 < \gamma < 1$, for $m > 3$ we have that $\gamma^* > 1$. Hence, $\gamma < \gamma^*$ always.

[15]Comparing (7.5) and (7.8) yields $\delta^* = \dfrac{\lambda}{\gamma}\left(\dfrac{\lambda n}{c}\right)^{\frac{\gamma}{1-\gamma}}\left[\left(\dfrac{3}{2}\right)^{\frac{\gamma}{1-\gamma}} - 1\right]$.

Table 7.1. A comparison of FC and HS under monopoly

Variable	Notation	FC	Effect	HS	Conditions
Frequency	f	f^d	<	f^h	non-binding capacity
Frequency	f	$2f^d$	=	f^h	if capacity is binding
Load Factor	$k/\bar{k}$	k^d	<	k^h	iff $\gamma < \gamma^*$
Fare to/from hub	$p_i,\ i = 1,2$	p_i^d	<	p_i^h	
Fare to non-hub	p_3	p_3^d	>	p_3^h	iff $\delta > \delta^*$

It is interesting to compare the results in Table 7.1 with those obtained by Borenstein (1989) and Morrison and Winston (1986). These authors found that following the US deregulation, fares on routes to and from hub cities dominated by a single airline, were much higher than their indirect counterparts. Table 7.1 indicates that the basis for these price differentials stems from the change in the network structure (i.e. from FC to HS) and *not* exclusively from the change in the regulatory regime.

Before investigating this issue, we ask under what conditions a monopoly airline firm would switch from a FC to HS. Comparing the profit levels (7.5) with (7.8) yields the following proposition:

Proposition 3

The monopoly airline will operate a HS network as long as passengers valuation of direct flights is less than a threshold value, i.e., $\delta < \delta^{**}$. Otherwise, it would operate a FC network.[16]

It should be emphasized, however, that in a regulated market, meeting the condition in Proposition 3 may be insufficient for airlines to switch to HS, if the regulatory agency prohibits them from abandoning direct, non-hub routes or increasing frequency on hub routes. As pointed out in the introduction section, this indeed was the case prior to the deregulation in the USA.

7.6 Conclusion

The purpose of this paper was to analyze the conditions under which airlines will design their equilibrium route network as a hub-and-spoke. A primary feature of

---

[16]Comparing the profit levels in (7.5) and (7.8) yields

$$\delta^{**} = \left(\frac{\lambda n}{c}\right)^{\frac{\gamma}{1-\gamma}} \left(\frac{1-\gamma}{\gamma}\right) \left[\left(\frac{3}{2^{\gamma}}\right)^{\frac{1}{1-\gamma}} - 3\right].$$

this analysis is the use of demand conditions, such as demand frequency elasticity and willingness to pay for direct (non-hub) flights, to explain the observed widespread use of this network type, the significant increase in route frequency (relative to a fully-connected network case), and the development of a fare structure which discriminates between hub-oriented and non-hub oriented flights. Other models in the literature have emphasized cost conditions such as network, density, and scope economies to explain these phenomena, mainly of the adoption of a hub-and-spoke network. In contrast, our model shows similar results on the basis of demand conditions only while using a cost function which is linear in output. We chose to model a monopoly market structure, since the monopoly airfare perfectly captures the tradeoff in passengers' surplus between flying indirectly via a hub and having a higher frequency service. Furthermore, we show that these results do not require a deregulated market regime since a monopoly airline, which is free to design its equilibrium network, will behave similarly. On the other hand, we have deliberately ignored the effect of fixed costs and airport charges since our purpose was to demonstrate the strategic use of network structure irrespective of these factors.

The principal result from the analysis is that an airline will choose to operate a hub-and-spoke network provided it can compensate passengers for their welfare loss, caused by transporting them to their final destination via the hub, by offering them higher flight frequencies on all hub routes. Adoption of such a strategy by a profit maximizing airline firm will increase its profits relative to the case when it chooses to operate a fully-connected network. This result will hold irrespective of the prevailing market regime (i.e. regulated vs. deregulated) as long as the airline is free to configure its route network to meet its strategic objectives.

Other results from the analysis pertain to the resultant fare structure and to aircraft capacity utilization. These results corroborate observed patterns in aviation markets such as higher fares per mile paid by passengers who begin or end their flight at a hub airport relative to fares charged for non-hub origin and destination flights.

The analysis carried out in this paper assumed away several important factors whose inclusion in this or an alternative model requires further analysis. First, we have considered two extreme types of network: a fully-connected and a hub-and-spoke. While the latter type is indeed the predominant one in contemporary aviation markets, nonetheless, we can also observe a significant number of non-hub routes. It is thus useful to develop a more general model which allows for such a 'mixed' network structure. Hub charges and hub capacity constraints, are key variables which influence the behaviour of airline firms relative to their choice of a network, level of service and fares. Since, in many cases airlines operate from secondary hubs it is desirable to investigate whether the above factors can explain this phenomenon or whether it is intrinsic to the strategic behavior of airlines. Lastly, The effect of competition, following deregulation, certainly requires a specific analysis. In particular, it is necessary to show whether, and under what conditions, the market will converge into an equilibrium solution when airlines compete relative to frequency, fares and market dominance.

At present we are unaware of a model which deals with all of these aspects in order to show which equilibrium network pattern will be selected by airlines.

Acknowledgements

We thank Neil Gandal, Val Lambson and Yossi Spiegel; as well as the participants in workshops given at Bellcore and Boston University for most valuable comments on an earlier draft.

References

Bailey EE, Graham D, Kaplan D (1985) Deregulating the Airlines. The MIT Press, Cambridge, MA.

Bailey EE, Williams JR (1988) Sources of Economic Rent in the Deregulated Airline Industry. Journal of Law and Economics 31:173-202

Barrett S (1992) Barriers to Constestability in the Deregulated European Aviation Market. Transportation Research A 26:159-165

Berechman J, Shy O (1991) Airline Deregulation and the Choice of Network. Tel Aviv University, The Sackler Institute of Economic Studies, Working Paper No. 2-91.

Berry S (1992) Estimation of a Model of Entry in the Airline Industry. Econometrica 60:889-917

Bittlingmayer G (1990) Efficiency and Entry in a Simple Airline Network. International Journal of Industrial Economics 8:245-257

Borenstein S (1989) Hubs and High Fares. Dominance and Market Power in the US Airline Industry. Rand Journal of Economics 20:344-365

Borenstein S (1992) The Evolution of US Airline Competition. Journal of Economic Perspectives 6:45-73

Brueckner JK, Spiller PT (1991) Competition and Mergers in Airline Networks. International Journal of Industrial Organization 9:323-342

Call GD, Keeler TE (1985) Airline Deregulation, Fares and Market Behavior. Some Empirical Evidence. In Daughety AF, Analytical Studies in Transportation Economics. Cambridge University Press.

Doganis R (1991) Flying off Course. Second Edition. London

Hansen M, Kanafani A (1990) Airline Hubbing and Airport Economics. Transportation Research 24: 217-230

Hendricks K, Piccione M, and Tan G (1992) The Economics of Hubs - The Case of Monopoly. Discussion Paper No. 92-09. The University of British Columbia, Department of Economics, Vancouver Canada

Hurdle GJ, Johnson RL, Joskow AS, Werden GJ, Williams MA (1989) Concentration, Potential Entry, and Performance in the Airline Industry. Journal of Industrial Economics 38:119-139

Johnson RL (1985) Networking and Market Entry in the Airline Industry. Journal of Transport Economics and Policy. 19:299-304

McShan S, Windle R (1989) The Implications of the Hub-and-Spoke Routing for Airline Costs and Competitiveness. The Logistics and Transportation Review 25:209-230

Morrison SA, Winston C (1986) Intercity Transportation Route Structure under Deregulation. Some Assessment Motivated by the Airline Experience. American Economic Review. Papers and Proceedings pp 57-65

Pucher J, Ioannides D, Hirschman I (1993) Passenger Transport in the United States and Europe. A Comparative Analysis of Public Sector Involvement. In: Banister D, Berechman J (Eds.) Transport in Unified Europe. North-Holland, Amsterdam

Reynolds AJ (1992) EC and US Air Traffic Patterns. A Comparative Spatial Analysis. Paper presented at the 32nd RSA European Congress, Brussels, August 25-28, 1992

Spiller P, (1989) A Note on Pricing of Hub-and-Spoke Network. Economics Letters 30:165-169

Standard and Poor (1992) Industry Survey. Aerospace and Air Transport. Standard and Poor Corporation, New York

Starr R, Stinchcombe M (1992) Efficient Transportation Routing and Natural Monopoly in the Airline Industry. An Economic Analysis of Hub-Spoke and Related Systems Department of Economics, University of California, San-Diego, Discussion Paper 92-25

Toh RS, Higgins RG (1985) The Impact of Hub and Spoke Network Centralization and Route Monopoly on Domestic Airline Profitability. Transportation Journal. Summer, 16-27

US DOT (1992) National Transport Statistics. Washington DC., Research and Special Programs Administration.

Appendix 7.1

A Round Trips Environment

Consider a single firm operating in an economy described in Figures 7.1c and 7.1d. In view of the problem solved by the airline firm operating one-way flights on a FC network, see (7.3), the problem solved by the airline firm operating round trips on a FC network is now given by

$$\max_{f} \pi^{d} = 6n\left[\delta + \frac{\lambda}{\gamma}(f)^{\gamma}\right] - 6cf$$

which yields the same frequency and prices as under the one-way environment given in (7.5).

The problem solved by the airline firm operating round trips on a HS network is now given by

$$\max_{f} \pi^{h} = 4n\delta + 6n\frac{\lambda}{\gamma}(f)^{\gamma} - 4cf$$

which again yields an identical solution to the problem solved in (7.7).

III

Public Sector Instruments

8 Public Investment Criteria for a Comprehensive Transport System Using an Interregional Input-Output Programming Model

Hirotada Kohno
University of Tsukuba

8.1 Introduction

This paper focuses on the derivation of the criteria to decide (i) where, (ii) how, and (iii) by what amount, public funds should be allocated to investment in transportation infrastructure such as railways, new trunk lines, highways, expressways, ports, airports etc. The methodology adopted is that of an extended interregional input-output programming model. This approach solves concurrently the problem of the measurement of the social benefits of private and public sector production, and the problem of derivation of the public investment criteria themselves.

The problem of how to allocate funds to public investments in infrastructure has been tackled by means of various valuation methods, such as (i) the computation of the Net Present Value or Benefit-Cost Ratio, (ii) general optimization subject to various constraints; (iii) a static or dynamic interregional input-output programming model, and (iv) a discrete type of interregional input-output programming model. The historical methodological development from (i) to (iv) has been one of progress toward more sophistication and more *overall* and *consistent* approaches. In this paper we shall use a *static* type of interregional input-output programming model (i.e. method (iii) above). The model is a generalized version of the Moses (1960) model.[1]

There is a great difficulty which awaits solution when one tries to solve the problem of the optimal allocation of infrastructure in a multi-regional comprehensive transport system. This difficulty is that the comprehensive transport system is formed by vehicles such as automobiles, trucks, regular trains, high speed trains, aeroplanes, ships, etc., of which some are used for both freight and passengers (e.g. trains and aeroplanes), some are used just for freight (e.g. container ships) and some are used exclusively for passenger transportation (e.g. passenger cars and high speed trains). Hence the analytical framework must

---

[1]An application of a discrete type of *dynamic* interregional input-output programming model for public investment can be found in Kohno and Higano (1995).

incorporate both freight services and passenger services. Consequently, the industrial input-output tables conventionally used for a freight flow analysis are not sufficient and the framework must also deal consistently with passenger services.

Of course, the business trip component of the passenger flows can be assumed to be proportional to the freight flows. Thus, the most important task is to formulate a model for final consumptive demand for transportation of the population. The task to create a column vector for final consumptive travel demand in an input-output framework has proved to be difficult. However, it is also difficult to find a suitable objective function for 'person trips' in the same way as we construct the programming model of the maximization of GNP (i.e. value added) through the commodity flows in the spatial economy. The reason is that the travel component of final consumptive demand increases utility directly rather than as an input. Hence, we use the concept of the marginal rate of substitution between goods (money) and leisure to adopt a methodology to derive the elements of the valuation row vector (see Kohno and Yoshida 1981a, pp 62-64).

Moreover, the model has the limitation of being comparatively static (with initial and terminal periods/points) rather than dynamic. Hence, dynamic processes are not analysed. This implies that we must assign a predetermined value to every element in the right hand side constraint column vector of the programming model; this implies, for example, in the analysis of production activity that the final demand by region at the terminal point is exogenously given.

Consequently, to analyse the flow of person trips in the same way we must find something suitable for the right hand side constraint, namely future (terminal point) total travel demand by mode (transport vehicle) for each region. Without this constraint, we cannot consider the person trip flows.

Hence, by means of these two *given* constraints (viz. the rate of valuation row vector and the total travel demand by mode for each region) we can introduce the optimization among alternative travel demand activities in the model. This builds into the model the degree of freedom of where people want to travel and what kind of travel mode they choose. Various alternative possibilities will exist, such as travelling to a tourist spot in the neighbourhood, to an attraction at a medium distance, or one at a longer distance. Moreover, trips may vary between being overnight, four to five nights, one week or a fortnight. The choice of the travel mode follows from the alternative travel possibilities.

8.2 Building Sightseeing Travel Activities in the Model

8.2.1 Measurement of Travel (Sightseeing) Consumption Activities

To put freight and passengers on an equal footing in the model, it is necessary to create a travel (sightseeing) consumption column vector, together with the existing industrial activities by sector (see also Kohno and Yoshida 1981a, pp 60-67).

It must be firstly recognised that the flows of goods and services in the industrial input-output table are presented in monetary form and that the passenger flow is not treated explicitly, but forms part of final consumption demand such as sightseeing travel.

However, even in the existing input-output table there are private passenger automobile activities, bus activities, railway activities, domestic aeroplane passenger activities, etc. as part of the production (column) and consumption (row) of the *transport services industries*. The output of a transport services industry shows the total freight charge which the industry concerned has made for the transport services provided in the year concerned (except for private passenger transportation by car). Output refers here not to its ordinary meaning, but to the shipment base between any origin and destination region.

If we convert the transport services output to one in person kilometres, it seems that we can get a passenger flow table. However, this will not be the case because the 'rate of valuation' row vector corresponding to these activities is the rate of value added of the transport services industry, which does not necessarily correspond to information on where, how many days, and by which mode, the population of each city in each region wants to travel in the future.

It is expected, for example, that the sum and composition of the valuation of a new trunk line in the future will be determined by the travel needs of the population in each big city in each region. Therefore, it is proposed in our study that by adding the new 'sightseeing consumption' column vector to the existing transport service column vector, we define this result to be the 'sightseeing travel service' column vector.

If we can construct such a sightseeing travel service column vector, we will be able to make these activities operate on a full scale either *within* the industrial input-output table, or *in tandem* with the table, provided that: (i) the future total sightseeing travel demand (within the neighbourhood or over a short/medium/long distance) has been forecasted exogenously; and (ii) the 'rate of valuation' column vector by vehicle mode of the sightseeing travel service activities has also been derived by a separate study.

It follows that by this method the transport service produced by the highway facilities, railway facilities etc. are consumed simultaneously by the sightseeing travel service demand and by the freight flow demand which is obtained from the original input-output table. Hence, by means of this approach we are able to deal with the freight and passenger flows simultaneously. The means through which freight and passengers are combined technically is the conversion of the number of passengers to tonnage, which in turn determines the capacity utilization of the transport facilities.

It is obvious that normally we will only be able to gain leisure utility from sightseeing travel when expenditure is incurred on various items such as suitable clothing, a bag, meals and souvenirs, accommodation and payments to the transport service industries. We cannot obtain leisure by paying out travel expenses to the transport service industries only. The transport service sector in the input-output table shows only the production (column) and consumption (row) of the transport services supplied by commercial carriers and private cars. All the items such as clothing, other personal effects, accommodation, meals, souvenirs, etc. except for the transportation payments to the transport service industries, constitute the sightseeing consumption expenditure vector, which is part of final consumption within the final demand column vector.

8.2.2 Specification of the Sightseeing Travel Consumption Activities by Transport Mode, Destination and Route

Specification of the Alternatives Regarding the Route Between Any Pair of Regions

The actual railway or highway network is very complicated. Many alternative transport routes exist from the origin to the destination in any pair of regions. As it is very tedious to include all these alternative routes in the model and derive the sightseeing travel consumption demand activities corresponding to these routes, we omit the alternative routes which are expected to be inefficient from our computational work. Inefficient here means that the transport route is unlikely to be selected where the travel time required will increase out of all proportion to the distance when the distance increases. For example, when a person travels from Hokkaido to Kyushu it is very unlikely that the route on the Japan Sea side will be selected, irrespective of whether travel will be by rail or road.

Taking this into account, the alternative transport routes can be specified. Two examples are as follows:

(1) Railway
Tohoku ↔ Hokuriku

(i) by way of Tokyo:

●——●——●——●——●
Sendai Tokyo Takasaki Toyama Kanazawa

(ii) by way of Niigata

●——●——●——●——●
Sendai Fukushima Niigata Toyama Kanazawa

(2) Highway
Hokkaido ↔ Hokuriku

(i) by way of Sendai, Niigata:

●——●——●——●——●
Sapporo Sendai Niigata Toyama Kanazawa

(ii) by way of Niigata:

●——●——●——●——●
Sapporo Akita Niigata Toyama Kanazawa

Alternative Combinations of Mixed Railway and New Trunk Line Passenger Services

The setting up of a mixed railway and new trunk line type of passenger service through any one route between any two regions and alternative combinations will be explained by considering travel between Hokkaido and Kyushu, where the existing new trunk line, newly proposed new trunk line and the railway are shown as - - - - - , ---------, and ——— respectively; and the transportation route is

The alternative combinations among railways and new trunk lines are as follows:

(1) Railway only

(2) Railway, new trunk line mixed type [I]

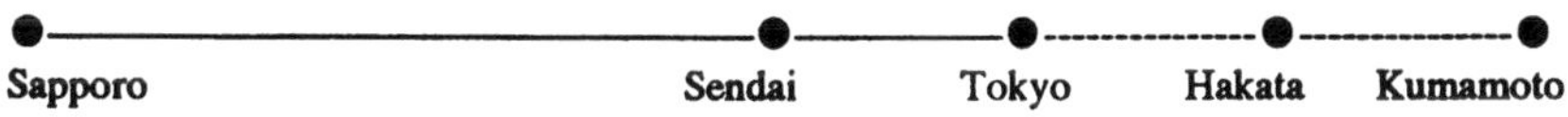

(3) Railway, new trunk line mixed type [II]

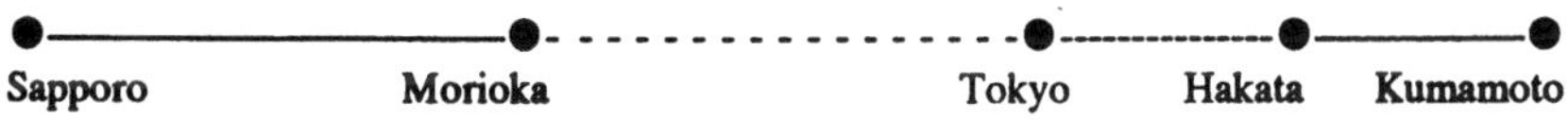

(4) Railway, new trunk line mixed type [III]

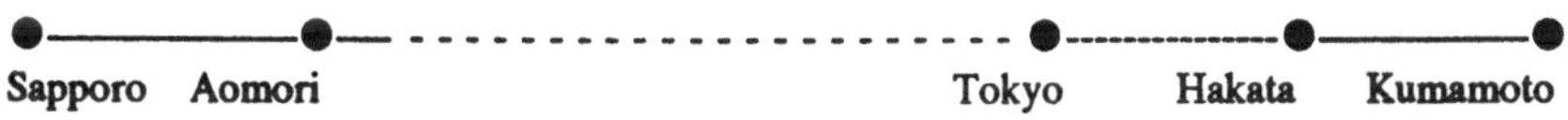

(5) Railway, new trunk line mixed type [IV]

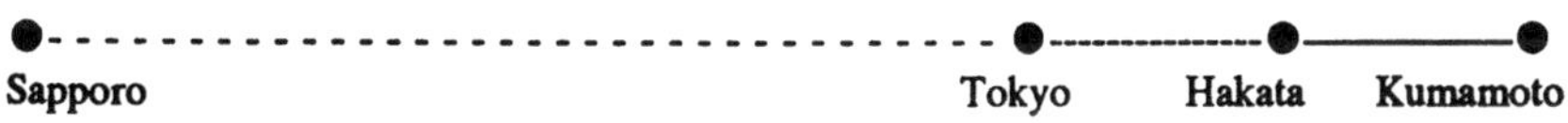

(6) Railway, new trunk line mixed type [V]

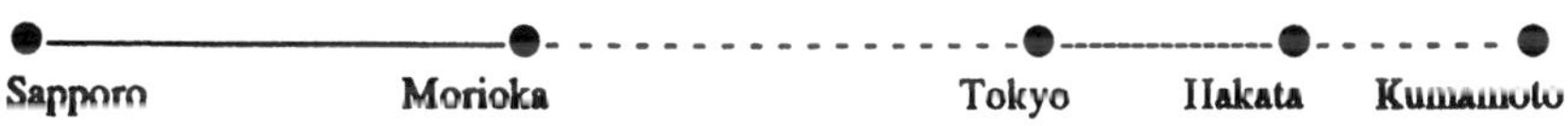

(7) Railway, new trunk line mixed type [VI]

(8) Railway, new trunk line mixed type [VII]

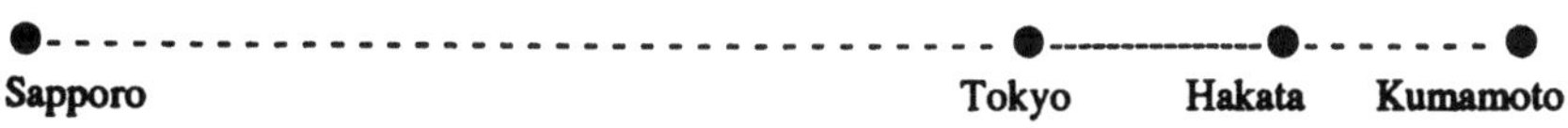

As can be seen above, there are eight alternative combinations for the railway and new trunk line mixed type passenger service between Hokkaido and Kyushu.

8.3 Formulation of the Model

The details of the model can be found in Kohno (in Japanese, 1975) and in Kohno (in English, forthcoming). The model is an interregional input-output programming model for the overall valuation of investments in social overhead capital. The linear objective function is

$$Z = (\epsilon_1,\epsilon_2,\ldots,\epsilon_m;\mu;\rho_r,\rho_n,\rho_h,\rho_e,\rho_s,\rho_f,\rho_g;0)[X,U,\Delta r,\Delta n,\Delta h,\Delta e,\Delta s,\Delta f,\Delta g,\Delta l]'. \tag{8.1}$$

The linear objective function represents a first approximation to public welfare by means of an amalgamation of output and investment components.

The problem of the overall valuation of investment in social overhead capital is one of *maximization* of the objective function subject to the following linear constraints (see also Kohno and Yoshida 1981a, pp 68-74):

$$\left[\begin{array}{c|c|cccccccc}
I-A(V)_T & & & & & & & & & \\ \hline
B_T & & & & & & & & & \\ \hline
R_T & & -I & & & & & & & \\
N_T & & & -I & & & & & & \\
H_T & & & & -I & & & & & \\
E_T & & & & & -I & & & & \\
S_T & & & & & & -I & & & \\
F_T & & & & & & & -I & & \\ \hline
G_T & Iu & & & & & & & -I & \\ \hline
 & & & & & c & & & & \\ \hline
L_T & & & & & & & & & -I \\ \hline
 & & & & & & & & & m \\ \hline
P_T & & & & & & & & & \\ \hline
D_T & & & & & & & & & \\ \hline
W & & & & & & & & &
\end{array}\right]
\begin{bmatrix} X \\ U \\ \Delta r \\ \Delta n \\ \Delta h \\ \Delta e \\ \Delta s \\ \Delta f \\ \Delta g \\ \Delta l \end{bmatrix}
\begin{array}{c} \geq \\ \leq \\ \leq \\ \leq \\ \leq \\ \leq \\ \leq \\ \leq \\ \leq \\ \leq \\ \leq \\ \leq \\ \leq \\ \leq \\ \leq \end{array}
\begin{bmatrix} Y^*_T \\ b^*_T \\ r^*_0 \\ n^*_0 \\ h^*_0 \\ e^*_0 \\ s^*_0 \\ f^*_0 \\ g^*_0 \\ C^*_{T-0} \\ l^*_{T0} \\ M^*_{T-0} \\ p^*_T \\ d^*_T \\ w^*_T \end{bmatrix},
\quad
\begin{bmatrix} X \\ U \\ \Delta r \\ \Delta n \\ \Delta h \\ \Delta e \\ \Delta s \\ \Delta f \\ \Delta g \\ \Delta l \end{bmatrix} \geq 0 \tag{8.2}$$

Note that the additional notation in Kohno and Yoshida (1981a), as compared with Kohno (1975) is as follows:

(1) The matrix $[I - A(V)_T]$, where $A = [A1 \mid A_2]$:

The A matrix represents the regional input-output coefficients matrix; A_1 corresponds to production activities so far and A_2 corresponds to sightseeing passenger travel activities investigated here; V is the input from the transport service which is specified by transport model, destination and transport route. This matrix is the largest and most important one in the model. The rigorous specification can be found in Kohno (1975) and Kohno (forthcoming).

(2) The matrix $[X] = [X_i^{pq}]$ where $[X]=[X_1 \mid X_2]$:

Here X_1 is the shipment variables column vector and X_2 is the travel (sightseeing) consumption variables column vector. These are the main activity level variables, which show the shipment (or travel consumption) of good i (or travel service) from origin region p to destination region q, where x represents the type of transport mode, y represents the type of transport route and z represents the type of production (or passenger travel). However, alternative production activities are not built-in this time.

(3) N_T, E_T, F_T: necessary transport service-shipment coefficients for the new trunk line, expressway and airport respectively.
(4) G_T: necessary public service-shipment coefficients.
(5) P_T: pollution effluent-shipment coefficients.
(6) D_T: travel (sightseeing) demand (number of persons) by distance zone - travel expenditure coefficients.
(7) I_u: consumption coefficient matrix of social overhead capital service, where all the diagonal elements of $[I_u]$ consist of ones and make a "unit matrix" composed of $+1$s; that is, here the amount as stock is correctly supplied as the flow at every moment.
(8) b_T^*: industrial production capacity constraint by region and by sector in the final period.
(9) n_0^*, e_0^*, f_0^*: transport service capacity column vector at the initial point at the right hand side.
(10) g_0^*: public service capacity column vector at the initial point at the right hand side.
(11) d_T^*: the maximum capacity for fowarding travel (sightseeing) demand to the destinations by distance zones, such as the neigbourhood area, the area at medium distance and the area at long distance (measured from a target point).
(12) p_T^*: pollution effluent absorption capacity column vector at the initial point at the right hand side.
(13) $[U]=[U_{g1},U_{g2},U_{g3},U_{g4}]$: consumption (level) variable column vector of social overhead capital service, where the subscripts g_1, g_2, g_3 and g_4 represent sewage treatment, public housing, public hospitals and public libraries respectively.
(14) Δn, Δe, Δf, Δg: additional variables of transport service capacity (Δn: new trunk line; Δe: expressway; Δf: airport) and of social overhead capital service (Δg)
(15) $(\epsilon) = (\epsilon_1, \epsilon_2,..., \epsilon_m)$: the rate of valuation row vector of industrial sectors and travel consumption sectors.

(16) $(\mu) = (\mu_1, \mu_2, \mu_3, \mu_4)$: rate of valuation for the service provided by social overhead capital; see (13) for the meaning of the subscripts corresponding to g_i (i=1,2,3,4).

(17) (r): additional variables of social overhead capital service for private production and the public sector, where $(\rho) = (\rho_r, \rho_n, \rho_h, \rho_e, \rho_s, \rho_f, \rho_g)$ and $(\rho_g) = (\rho_{g1}, \rho_{g2}, \rho_{g3}, \rho_{g4})$. For any other notation, see Kohno (forthcoming).

8.4 Overall Valuation Criteria for Constructing Social Overhead Capital

8.4.1 Meaning of the Overall Valuation

How should we determine investment in transportation infrastructure such as railways (r), new trunk lines (n), highways (h), expressways (e), ports (s), airports (f), etc. and the investment in public facilities (g) such as sewerage, public housing, public hospitals, public libraries, etc? The objective function for this problem is neither maximization of GNP for any specific region, nor maximisation of earnings for any enterprise, but - instead - the 'overall' and 'consistent optimization of the aggregate level of welfare over the whole nation, which is the sum of the welfare of each citizen in every region. Whatever objective function will be formulated other than the one just mentioned, it may without fail lead to contradictions, and it may be inconsistent with other separate objectives (see Kohno and Yoshida 1981a, pp 74-76).

Hence, to achieve an 'overall valuation', we must define an objective function of 'optimisation of the welfare level index over the whole nation'. The overall valuation refers to having the industrial input-output relationships built into the model. The objective function must be consistent in three ways. First, all the industries must be built in, globally without leakage and overlap. Second, it must be consistent in the sense that industrial output and capital are determined concurrently with the construction (allocation) of social overhead capital facilities. Finally, it must be consistent in the sense that the demand and supply for freight transport services are determined simultaneously with the level of each type of industrial activity.

The interregional input-output programming model is the best model or technique to meet all these requirements.

Then, to derive the optimal level of social overhead capital involves adding the optimal investment in social overhead capital $[\Delta r, \Delta n, \Delta h, \Delta e, \Delta s, \Delta r, \Delta g]$ to the constraint column vector of social overhead capital in the initial period $[\overset{*}{r}_0, \overset{*}{n}_0, \overset{*}{h}_0, \overset{*}{e}_0, \overset{*}{s}_0, \overset{*}{f}_0, \overset{*}{g}_0]$, which is part of the right hand-side constraint column vector $\left[\overset{*}{Y}_T, \overset{*}{b}_T, \overset{*}{r}_0, \overset{*}{n}_0, \overset{*}{h}_0, \overset{*}{e}_0, \overset{*}{s}_0, \overset{*}{f}_0, \overset{*}{g}_0, \overset{*}{C}_{T-0}, \overset{*}{l}_{T0}, \overset{*}{M}_{T-0}, \overset{*}{p}_T, \overset{*}{d}_T, \overset{*}{w}_T\right]$ and derive equation (8.3) for the final period as follows:

$$\begin{bmatrix} Y_T^* \\ b_T^* \\ r_0^* + \Delta r \\ n_0^* + \Delta n \\ h_0^* + \Delta h \\ e_0^* + \Delta e \\ s_0^* + \Delta s \\ f_0^* + \Delta f \\ g_0^* + \Delta g \\ C_{t-0}^* \\ l_{t0}^* \\ M_{T-0}^* \\ p_T^* \\ d_T^* \\ w_T^* \end{bmatrix} = \begin{bmatrix} Y_T^* \\ B_T^* \\ r_t^* \\ n_y^* \\ h_t^* \\ e_t^* \\ s_t^* \\ f_t^* \\ g_t^* \\ c_{t-0}^* \\ l_{t0}^* \\ m_{t-0}^* \\ p_t^* \\ d_t^* \\ w_t^* \end{bmatrix} \tag{8.3}$$

It should be noted that there are several ways of setting up the vector of required additional capital $[\Delta r, \Delta n, \Delta h, \Delta e, \Delta s, \Delta r, \Delta g]$. Usually, where an optimisation of social overhead capital is attempted, various patterns of the allocation of this social overhead capital correspond to various visions:

$$\begin{matrix} \text{vision 1} & \text{vision 2} & \cdots & \text{vision u} \\ \begin{bmatrix} \Delta r_1 \\ \Delta n_1 \\ \Delta h_1 \\ \Delta e_1 \\ \Delta s_1 \\ \Delta f_1 \\ --- \\ \Delta g_1 \end{bmatrix} & \begin{bmatrix} \Delta r_2 \\ \Delta n_2 \\ \Delta h_2 \\ \Delta e_2 \\ \Delta s_2 \\ \Delta f_2 \\ --- \\ \Delta g_2 \end{bmatrix} & & \begin{bmatrix} \Delta r_u \\ \Delta n_u \\ \Delta h_u \\ \Delta e_u \\ \Delta s_u \\ \Delta f_u \\ --- \\ \Delta g_u \end{bmatrix} \end{matrix} \tag{8.4}$$

such as 'development of underdeveloped regions' (vision 1), 'removal of congestion in large cities" (vision 2), "creation of garden cities' (vision 3), etc.

Provided that every optimization model corresponding to such visions can be solved, it follows that the resulting values of the objective function form a solution vector $(Z_1, Z_2, \ldots, Z_u)$ and the *optimum optimorum* among the solutions will be

$$Z_i^{*} = Max\ [Z_1,Z_2,\ldots,Z_u]\ i = 1,2,\ldots,u.$$

However, the selection of visions is rather arbitrary, such as is the case frequently with the policy formulation of the Japanese government, and there is then a possibility that what is truly optimal is undetermined.

This defect represents a problem to be solved. So, if we adopt in the model developed here a formulation in which the target of the allocation of the transport capital among various modes of facilities itself is endogenous, then we will derive the socially optimal 'allocation vector' without fail.

8.4.2 Overall Valuation Criteria

The resource allocation problem in the format of the previous section, in which the additional amount of social overhead capital is added to each component of the constraint column vector in the initial period, will usually be solved by making the imputed price to the resource constraint to be Merkmal.

Such an imputed price is a unit price of the resource in the case where the optimum quantities in the objective function are allocated to every contributing resource, without residual and according to their contribution. From another point of view, it indicates the rate of change to the objective function resulting from relaxing or tightening any constraint element in the right hand side column vector of resource constraints by one unit. That is,

$$\frac{\partial Z}{\partial r} = \Theta_r,\ \frac{\partial Z}{\partial n} = \Theta_n,\ \frac{\partial Z}{\partial h} = \Theta_h,\ \frac{\partial Z}{\partial e} = \Theta_e,\ \frac{\partial Z}{\partial s} = \Theta_s,\ \frac{\partial Z}{\partial f} = \Theta_f,\ \frac{\partial Z}{\partial g} = \Theta_g,$$

where $\Theta_r,\ldots$ are imputed prices for the amount of railway facilities in terms of the stock, etc. Of course the imputed price is zero when the corresponding resource constraint is not binding and positive where it is binding. We can therefore judge the relative importance of a resource constraint by the magnitude of the imputed price.

However, as the imputed price corresponds to a one unit adjustment of each constraint, it follows that when a large number of units is added to the existing facilities, it is not certain that the imputed price will still indicate the importance of the facilities.

Moreover, given that the units of the various resources are different in physical terms, it is also very difficult to make a direct comparison of imputed prices of resources.

In contrast, in the overall valuation model proposed in this paper, the increment to the physical resource constraint is carried out in a formulation in which the constraint in terms of total funds is converted to a constraint in terms of physical facilities. A 'physical ratio' such as 'the coefficient of construction costs for the required transport facilities and transport volume' is used. Such a ratio has a dimension of, for example, one million yen per one million vehicle kilometres (Kohno 1975, pp 63-64).

Next, if we carry out a normalisation by dividing each imputed price of Θ_r, Θ_n, Θ_h, ... etc. by the corresponding 'physical ratio' we do not derive an imputed price per facility providing one million vehicle kilometres, but one per *facility of one million yen,* as follows:

$$\Theta_r^* = \frac{\Theta_r}{c_r},\ \Theta_n^* = \frac{\Theta_n}{c_n},\ \Theta_h^* = \frac{\Theta_h}{c_h},\ \Theta_e^* = \frac{\Theta_e}{c_e},\ \Theta_s^* = \frac{\Theta_s}{c_s},\ \Theta_f^* = \frac{\Theta_f}{c_f},\ \Theta_g^* = \frac{\Theta_g}{c_g}\ .$$

Using these modified shadow prices, we can derive by solving the model corresponding to each vision, an imputed price vector which satisfies for the ith vision an inequality such as

$$\Theta_{ri}^* > \Theta_{ni}^* > \Theta_{hi}^* > \Theta_{ei}^* \tag{8.5}$$

can be obtained. This implies that when we construct the various facilities according to the order of magnitude of the imputed price, an optimal allocation can be derived, at least as far as the marginal increment in the facilities is concerned. It follows that we really have obtained the 'public investment criteria' in terms of the ordinary meaning.

When the public investment is a large indivisible unit, there is a possibility that the *ex post facto* price may be drastically different from the imputed price before the investment took place, which is a disadvantage when the imputed price for investment choice is made Merkmal.

However, as the joint optimisation of the economic activity sector and the closely linked social overhead capital supply sector is executed thoroughly, the total funds given as one component of the right hand side constraint will be allocated by our overall valuation without residual and in order of magnitude of the rate of urgency. The modified imputed price will then be:

$$\frac{\Theta_r}{c_r} = \frac{\Theta_n}{c_n} = \frac{\Theta_h}{c_h} = \frac{\Theta_e}{c_e} = \frac{\Theta_e}{c_e} = \frac{\Theta_s}{c_s} = \frac{\Theta_f}{c_f} = \frac{\Theta_g}{c_g}$$

that is,

$$\Theta_r^* = \Theta_n^* = \Theta_h^* = \Theta_e^* - \Theta_s^* = \Theta_f^* = \Theta_g^*\ , \tag{8.6}$$

where the weighted imputed price for each type of social overhead capital has been equalised. Hence, our model of the 'public investment allocation framework' has the desirable feature that the public investment funds can be allocated such that the marginal rate of importance of all facilities, when they are utilised entirely, are equalised. Otherwise, we will be able to improve the optimum value of the model by reallocation.

As stated above, we are able to determine our social overhead capital augmentation programme automatically with our overall valuation model by maximising the value index, which is a linear combination of the activity column vector and the rate of valuation row vector. All the direct economic effects and the indirect effects brought about by the social overhead capital augmentation programme will be included. Hence, it follows that equation (8.6) turns out to be the overall valuation criterion for the social overhead capital augmentation programme which we have been pursuing.

In the case where equation (8.6) has been realised, the value of the endogenous investment vector in the model, i.e.

$$\left[\Delta \tilde{r}, \Delta \tilde{n}, \Delta \tilde{h}, \Delta \tilde{e}, \Delta \tilde{s}, \Delta \tilde{f}, \Delta \tilde{g}\right]' \tag{8.7}$$

turns out to be the *optimum additional amount of each type of social overhead capital construction.*

It goes without saying that what is really meant by these $\Delta\tilde{r}$.., etc. are those disaggregated by region, by link between any two interchanges, by route, etc.

Multiplying the column vector (8.7) by the coefficient row vector of the construction cost for the necessary social overhead capital facilities $c = (c_r, c_n, c_h, c_e, c_s, c_f, c_g)$ (on the left hand side of (8.7)), then

$$(c_r\ c_n\ c_h\ c_e\ c_s\ c_f\ c_g)\begin{bmatrix}\Delta \tilde{r}\\ \Delta \tilde{n}\\ \Delta \tilde{h}\\ \Delta \tilde{e}\\ \Delta \tilde{s}\\ \Delta \tilde{f}\\ \Delta \tilde{g}\end{bmatrix} = \left[C^{\bullet}_{T-0}\right]$$

can be obtained which, needless to say, coincides with the total public investment funds given. That is,

$$\left[c_r.\Delta \tilde{r}, c_n.\Delta \tilde{n}, c_h.\Delta \tilde{h}, c_e.\Delta \tilde{e}, c_s.\Delta \tilde{s}, c_f.\Delta \tilde{f}, c_g.\Delta \tilde{g},\right]' \tag{8.8}$$

turns out to be the vector of the optimal investment in social overhead capital, such as railways, new trunk lines, highways, expressways, ports, airports and public facilities (public parks, libraries, hospitals, housing, etc.), which therefore provides symbolically a direct answer to the question posed in this paper.

8.5 An Application of the Model to the Augmentation of the Japanese Comprehensive Transport System

This empirical application investigated, and derived public investment criteria, to determine within the more comprehensive framework whether five proposed new trunk lines in relatively rural areas (viz. Hakodate-Sapporo, Morioka-Aomori, Nagaoka-Maibara, Tosu-Nagasaki and Hakata-Kagoshima) should be constructed. Our extended model was applied for this purpose. Hence, we provided a general equilibrium-like generalisation of a case by case benefit-cost analysis for the individual projects.

We were also able by means of this technology to compare the construction of an airport in the rural area with a new trunk line in the same or another area. We derived our public investment criteria regarding these project proposals based on the overall and consistent analytical framework.

For the measurement of the rate of valuation (row vector) by transport mode and by distance zone we applied our submodel. This model was developed to derive the 'rate of valuation for travel service choice' consistently with the coefficient of valuation (rate of value added) corresponding to the G.N.P. from the marginal rate of commodity substitution. This rate of substitution is not the one between one commodity and another, but between income and transport service, income and the services provided by the social overhead capital, etc., based on the theory of Keeney (1974). A related investigation of the marginal rate of income to noise pollution substitution has also been fairly systematically carried out (see Kohno and Yoshida 1981b).

To save space, only the final results of the optimal allocation of public investment funds among the comprehensive transport system are presented in Table 1. These are as yet the most detailed and comprehensive results derived from the normative model. However, it must be recalled that this is a comparative static model. The sections between any pairs of interchange such as Sapporo-Kitahiroshima are the current real ones. Socio-economic measurements such as in this example are useful to any real policy making situation as policy-supporting prescriptions.

Table 8.1. Optimal allocation of public investment

Part A Public investment in transport facilities

Row Number	Transport Facilities and link (Section)	Distance (km)	The derived optimum amount of public investment (million yen)	The imputed price of the constraint row
	Interregional Transport Facilities			
	railway			
1	Sapporo-Hakodate	286.3	236088	0.0559
2	Hakodate-Aomori	54.0	41436	0.0099
3	Aomiri-Morioka	203.9	122458	0.0336
4	Morioka-Sendai	183.5	123538	0.0336
5	Sendai-Fukushima	79.0	31567	0.0140
6	Fukushima-Ueno	269.3	124608	0.0559
7	Tokyo-Nagoya	366.0	9919	0.0559
8	Aomori-Nagaoka	505.6	47738	0.0839
9	Nagaoka-Toyama	190.8	23868	0.0419
10	Toyama-Kanazawa	59.5	6555	0.0112
11	Kanazawa-Tsuruga	130.7	0	0.0000
12	Tsuruga-Maibara	45.9	0	0.0000
13	Tokyo-Takasaki	105.0	0	0.0000
14	Takasaki-Niigata	228.9	0	0.0000
15	Nagoya-Maibara	79.9	1798	0.0140
16	Maibara-Osaka	110.5	2697	0.0210
17	Osaka-OPkayama	176.5	5795	0.0336
18	Okayama-Uno	32.9	773	0.0060
19	Uno-Takamatsu	18.0	352	0.0033
20	Takamatsu-Ikeda	76.5	1801	0.0140
21	Okayama-Hiroshima	161.9	2189	0.0280
22	Hiroshima-Hakata	281.7	0	0.0000
23	Hakata-Tosu	28.6	0	0.0000
24	Tosu-Nagasaki	125.3	0	0.0000
25	Tosu-Kumamoto	89.8	0	0.0000
26	Kumamoto-Kagoshima	201.9	0	0.0000
	new trunk line			
27	Sapporo-Hakodate	316.0	20948	0.0117
28	Hakodate-Aomori	54.0	3572	0.0020
29	Aomori-Morioka	170.0	8990	0.0063
30	Morioka-Sendai	171.0	0	0.0000
31	Sendai-Fukushima	70.0	0	0.0000
32	Fukushima-Uno	255.0	0	0.0000
33	Tokyo-Nagoya	342.0	865188	0.0126
34	Niigata-Toyama	230.2	0	0.0000
35	Toyama-Kanazawa	53.9	0	0.0000
36	Kanazawa-Tsuruga	118.4	0	0.0000

Table 8.1. Optimal allocation of public investment (continued)

37	Tsuruga-Osaka	110.0	0	0.0000
38	Tokyo-Takasaki	108.0	0	0.0000
39	Takasaki-Toyama	302.1	0	0.0000
40	Nagoya-Maibara	66.2	105231	0.0043
41	Maibara-Osaka	107.1	172516	0.0070
42	Osaka-Okayama	161.0	0	0.0000
43	Okayama-Sakaide	49.2	0	0.0000
44	Sakaide-Ikeda	56.9	0	0.0000
45	Okayama-Hiroshima	145.0	0	0.0000
46	Hiroshima-Hakata	248.0	0	0.0000
47	Hakata-Tosu	28.6	0	0.0000
48	Tosu-Nagasaki	120.0	0	0.0000
49	Tosu-Kumamoto	71.0	0	0.0000
50	Kumamoto-Kagoshima	170.0	0	0.0000
	ordinary road *1			
51	Sapporo-Kitahiroshima	12.0	512	0.0003
52	Kitahiroshima-Tomakomaihigashi	49.0	0	0.0000
53	Tomakomaihigashi-Toya	83.0	3530	0.0023
54	Toya-Hakodate	153.0	61840	0.0042
55	Hakodate-Aomori	113.0	45806	0.0031
56	Aomori-Ikarigaseki	44.0	0	0.0000
57	Ikarigaseki-Ashiro	64.0	2738	0.0018
58	Ashiro-Sendai	238.0	0	0.0000
59	Sendai-Tokyo	326.0	57741	0.00910
60	Aomori-Niigata	465.0	0	0.0000
61	Sendai-Niigata	229.0	0	0.0000
62	Niigata-Nagaoka	62.0	0	0.0000
63	Tokyo-Tsukiyono	135.0	20	0.0037
64	Tsukiyono-Yuzawa	36.0	1498	0.0010
65	Yuzawa-Nagoaka	78.0	12	0.0021
66	Nagaoka-Kakizaki	47.0	7	0.0013
67	Kakizaki-Asahira	91.0	3780	0.0025
68	Asahi-Kanazawa	102.0	15	0.0028
69	Kanazawa-Maibara	181.0	0	0.0000
70	Tokyo-Nagoya	347.0	0	0.0000
71	Nagoya-Maibara	59.0	0	0.0000
72	Maibara-Osaka	110.0	0	0.0000
73	Osaka-Himeji	85.0	28700	0.0023
74	Himeji-Bizen	53.0	0	0.0000
75	Bizen-Okayama	36.0	12180	0.0010
76	Ochiai-Okayama	60.0	10711	0.0016
77	Okayama-Kurashiki	25.0	4464	0.0007
78	Kurashiki-Kojima	23.9	6299	0.0007
79	Kojima-Sakaide	18.4	4826	0.0005
80	Sakaide-Kawanoe	47.0	12400	0.0013
81	Kurashiki-Fukuyama	48.0	4075	0.0013
82	Fukuyama-Shiwa	65.0	33760	0.0018
83	Shiwa-Hiroshima	37.0	3138	0.0010

Table 8.1. Optimal allocation of public investment (continued)

84	Hiroshima-Hatsukaichi	25.0	2126	0.0007
85	Hatsukaichi-Iwakuni	20.0	7126	0.0006
86	Iwakuni-Yuda	87.0	7422	0.0024
87	Yuda-Kitakyusya	72.0	6136	0.0020
88	Kitakyusyu-Wakamiya	44.0	3739	0.0014
89	Wakamiya-Tosu	52.0	4428	0.0014
90	Tosu-Kanzaki	13.0	0	0.0000
91	Kanzaki-Oshmura	80.0	0	0.0000
92	Oshmura-Nagasaki	18.0	0	0.0000
93	Tosu-Kumamoto	128.0	10824	0.0035
94	Kumamoto-Yatsushiro	43.0	303	0.0012
95	Yatsushiro-Ebino	63.0	17575	0.0017
96	Ebino-Kagoshimakita	71.0	499	0.0020
97	Kagoshimakita-Kagoshima expressway	5.0	6625	0.0007
98	Sapporo-Kitahiroshima	12.0	4381	0.0003
99	Kitahiroshima-Tomakomaihigashi	49.0	18669	0.0014
100	Tomakomaihigashi-Toya	83.0	30238	0.0023
101	Aomori-Ikarigaseki	44.0	11773	0.0008
102	Ikarigaseki-Ashiro	64.0	16568	0.0012
103	Ashiro-Sendai	238.0	63074	0.0043
104	Sendai-Tokyo	326.0	602318	0.0060
105	Niigata-Nagaoka	62.0	0	0.0000
106	Tokyo-Tsukiyono	135.0	296289	0.0062
107	Tsukiyono-Yuzawa	36.0	78344	0.0016
108	Yuzawa-Nagaoka	78.0	170722	0.0036
109	Nagaoka-Kakizaki	47.0	103334	0.0022
110	Kakizaki-Asahi	91.0	182140	0.0033
111	Asahi-Kanazawa	102.0	204395	0.0037
112	Kanazawa-Maibara	181.0	13130	0.0066
113	Tokyo-Nagoya	347.0	880316	0.0161
114	Nagoya-Maibara	59.0	991838	0.0027
115	Maibara-Osaka	110.0	1349935	0.0051
116	Osaka-Achiai	151.0	1172776	0.0055
117	Ochiai-Hiroshima	181.0	1401245	0.0066
118	Hiroshima-Yuda	149.0	1726048	0.0054
119	Himeji-Bizen	53.0	9105	0.0019
120	Okayama-Kurashiki	25.0	4301	0.0009
121	Kurashiki-Kojima	23.9	6337	0.0009
122	Kojima-Sakaide	18.4	4878	0.0007
123	Sakaide-Kawanoe	47.0	12609	0.0017
124	Kurashiki-Fukuyama	48.0	14636	0.0018
125	Shiwa-Hiroshima	37.0	11275	0.0014
126	Hiroshima-Hatsukaichi	25.0	2939	0.0009
127	Iwakuni-Yuda	87.0	10751	0.0032
128	Yuda-Kitakyusyu	72.0	851445	0.0026
129	Kitakyusyu-Wakamiya	44.0	519873	0.0016

Table 8.1. Optimal allocation of public investment (continued)

130	Wakamiya-Tosu	52.0	610110	0.0047
131	Tosu-Kanzaki	13.0	662	0.0001
132	Kanzaki-Ohmura	80.0	4074	0.0004
133	Ohmura-Magasaki	18.0	916	0.0001
134	Tosu-Kumamoto	128.0	1491899	0.0054
135	Kumamoto-Yatsushiro	43.0	8224	0.0016
136	Yatsushiro-Ebino	63.0	4519	0.0023
137	Ebino-Kagoshimakita	71.0	13602	0.0026
138	Kagoshimakita-Kagoshima	5.0	353	0.0002
	airport			
139	Hokkaido		51357	0.0671
140	Tohuku		47224	0.0671
141	Kanto		339959	0.0671
142	Chukyo		40266	0.0671
143	Hokuriku		9334	0.0671
144	Kinki		279109	0.0671
145	Chugoku		2054	0.0671
146	Shikoku		5042	0.0671
147	Kyusyu		65102	0.0671
148	Okimnawa		99185	0.0671
	port			
149	Hokkaido		1491670	0.0556
150	Tohoku		2188787	0.5557
151	Kanto		0	0.0000
152	Chukyo		2509791	0.5557
153	Hokuriku		0	0.0000
154	Kinki		2485973	0.0556
155	Chugokiu		0	0.0000
156	Shikoku		0	0.0000
157	Kyusyu		946345	0.0556
158	Okinawa		260171	0.0556
	Intraregional transport facilities			
	railway			
159	Hokkaido		0	0.0000
160	Tohoku		38058	0.0052
161	Kanto		81660	0.0056
162	Chukyo		69173	0.0034
163	Hokuriku		69170	0.0060
164	Kinki		9219	0.0047
165	Chugoku		15295	0.0056
166	Shikoku		42919	0.0052
167	Kyusyu		19641	0.0056

Table 8.1. Optimal allocation of public investment (continued)

	new trunk line			
168	Hokkaido		0	0.0000
169	Tohoku		51700	0.0033
170	Kanot		0	0.0000
171	Chukyo		0	0.0000
172	Hokuriku		0	0.0000
173	Kinki		0	0.0000
174	Chugoku		0	0.0000
175	Shikoku		0	0.0000
176	Kyusyu		16585	0.0036
	ordinary road			
177	Hokkaido		856404	0.0015
178	Tohoku		784964	0.0016
179	Kanto		2106160	0.0016
180	Chukyo		341290	0.0016
181	Hokuriku		88846	0.0016
182	Kinki		929691	0.0016
183	Chugoku		265674	0.0017
184	Shikoku		302482	0.0015
185	Kyusyu		927791	0.0016
186	Okinawa		167547	0.0018
	expressway			
187	Hokkaido		26954	0.0001
188	Tohoku		92300	0.0003
189	Kanto		980705	0.0011
190	Chukyo		98382	0.0008
191	Hokuriku		40799	0.0011
192	Kinki		385934	0.0009
193	Chugoku		154698	0.0011
194	Shikoku		23082	0.0003
195	Kyusyu		253651	0.0007
196	Okinawa		0	0.0000
	National Railways electric train service			
197	Kanto		499092	0.0280
198	Kinki		97648	0.0280

Notes

*1 For convenience, it is assumed that where ordinary roads and expressways serve the section, the distance is the same.

*2 As each investment fund is applied to three construction phases, of which the expenditures are summed in the table, the imputed price for the funding constraint cannot be given.

The information on the total public investment funds row constraint at the upper limit is as follows:

1. The maximum value of the optimum solution is 73.5 trillion yen
2. Slack activity: 0.0
3. Dual activity (imputed price): 0.08387

Table 8.1. Optimal allocation of public investment

Part B Investment in social overhead capital

Row number	Region	Facilities	The derived optimum amount of public investment (million yen) *2
	Hokkaido		
201		environmental sanitation	747760
202		public rental housing	396800
203		public welfare service facilities	230368
204		education service facilties	410839
	Tohuku		
205		environmental sanitation	477479
206		public rental housing	732840
207		public welfare service facilities	397992
208		education service facilties	774741
209		environmental sanitation	3517351
210		public rental housing	2589740
211		public welfare service facilities	8903644
212		education service facilties	319308
213		environmental sanitation	1222629
214		public rental housing	671770
215		public welfare service facilities	278208
216		education service facilties	766277
217		environmental sanitation	144595
218		public rental housing	207080
219		public welfare service facilities	123464
220		education service facilties	441694
221		environmental sanitation	2291066
222		public rental housing	4322640
223		public welfare service facilities	602976
224		education service facilties	983994
225		environmental sanitation	329338
226		public rental housing	537540
227		public welfare service facilities	278024
228		education service facilties	623832
229		environmental sanitation	146423
230		public rental housing	286440
231		public welfare service facilities	186760
232		education service facilties	839169
233		environmental sanitation	696016
234		public rental housing	777480
235		public welfare service facilities	539856
236		education service facilties	802791
237		environmental sanitation	93470
238		public rental housing	166600
239		public welfare service facilities	21896
240		education service facilties	3253

References

Keeney RL (1974) Multiplicative Utility Functions. Operations Research 22:22-34

Kohno H (forthcoming) The Optimum Allocation of Public Investment Using an Interregional Input-Output Model. In: Button K, Nijkamp P Kohno H (Eds) Transportation and Land Use (the 5th Volume in Classics in Regional Science), Edward Elgar Publishing

Kohno H (1975) The Optimum Allocation of Public Investment Using an Interregional Input-Output Model. The Journal of Economics, The Society of Economics. University of Tokyo 41:61-82 (in Japanese)

Kohno H, Higano Y (forthcoming) Dynamic Takeoff-Accelerating Effects of the Asian Expressway Network on China. In: Chatterji M (Ed) Perspectives for the Future of Regional Science. Macmillan, London

Kohno H, Mitomo H (1982) The Optimum Allocation of Public Investment in the Trunk Highway Network. Road Transportation Economy 19, Economic Research Association (in Japanese)

Kohno H, Yoshida M (1981a) The Simultaneous Optimum Allocation of Public Investment in the Interurban Comprehensive Transport System and the Regional Living Environmental Facilities. Studies in Regional Science, Japan Section of the Regional Science Association, 11:59-79 (in Japanese)

Kohno H, Yoshida M (1981b) The Theory and Measurement of the Marginal Rate of Valuation of a Public Nuisance. Papers of the Regional Science Association 46:45-59

Lefeber L (1958) Allocation in Space. Production, Transport and Industrial Location. North-Holland, Amsterdam

Ministry of International Trade and Industry (1975) 1970 Interregional Input-Output Tables. Report on the Research Results. The Ministerial Secretariat, Tokyo

Ministry of International Trade and Industry (1980) 1975 Interregional Input-Output Tables. Tentative Version. The Ministerial Secretariat, Tokyo

Moses LN (1960) A General Equilibrium Model of Production, Interregional Trade and the Location of Industry. Review of Economics and Statistics 42:373-397

9 Spatial Price Equilibrium Models with Discriminatory Ad Valorem Tariffs: Formulation and Comparative Computation Using Variational Inequalities

Anna Nagurney
University of Massachusetts

Charles F. Nicholson and Phillip M. Bishop
Cornell University

9.1 Introduction

Ad valorem tariffs are a trade policy instrument used by nearly all nations to restrict imports of a wide variety of commodities. Recent emphasis on tariffication as a means of trade liberalization underscores the importance of appropriately incorporating discriminatory ad valorem tariffs (tariffs imposed by the importing country that vary according to the exporting country) into spatial economic models. Both the recently-approved North American Free Trade Agreement (NAFTA) and the General Agreement on Tariffs and Trade (GATT) negotiations specify tariffication as a mechanism to lower trade barriers.

Under NAFTA, for example, tariffication involves the establishment of Tariff Rate Quotas (TRQs) in which no tariff is applied to a quota amount and a discriminatory ad valorem tariff is assessed on over-quota imports (NAFTA, 1992). The 'above-quota' tariff is lowered over a specified period and the quota is simultaneously increased. Ad valorem tariffs on above-quota imports of certain commodities are initially greater than 100 per cent under NAFTA provisions. Recent GATT negotiations have focused on proposals to convert all forms of agricultural protection to equivalent ad valorem and (or) unit tariffs (see, e.g. Institute for International Legal Information 1992). Thus, discriminatory ad valorem tariffs represent important restrictions on trade patterns while simultaneously serving as a mechanism for trade liberalization.

Spatial price equilibrium (SPE) models are commonly-used tools for spatial analysis of commodity trade (see Judge and Takayama 1973; Labys 1989; Thompson 1989; Hertel 1990; Labys et al. 1991). Most early SPE models specified linear, separable supply and demand functions. Non-discriminatory ad valorem tariffs, that is ad valorem tariffs that do not vary by exporting country, have been incorporated in SPE models by modification of linear, separable excess supply and demand price functions (Cramer et al. 1991). However, the aggregation of countries into excess price functions implies that this method cannot be applied to analyze the impacts of discriminatory ad valorem tariffs.

Introduction of discriminatory ad valorem tariffs into SPE will typically violate the conditions which allow the spatial price equilibrium conditions to be expressed as the first-order conditions of an optimization problem. Takayama and Judge (1971) recognized this fact and provided a linear complementarity formulation of the SPE with linear supply and demand functions, along with a numerical example. Holland and Sharples (1984) proposed a fixed point algorithm, whereas Chang, et al. (1991) proposed a heuristic method to solve SPE models with discriminatory ad valorem tariffs. Most recent SPE models and computable general equilibrium models have incorporated nonlinear supply and demand functions (often constant elasticity of substitution, or CES, functions) to better represent the underlying economic phenomena (see Hertel 1990).

In this paper we apply the methodology of the theory of variational inequalities for the formulation and computation of spatial price equilibrium models with discriminatory ad valorem tariffs, a widely-used trade policy instrument. A variety of medium-scale and large-scale problems relevant for empirical modelling of international commodity trade are then solved on both serial and massively parallel architectures to illustrate the performance of the proposed computational procedure, for which convergence is also established.

In Section 9.2 we develop the model with fixed per unit transportation costs, state the equilibrium conditions and derive the variational inequality formulation. In Section 9.3 we propose an algorithm, the modified projection method (see Korpelevich 1977), which in the context of our model takes on a very simple computational form and provide conditions for convergence. In particular, at each iteration, the updated commodity shipment pattern can be computed via a closed-form expression. Moreover, each of the commodity shipment updates between a pair of markets can be computed simultaneously. Hence, this computational scheme unveils the opportunity for the (massively) parallel computation of the commodity shipments.

In Section 9.4 we present and solve representative SPE problems with discriminatory ad valorem tariffs of sizes relevant for empirical modelling of international commodity trade. In particular, we first consider medium-scale problems ranging in size from ten supply markets and ten demand markets (one hundred commodity shipment variables) to ninety supply markets and ninety demand markets (eight thousand one hundred commodity shipment variables). We then consider large-scale problems ranging in size from one hundred supply markets and one hundred demand markets (ten thousand shipment variables) to five hundred supply markets and five hundred demand markets (two hundred and fifty thousand shipment variables).

For the solution of the SPE problems we utilize both a serial implementation of the algorithm on the IBM ES/9000 and a massively parallel implementation of the algorithm on the Thinking Machines CM-2 (Connection Machine) with 32K (32,768) processors. Our goal in conducting the computational work on two different computer architectures is to gain insight as to when a switch to a parallel architecture becomes advantageous in terms of the size of the problem.

In Section 9.5 we present a summary and our conclusions and also suggest directions for future research.

9.2 The Spatial Market Model with Discriminatory Ad Valorem Tariffs

In this section we introduce the spatial market model with ad valorem tariffs. We state the governing equilibrium conditions and then derive the variational inequality formulation. The model in the absence of tariffs collapses to the well-known spatial price equilibrium models pioneered by Samuelson (1952) and Takayama and Judge (1971).

We consider m supply markets involved in the production of a homogeneous commodity and n demand markets. We denote a typical supply market by i and a typical demand market by j. Let s_i denote the supply at supply market i and d_j the demand at demand market j. We group the supplies into a column vector $s \in R^m$ and the demands into a column vector $d \in R^n$. Let Q_{ij} denote the nonnegative commodity shipment between supply and demand market pair (i, j), and group the commodity shipments into a column vector $Q \in R^{mn}$.

The commodity shipments and the supplies and demands must satisfy the following conservation of flow equations:

$$s_i = \sum_{j=1}^{n} Q_{ij}\ ,\ i = 1,\ldots,m. \tag{9.1}$$

or, equivalently, in vector form

$$s = AQ,$$

where A is the $(m \times mn)$ node-arc incidence matrix with components $(A)_{ij} = 1$, for $j = (i - 1)n + 1,\ldots,in$, and 0, otherwise,

$$d_j = \sum_{i=1}^{m} Q_{ij}\ ,\ j = 1,\ldots,n, \tag{9.2}$$

or, equivalently, in vector form

$$d = BQ,$$

where B is the $(n \times mn)$ node-arc incidence matrix such that $B = (I_n | I_n | \ldots | I_n)$, where I_n denotes the $n \times n$ identity matrix.

Hence, the supply at each supply market must be equal to the sum of the commodity shipments from that market to all the demand markets, and the demand at each demand market must be equal to the sum of the commodity shipments from all supply markets to each demand market.

We now describe the price and cost structure. Let π_i denote the supply price at supply market i and ρ_j the demand price at demand market j. We group the supply prices into a row vector $\pi \in R^m$ and the demand prices into a row vector $\rho \in R^n$. The transportation cost associated with shipping the commodity between supply market i and demand market j is denoted by c_{ij}. We group the transportation costs into a row vector $c \in R^{mn}$.

We assume that the supply price at a supply market may, in general, depend upon the supplies of the commodity at every supply market, that is,

$$\pi = \pi(s). \tag{9.3}$$

Similarly, the demand price at a demand market may, in general, depend upon the demands for the commodity at every demand market, that is,

$$\rho = \rho(d). \tag{9.4}$$

These static supply and demand price functions assume perfect certainty and ignore seasonal variation in production and consumption.

The per unit transportation cost, in turn, associated with shipping the commodity between a pair of supply and demand markets is assumed to be fixed, that is, it is independent of the volume of commodity shipments, where the fixed per unit cost is denoted by $\bar{c}_{ij}$, and the associated mn-dimensional vector by $\bar{c}$. Hence, we have that

$$c = \bar{c}. \tag{9.5}$$

Note that other fixed per unit transfer costs and per unit tariffs can be readily incorporated in the fixed $\bar{c}$ function.

In the absence of policy interventions and, under the assumption of perfect competition, the well-known spatial price equilibrium conditions (see Samuelson 1952; Takayama and Judge 1971) are stated as follows: For all pairs of supply and demand markets (i, j); $i = 1,\dots,m$; $j = 1,\dots,n$, a commodity supply, shipment and demand pattern (s^*, Q^*, d^*) satisfying constraints (9.1) and (9.2) is in equilibrium if

$$\pi_i(s^*) + \bar{c}_{ij} \begin{cases} = \rho_j\,(d^*), & \textit{if}\;\; Q_{ij}^* > 0 \\ \geq \rho_j\,(d^*), & \textit{if}\;\; Q_{ij}^* = 0. \end{cases} \tag{9.6}$$

In other words, in equilibrium, if a positive amount of the commodity is shipped between a pair of supply and demand markets, then the supply price at the supply market plus the cost of transportation must be equal to the demand price at the demand market. There will be no shipment of the commodity between a pair of markets if the sum of the supply price and transportation cost exceeds the demand price.

As established in Florian and Los (1982) (see also, Dafermos 1980; Friesz et al. 1984), the above equilibrium conditions can be formulated as a variational inequality problem as stated in the subsequent Theorem.

Theorem 1

A pattern (s^*, Q^*, d^*) satisfying (9.1) and (9.2) is a spatial price equilibrium if and only if it satisfies the variational inequality

$$\pi(s^*) \cdot (S - s^*) + \bar{c} \cdot (Q - Q^*) - \rho(d^*) \cdot (d - d^*) \geq 0, \quad (9.7)$$

for all (s, Q, d) satisfying (9.1) and (9.2).

We now introduce discriminatory ad valorem tariffs into the above model. Let τ_{ij} denote the ad valorem tariff, assumed positive, and applied to imports by demand market j from supply market i. The incorporation of ad valorem tariffs modifies the spatial price equilibrium conditions as follows: For all pairs of supply and demand markets (i, j); $i = 1,\ldots,m$; $j = 1,\ldots,n$, a commodity supply, shipment and demand pattern (s^*, Q^*, d^*) satisfying (9.1) and (9.2) is said to be in equilibrium if

$$(\pi_i(s^*) + \bar{c}_{ij}) \cdot (1 + \tau_{ij}) \begin{cases} = \rho_j(d^*), & \text{if } Q_{ij}^* > 0 \\ \geq \rho_j(d^*), & \text{if } Q_{ij}^* = 0. \end{cases} \quad (9.8)$$

Hence, in equilibrium, if a positive amount of the commodity is shipped between a pair of supply and demand markets, then the effective supply price plus transportation cost after the imposition of ad valorem tariffs must be equal to the demand price at the demand market. If there is no commodity shipment between a pair of supply and demand markets, then the effective supply price plus transportation cost can exceed the demand price.

In view of constraints (9.1) and (9.2), we define the functions $\hat{\pi}_i(Q) \equiv \pi_i(AQ)$, $i = 1,\ldots,m$, and the functions $\hat{\rho}_j(Q) = \rho_j(BQ)$, $j = 1,\ldots,n$. We are now ready to derive the variational inequality formulation of the equilibrium conditions governing the spatial market model with ad valorem tariffs. In particular, we state the following.

Theorem 2

A commodity shipment $Q^* \in R_+^{mn}$ is an equilibrium pattern, that is, it satisfies conditions (9.8) if and only if it satisfies the variational inequality problem

$$\sum_{i=1}^{m}\sum_{j=1}^{n} ((\hat{\pi}_i(Q^*) + \bar{c}_{ij}) \cdot (1 + \tau_{ij}) - \hat{\rho}_j(Q^*)) \cdot (Q_{ij} - Q_{ij}^*) \geq 0, \quad \forall Q \in R_+^{mn}. \quad (9.9)$$

Proof:

We first establish that if a pattern Q^* satisfies equilibrium conditions (9.8) then it also satisfies variational inequality (9.9).

For fixed market pair (i, j), (9.8) implies that

$$((\pi_i(s^*) + \bar{c}_{ij}) \cdot (1 + \tau_{ij}) - \rho_j(d^*)) \cdot (Q_{ij} - Q_{ij}^*) \geq 0, \quad \forall Q_{ij} \geq 0,$$

or, equivalently,

$$((\hat{\pi}_i(Q^*) + \bar{c}_{ij}) \cdot (1 + \tau_{ij}) - \hat{\rho}_j(Q^*)) \cdot (Q_{ij} - Q_{ij}^*) \geq 0, \quad \forall Q_{ij} \geq 0. \tag{9.10}$$

Summing (9.10) over all market pairs (i, j), we obtain variational inequality (9.9).

We now show that any solution to variational inequality (9.9) also satisfies equilibrium conditions (9.8).

In variational inequality (9.9), set $Q_{ij} = Q_{ij}^*$, for all market pairs $(i, j) \neq (k, l)$. Then (9.9) reduces to:

$$((\hat{\pi}_k(Q^*) + \bar{c}_{kl}) \cdot (1 + \tau_{kl}) - \hat{\rho}_l(Q^*)) \cdot (Q_{kl} - Q_{kl}^*) \geq 0, \quad \forall Q_{kl} \geq 0, \tag{9.11}$$

or,

$$((\pi_k(s^*) + \bar{c}_{kl}) \cdot (1 + \tau_{kl}) - \rho_l(d^*)) \cdot (Q_{kl} - Q_{kl}^*) \geq 0, \quad \forall Q_{kl} \geq 0, \tag{9.12}$$

which implies equilibrium conditions (9.8) for market pair (k, l). Since this statement is independent of how we select the pair, the conditions hold for all market pairs.

Takayama and Judge (1971) proposed a linear complementarity formulation of the spatial price equilibrium conditions in the presence of discriminatory ad valorem tariffs for a model with linear supply and demand price functions. Indeed, even in the simplest case of linear and separable supply and demand price functions the model with tariffs (unlike the classical SPE model) could not be reformulated as a *QP* problem.

Variational inequality (9.9) can be put into standard form (see Nagurney 1993). Define

$$F_{ij}(Q) \equiv ((\hat{\pi}_i(Q) + \bar{c}_{ij}) \cdot (1 + \tau_{ij}) - \hat{\rho}_j(Q)), \quad \forall i, j, \tag{9.13}$$

and let $F(Q) \in R^{mn}$ be the row vector with (i, j)-th component $F_{ij}(Q)$. Then variational inequality (9.9) may be expressed in standard form:

Determine $Q^* \in K$, such that

$$F(Q^*) \cdot (Q - Q^*) \geq 0, \quad \forall Q \in K, \tag{9.14}$$

where the feasible set $K \equiv \{Q \mid Q \in R_+^{mn}\}$.

We now discuss certain qualitative properties of the function $F(Q)$ which will be useful in establishing convergence of the computational procedure. In particular, we give conditions that guarantee monotonicity of the function F, and the Lipschitz continuity.

Lemma 1

Assume that $\pi(s)$ and $-\rho(d)$ are each strongly monotone in s and d, respectively, i.e;

$$\begin{aligned} \left[\pi(s^1) - \pi(s^2)\right] \cdot \left[s^1 - s^2\right] &\geq \alpha \| s^1 - s^2 \|^2, \quad \forall s^1, s^2, \\ -\left[\rho(d^1) - \rho(d^2)\right] \cdot \left[d^1 - d^2\right] &\geq \beta \| d^1 - d^2 \|^2, \quad \forall d^1, d^2, \end{aligned} \tag{9.15}$$

for some $\alpha, \beta > 0$. Then $F(Q)$ with components $F_{ij}(Q)$ defined in (9.13) is monotone, i.e.

$$\left[F(Q^1) - F(Q^2)\right] \cdot \left[Q^1 - Q^2\right] \geq 0, \quad \forall Q^1, Q^2 \in K. \tag{9.16}$$

Proof:

We use the fact that if $\nabla F(Q) = \begin{bmatrix} \frac{\partial F_{11}}{\partial Q_{11}} & \cdots & \frac{\partial F_{11}}{\partial Q_{mn}} \\ \vdots & & \vdots \\ \frac{\partial F_{mn}}{\partial Q_{11}} & \cdots & \frac{\partial F_{mn}}{\partial Q_{mn}} \end{bmatrix}$ is positive semidefinite for all $Q \in K$, then $F(Q)$ is monotone.

$$\nabla F(Q) = \begin{bmatrix} \frac{\partial \hat{\pi}_i(Q)}{\partial Q_{11}}(1+\tau_{11}) - \frac{\partial \hat{\rho}_1(Q)}{\partial Q_{11}} & \cdots & \frac{\partial \hat{\pi}_1(Q)}{\partial Q_{mn}}(1+\tau_{11}) - \frac{\partial \hat{\rho}_i(Q)}{\partial Q_{mn}} \\ \vdots & & \vdots \\ \frac{\partial \hat{\pi}_m(Q)}{\partial Q_{11}}(1+\tau_{mn}) - \frac{\partial \hat{\rho}_n(Q)}{\partial Q_{11}} & \cdots & \frac{\partial \hat{\pi}_m(Q)}{\partial Q_{mn}}(1+\tau_{mn}) - \frac{\partial \hat{\rho}_n(Q)}{\partial Q_{mn}} \end{bmatrix}$$

$$= A^T \left\{\frac{\partial \pi}{\partial s}\right\} A + A^T \left\{\frac{\partial \pi}{\partial s}\right\} A\tau - B^T \left\{\frac{\partial \rho}{\partial d}\right\} B \tag{9.17}$$

where τ is the $(mn \times mn)$-dimensional matrix with diagonal components:

$$\{\tau_{11},\ldots,\tau_{1n},\ldots,\tau_{m1},\ldots,\tau_{mn}\}.$$

Under the assumption of strong monotonicity of π and $-\rho$, $\left[\frac{\partial \pi}{\partial s}\right]$ and $-\left[\frac{\partial \rho}{\partial d}\right]$ are positive definite, and, hence, $A^T\left[\frac{\partial \pi}{\partial s}\right]A$ is positive semidefinite, as is $-B^T\left[\frac{\partial \rho}{\partial d}\right]B$. Further, because τ is a positive definite matrix, we can conclude that $A^T\left[\frac{\partial \pi}{\partial s}\right]A\tau$ is positive semidefinite. Moreover, since the sum of positive semidefinite matrices is a positive semidefinite matrix, it follows that $\nabla F(Q)$ is positive semidefinite and, consequently, $F(Q)$ is monotone.

The proof is complete.

Note that the strong monotonicity condition holds in the case of linear, separable supply and demand price functions where

$$\pi_i(s_i) = r_i s_i + t_i\ , \quad i = 1,\ldots,m, \tag{9.18}$$

$$\rho_j(d_j) = -m_j d_j + q_j\ , \quad j = 1,\ldots,n, \tag{9.19}$$

with $r_i,\ m_j > 0,\ \forall_{i,j}$.

Lemma 2

$F(Q)$ is Lipschitz continuous on K, that is, there exists an $L > 0$, such that

$$\|F(Q^1) - F(Q^2)\| \leq L\|Q^1 - Q^2\|, \quad \forall Q^1,Q^2 \in K, \tag{9.20}$$

under the assumption that $F_{ij}(Q)$ has bounded first order derivatives for all $Q \in K$.

Proof:

Several relationships are first presented, the proofs of which immediately follow.

$$\|F(Q^1) - F(Q^2)\|^2 = \sum_{i=1}^{m}\sum_{j=1}^{n} [F_{ij}(Q^1) - F_{ij}(Q^2)]^2 \tag{9.21}$$

$$= \sum_{i=1}^{m}\sum_{j=1}^{n} [\nabla F_{ij}(\tilde{Q}^{ij}) \cdot (Q^1 - Q^2)]^2 \tag{9.22}$$

$$\leq \sum_{i=1}^{m}\sum_{j=1}^{n} \|\nabla F_{ij}(\tilde{Q}^{ij})\|^2 \cdot \|Q^1 - Q^2\|^2 \tag{9.23}$$

$$\leq \sum_{i=1}^{m}\sum_{j=1}^{n} L_{ij}^2 \|Q^1 - Q^2\|^2 \leq L^2 \|Q^1 - Q^2\|^2. \tag{9.24}$$

Since $F(Q)$ is differentiable, applying the Mean Value Theorem to each component $F_{ij}(Q)$ of $F(Q)$, on knows that

$$(\tilde{Q}^{ij}) = \theta_{ij}(Q^1) + (1 - \theta_{ij})(Q^2), \quad 0 < \theta_{ij} < 1,$$

and, hence, $\tilde{Q}^{ij} \in K$, $i = 1,\ldots,m$; $j = 1,\ldots n$, so that (9.21) equals (9.22). Applying the Schwartz inequality, one then obtains (9.23) from (9.22).

Also, since each $F_{ij}(\tilde{Q}^{ij})$ is assumed to have bounded first order derivatives for all the variables, we can conclude that $\nabla F_{ij}(\tilde{Q}^{ij})$ is bounded for all i, j. Therefore there exist $L_{ij} > 0$, $i = 1,\ldots m$; $j = 1,\ldots,n$, such that $\|\nabla F_{ij}(\tilde{Q}^{ij})\| \leq L_{ij}$, $\forall i, j$, and $\forall Q \in K$. Hence, the first inequality in (9.24) is obtained from (9.23).

Let now

$$L = \max_{\{1 \leq \ldots \leq m; 1 \leq j \leq n\}} \{L_{ij}\},$$

from which the remaining inequality in (9.24) follows.

It also follows that $F(Q)$ will have bounded first order derivatives if the supply and demand price functions have bounded first order derivatives and the tariff rates are all finite.

9.3 The Computational Procedure

In this section we propose the computational procedure for the solution of variational inequality (9.9), or equivalently, variational inequality (9.14), governing the spatial price equilibrium model with discriminatory ad valorem tariffs. We also establish convergence.

The algorithm that we propose is the modified projection method of Korpelevich (1977) which, as we shall show, resolves the variational inequality problem under consideration here into subproblems that are very simple for computational purposes. Indeed, we obtain a closed form expression for the determination of the commodity shipments at each iteration. Moreover, since each of the commodity shipments between a pair of supply and demand markets can be evaluated separately and simultaneously at any iteration, this algorithmic scheme enables one to exploit the availability of (massively, parallel computer architectures. Finally, this algorithm, unlike the algorithms proposed in Takayama and Judge (1971) and Cramer et al. (1991), can also handle spatial price models with nonlinear supply and demand price functions.

The statement of the algorithm follows.

The Modified Projection Method

Step 0: Initialization

Start with a $Q^0 \in K$. Set $k := 1$ and select γ, such that $0 < \gamma <, \frac{1}{L}$ where L is the Lipschitz constant in (9.20).

Step 1: Construction and Computation

Compute $\bar{Q}^{k-1}$ by solving the variational inequality subproblem:

$$[\bar{Q}^{k-1} + (\gamma F(Q^{k-1})^T - Q^{k-1})]\cdot[Q' - \bar{Q}^{k-1}] \geq 0, \quad \forall Q' \in K. \tag{9.25}$$

Step 2: Adaptation

Compute Q^k by solving the variational inequality subproblem:

$$[Q^k + (\gamma F(\bar{Q}^{k-1})^T - Q^{k-1})]^T \cdot [Q' - Q^k] \geq 0, \quad \forall Q' \in K. \tag{9.26}$$

Step 3: Convergence Verification

If $|Q^k - Q^{k-1}| \leq \epsilon$, for $\epsilon > 0$, a prespecified tolerance, then, stop; otherwise set $k := k + 1$ and go to Step 1.

We now provide the closed form expressions for the solution of problems (9.25) and (9.26). In particular, we have that (9.25) can be solved as: For all supply and demand market pairs (i, j), $i = 1,\ldots,m$; $j = 1,\ldots,n$, set

$$\bar{Q}_{ij}^{k-1} = \max\left\{0, \gamma((-\pi_i(\bar{s}^{k-1}) - \bar{c}_{ij})(1 + \tau_{ij}) + \rho_j(d^{k-1})) + Q_{ij}^{k-1})\right\}. \tag{9.27}$$

(9.26), in turn, can be solved as: For all supply and demand market pairs (i, j), $i = 1,\ldots,m; j = 1,\ldots,n$, set

$$Q_{ij}^{k} = \max\left\{0, \gamma((-\pi_i(\bar{s}^{k-1}) - \bar{c}_{ij})(1 + \tau_{ij}) + \rho_j(\bar{d}^{k-1})) + Q_{ij}^{k-1}\right\}. \tag{9.28}$$

In view of expressions (9.27) and (9.28), one sees that all of the *mn* commodity shipments can be solved simultaneously at each iteration. Hence, an 'ideal' computer architecture for the solution of such problems may be one in which there are as many processors as there are pairs of markets. We investigate this issue in the subsequent section.

The convergence results are now given.

Theorem 3

Assume that $\pi(s)$ and $-\rho(d)$ are each strongly monotone in s and d, respectively, and have bounded first order derivatives. Also, assume that the tariff rates τ_{ij} are finite for all i, j. Then the modified projection method converges to the solution of variational inequality (9.9), or, equivalently, (9.14), provided that a solution exists.

Proof:

As established in Korpelevich (1977), the modified projection method is guaranteed to converge if the function $F(\cdot)$ that enters the variational inequality problem is monotone and Lipschitz continuous. Monotonicity follows from Lemma 1, and Lipschitz continuity from Lemma 2, under the above assumptions.

The proof is complete.

It is important to note that the modified projection method can also be applied to compute the solution to the spatial price equilibrium model without ad valorem tariffs, governed by variational inequality (9.7). Indeed, convergence will be guaranteed under the same conditions on the supply and demand price functions as imposed above.

Computational procedures to date for the solution of spatial price equilibrium problems and based on variational inequality theory have focused on models that satisfy the strong monotonicity condition (rather than the monotonicity condition that is appropriate here). In such models the transportation cost associated with shipping a commodity between a pair of supply and demand markets is assumed to be an increasing function of that commodity shipment. For a variety of such computational procedures, we refer the reader to Nagurney (1995) and the references therein.

In the subsequent section we apply the modified projection method for the solution of spatial price equilibrium problems with and without discriminatory ad valorem tariffs on two distinct computer architectures.

9.4 Numerical Results on Serial and Parallel Architectures

In this section we present our numerical results for the implementations of the modified projection method on two distinct architectures, the IBM ES/9000, when the algorithm is implemented in FORTRAN 77, compiled, and executed using a single processor, and the massively parallel architecture, the Thinking Machines CM-2 (Connection Machine), when the algorithm is implemented in CM FORTRAN and executed on 8K (9,192), 16K (16,384) and 32K (32,768) processors. We consider the solution of medium-scale and large-scale spatial price equilibrium problems with discriminatory ad valorem tariffs. For completeness, we also present numerical results for the problems without ad valorem tariffs.

To date, only spatial price equilibrium problems without any policy interventions have been solved on a massively parallel architecture (see Nagurney, Takayama and Zhang, 1993). Therein an entirely distinct methodology from the one proposed here was utilized, based on dynamical systems.

Specifically, we consider spatial price equilibrium problems in which the supply price and demand price functions are asymmetric and the transportation cost functions are fixed. Of course, the algorithm, as noted in Section 9.3, can also handle nonlinear supply and demand price functions. In particular, we consider a linear, asymmetric supply price function for each supply market i, see (9.3), of the form

$$\pi_i(s) = \sum_j r_{ij}\, s_j + t_i\,, \tag{9.29}$$

where the cross-terms are such that $r_{ij} \neq r_{ji}$, for $j \neq i$, and a linear, asymmetric demand price function for each demand market j, see (9.4), of the form

$$\rho_j(d) = -\sum_k m_{jk} d_k + q_j, \tag{9.30}$$

where the cross-terms are such that $m_{jk} \neq m_{kj}$, for $j \neq k$. The per unit transportation cost function associated with each market pair (i, j), see (9.5), is fixed, where

$$\bar{c}_{ij} = h_{ij} \tag{9.31}$$

In subsection 9.4.1 we discuss our implementation of the modified projection method on the massively parallel CM-2 architecture and then in Subsection 9.4.2 we present our numerical work and computational comparisons. A discussion of the specific ranges of the function parameter is (9.29), (9.30), and (9.31) used in the computations is given in the numerical section 9.4.2.

9.4.1 Implementation of the Modified Projection Method on the CM-2

The CM-2 is a distributed SIMD (Single-Instruction Multiple Data) massively parallel processing system. The mode of computation is data level parallelism, that is, all processors execute identical operations. The CM-2 system that we utilized had 32K processors in its configuration and a SUN workstation served as its front end. The language used for the parallel implementation was CM FORTRAN.

The CM FORTRAN code for the implementation of the modified projection method for the model consisted of an input and setup routine and a computation routine to implement the iterative steps (9.27) and (9.28). The crucial feature in the design of the program was the construction of the data structures to take advantage of the data level parallelism and computation. We first constructed the array H, of dimension $m \times n$, to store the elements $\{h_{ij}\}$. We then constructed the array 'tau' to store the tariff rates, with the (i, j)-th component equal to τ_{ij}.

The supply price coefficients $\{\tau_{ij}\}$ were stored in an $m \times m$ array SC, and the demand price coefficients $\{-m_{jk}\}$ were stored in an $n \times n$ array DC. We also introduced additional arrays SP and DP to denote, respectively, the supply prices and the demand prices at a given iteration, where the i-th row of SP consisted of the identical elements $\{\pi_i\}$ and the j-th column of DP consisted of the identical elements $\{\rho_j\}$. To compute the supply prices, we used the *spread* command to spread the supplies and then multiplied the resulting matrix with the SC matrix. Specifically, the *spread* command makes multiple copies of a vector along columns or along the rows to create a 2-dimensional array. We then used the *sum* command to add the elements of each row of the resulting arrays and added the resulting vector to the vector containing the fixed supply price terms $\{t_i\}$, see (9.29). The result was then *spread* to create the supply prices SP at the particular iteration. The demand prices were obtained in an analogous fashion. The array QO was used to store the values of Q from the previous iteration and was used for convergence purposes.

We now present the critical steps in the CM FORTRAN computation section.

Implementation of the Modified Projection Method

```
 Do while (err.ge.ε)
c     Initialization
 1.  QO(:,:)=Q(:,:)
c     Construction and Computation
 2.  construct SP and DP
 3.  temp(:,:)=Q(:,:)+γ(DP(:,:)-(H(:,:)+SP(:,:))*(1+tau))
 4.  Q(:,:)=temp(:,:)
 5.  where(temp(:,:).lt.0.) Q(:,:)=0.
```

```
c    Adaptation

 6.  update SP and DP with new Q

 7.  temp(:,:)=QO(:,:)+γ(DP(:,:)-(H(:,:)+SP(:,:))*(1+tau))

 8.  Q(:,:)=temp(:,:)

 9.  where(temp(:,:).lt.0.)Q(:,:)=0.

10.  err=maxval(abs(Q-QO))]

11.  update supplies and demands

end do
```

Hence, from Step 3 above it can be seen that element (i, j) of the array 'temp' contains at the k-th iteration in the Construction and Computation Step the value of: $\gamma(\rho_j(d^{k-1}) - (\bar{c}_{ij} + \pi_i(s^{k-1}))(1 + \tau_{ij})) + Q_{ij}^{k-1}$ (see (9.27)). In the Adaptation Step 7 above, on the other hand, it can be seen that element (i, j) of the array 'temp' now contains at the k-th iteration the value of: $\gamma(\rho_j(\bar{d}^{k-1}) - (\bar{c}_{ij} + \pi_i(\bar{s}^{k-1}))(1 + \tau_{ij})) + Q_{ij}^{k-1}$ (see (9.28)).

All the variables above followed by a '(:,:)' are 2-dimensional arrays. Therefore, the array 'temp' contains not a single value but $m \times n$ values, with position (i, j) holding the tentative value of the Q_{ij}.

Q is updated by using a mask in Steps 5 and 9, where the (i, j)-th element is set to zero if the value of temp (i, j) is negative. What is important to note is that, at each iteration, all of the Q_{ij}'s, for $i = 1,\ldots,m$; $j = 1,\ldots,n$, are computed and updated *simultaneously*. This is not possible when the algorithm is implemented on a serial architecture with consequences that shall be highlighted in the numerical section 9.4.2.

Note that the above code can be easily adapted to solve spatial price equilibrium problems without ad valorem tariffs, but with fixed unit transportation costs, by simply removing the (1+tau) expression.

9.4.2 Numerical Results

We now turn to the presentation of the numerical results. We report the results on two different sets of examples. The first set consists of medium-scale problems ranging in size from ten supply markets and ten demand markets to ninety supply markets and ninety demand markets, that is, with one hundred to eight thousand one hundred commodity shipment variables. The second set of problems consists of large-scale problems ranging in size from one hundred supply markets and one hundred demand markets to five hundred supply markets and five hundred demand markets, that is, with ten thousand to two hundred and fifty thousand commodity shipment variables.

The numerical results for both the medium-scale and the large-scale problems are reported for both the serial implementation of the algorithm in FORTRAN 77 on the IBM ES/9000 and the parallel implementation in CM FORTRAN presented above on the CM-2 architecture.

The data, see (9.29), (9.30) and (9.31), were generated randomly and uniformly in the ranges: $r_{ii} \in [10,30]$, $t_i \in [10,100]$, $-m_{ij} \in [-5, -55]$, $q_j \in [50,5000]$, and $h_{ij} \in [.05,.50]$, for all $i = 1,\ldots,m; j = 1,\ldots n$. The off-diagonal terms in the supply and demand price functions were generated to ensure strict diagonal dominance, thus guaranteeing strong monotonicity of the supply and demand price functions. Each asymmetric function had five terms. Spatial price equilibrium problems with functions in identical ranges were previously solved in Nagurney, Takayama and Zhang (1995). The focus therein, however, was on problems that satisfied stronger conditions for convergence.

The tariff rates, in turn, were generated randomly and uniformly in the range $\tau_{ij} \in [0,2]$, for all i, j. We selected this range for consistency with the range of ad valorem tariffs specified under NAFTA and GATT.

We set the convergence tolerance $\epsilon = .01$, and set $\gamma = .0001$ for all the numerical examples. Also, we initialized the algorithm for each example with $Q^0_{ij} = 0$ for all i, j.

We measured the computational accuracy of the modified projection method in terms of the average error, where

$$\text{Average Error} = \frac{100}{p} \sum_{ij} \frac{|(\pi_i + \bar{c}_{ij})(1 + \tau_{ij}) - \rho_j|}{(\pi_i + \bar{c}_{ij})(1 + \tau_{ij})}$$

where p = the number of supply and demand market pairs (i, j) such that $Q_{ij} > 0$, and the maximum error, where

$$\text{Maximum Error} = \max_{ij} \frac{100|(\pi_i + \bar{c}_{ij})(1 + \tau_{ij}) - \rho_j|}{(\pi_i + \bar{c}_{ij})(1 + \tau_{ij})}$$

for all (i, j) such that $Q_{ij} > 0$ in the case of the model with tariffs. The measures in the case of the model without tariffs would have the term: $(1 + \tau_{ij})$ removed from all the above expressions. Such measures of accuracy (for the case of spatial price models without discriminatory ad valorem tariffs) had been used earlier by Friesz, Harker and Tobin (1984) and Nagurney (1987) and more recently, by Nagurney, Takayama and Zhang (1995).

The serial implementation of the modified projection method on the IBM ES/9000 yielded the same number of iterations as had been obtained on the CM-2 for each example. We implemented the modified projection method on a serial architecture for several reasons. First, for completeness, second, for illustrative purposes, third, for validation purposes and, fourth, because such a (serial) computer architecture may be the only type available for certain practitioners.

9.4.2.1 Numerical Results for Medium-Scale Problems

We report the CPU times (without input/output) for medium-scale problems in Table 9.1. For each of the five medium-scale problems we report the time required for the algorithm on the serial and on the parallel architecture for a given problem without any tariffs, and then with the inclusion of the tariffs. For the medium-scale problems we only utilized 8K (8,192) processors of the CM-2, since in this set the greatest number of unknowns was 8,100.

Table 9.1. Numerical results for medium-scale spatial price equilibrium problems

CPU Times in Seconds

	IBM ES/9000 (serial)		CM-2 (8K) (parallel)	
Example $m \times n$	Without Tariffs	With Tariffs	Without Tariffs	With Tariffs
10×10	.31	1.56	10.41	47.25
30×30	3.41	14.70	15.95	63.54
50×50	28.76	52.65	12.06	91.45
70×70	30.63	65.86	17.59	87.34
90×90	6.58	147.55	4.85	88.08

The first example. 10×10, consisting of ten supply markets and ten demand markets, required 533 iterations for convergence in the absence of tariffs and 2,427 iterations when the tariffs were added. The average error was 2.42 and the maximum error 6.24 for the without-tariff case and, respectively, 1.39 and 4.67 for the with-tariff case. The second example, 30×30, consisting of thirty supply markets and thirty demand markets, required 759 iterations for convergence in the absence of tariffs, and 3,021 iterations in their presence. The average error was equal to .47 and the maximum to 7.76 for the without-tariff case and, respectively to 2.15 and 8.77 for the with-tariff case.

The third example, 50×50, consisting of fifty supply markets and fifty demand markets required 553 iterations for convergence in the absence of tariffs and 3,917 in the presence of tariffs. The average error for the without-tariff case was .117 and the maximum was 5.45; the analogous values for the with-tariff case were 1.13 and 4.78, respectively. The fourth example, 70×70, consisting of seventy supply markets and seventy demand markets, required 557 iterations for convergence in the absence of tariffs and 2,558 iterations in their presence. The average error for the latter problem was 1.75 and the maximum was 7.10. The final example, 90×90, consisting of ninety supply markets and ninety demand markets, required 171 iterations for convergence in the absence of tariffs and 2,860 iterations in the presence of tariffs. The average error for the former problem was .17 and the maximum error was 5.70; the average error for the latter problem was 1.345 and the maximum was 6.41.

Note that the largest problem in this set, in the case of no tariffs, consisting of 90 supply markets and 90 demand markets, required only 4.85 CPU seconds on the CM-2 and 6.58 CPU seconds on the IBM ES/9000, whereas the next largest problem consisting of 70 supply markets and required 17.59 CPU seconds on the CM-2 and 30.63 seconds on the ES/9000. This is due to the fact that the largest problem only required 171 iterations in the absence of tariffs whereas the next larger one required 557 iterations. Most likely the initial conditions were closer to the equilibrium solution for the larger problem.

9.4.2.2 Numerical Results for Large-Scale Problems

Table 9.2. Numerical results for large-scale spatial price equilibrium problems

CPU Times in Seconds

	IBM ES/9000		CM-2 (8K)		CM-2 (16K)		CM-2 (32k)	
Example $m \times n$	Without Tariffs	With Tariffs	Without Tariffs	With Tariffs	Without Tariffs	With Tariffs	Without Tariffs	With Tariffs
100×100	8.8	241.50	7.58	186.72	5.22	130.90	--	--
200×200	53.83	632.08	17.23	178.72	14.96	155.19	10.89	112.89
300×300	107.94	>900	21.28	239.96	14.04	158.19	10.05	113.07
400×400	246.31	>900	38.23	523.15	24.92	340.64	17.43	238.29
500×500	880.47	>900	158.82	657.48	120.78	499.26	68.88	284.58

In the second set of examples, reported in Table 9.2, we solved problems on a scale larger than those considered in the previous subsection. The first example in this set, 100×100, consisting of one hundred supply markets and one hundred demand markets, required 185 iterations for convergence in the absence of tariffs and 4,611 iterations in the presence of tariffs. The average error for the without-tariff case was .237 and the maximum was 5.95; the average error for the with-tariff case was 1.10 and the maximum was 5.73. The second example in this set, 200×200, consisting of two hundred supply markets and two hundred demand markets, required 286 iterations for the without-tariff case and 2,951 iterations for the with-tariff case. The average error for the former problem was .35 and the maximum was 5.41. The average error for the latter problem was 1.35 and the maximum was 5.73.

The third example in Table 9.2, 300×300, consisting of three hundred supply markets and three hundred demand markets, required 250 iterations for convergence for the without-tariff case and 2,796 iterations for the with-tariff case. The former problem had an average error of .636 and a maximum error of 3.8; the latter problem had an average error of 1.75 and a maximum error of 5.99. The fourth example in Table 9.2, 400 × 400, consisting of four hundred supply markets and four hundred demand markets, required 305 iterations for the without-tariff case, and 4,140 iterations for the with-tariff case. The average error for the problem without tariffs was 1.17 and the maximum error was 4.08 and the average error for the problem with tariffs was .93 and the maximum error was 4.07. The final problem in this set, 500×500, consisting of five hundred supply markets and five hundred demand markets, required 686 iterations for convergence for the problem without tariffs, and 2,825 iterations for convergence for the problem with tariffs. The average error for the former problem was 1.033, whereas the maximum error was 4.05. The average error for the latter problem was 2.5, whereas the maximum error was 4.0.

9.4.3 Discussion

We first discuss the results for medium-scale problems and then those for large-scale problems reported above. Clearly, the addition of tariffs to any given problem increases both the number of iterations and the required CPU time. Interestingly, for all the problems with tariffs, except for the largest one in this set, 90×90, the modified projection method implemented on the serial architecture outperformed the algorithm implemented on the massively parallel architecture. One has to understand, however, that the CM-2 is accessed only according to 8K, 16K and 32K processors in the utilized configuration and, hence, the full features of the architecture are not exploited until approximately 8K variables are encountered for a problem. Nevertheless, spatial price equilibrium problems with fixed transportation costs, with and without tariffs, for problems as large as four thousand and nine hundred variables, were solved serially in approximately 60 seconds of CPU time or less. This is reasonable from a practical standpoint and allows one to conduct simulations of alternative tariff policy schemes.

For the large-scale problems, the addition of tariffs to any given problem also increased both the number of iterations required for convergence and the CPU time. It was apparent that the use of a massively parallel architecture for the large-scale realized substantial savings in CPU time over the time required on the serial architecture. For example, in the smallest problem in this set, 100 × 100, and without tariffs, the time on the IBM ES/9000 was 8.8 seconds, whereas the time used 16K processors of the CM-2 was 5.22 seconds. (We did not solve this problem on 32K processors since there were only 10,000 variables in this size of problem.) In the next largest problem in this set, 200×200, the time on the ES/9000 for the problem without tariffs was 53.83, whereas the same problem was solved in only 10.89 seconds using 32K processors of the CM-2, a five-fold improvement. This improvement in relative performance increased as the size of the problem increased, with the result that the largest problem in this set, 500×500, required 880.47 seconds on the

ES/9000 and less than a tenth of that time, 68.88 seconds, when 32K processors of the CM-2 were utilized.

In terms of the large-scale problems with tariffs, the superiority of the parallel implementation of the algorithm on the massively parallel architecture over the serial architecture was readily apparent. For example, the smallest problem in this set, 100×100, required 241.50 seconds on the ES/9000 and only 130.90 seconds when 16K processors of the CM-2 were utilized. The next larger problem, 200×200, required 632.08 seconds on the ES/9000 and only 112.89 seconds using 32K processors of the CM-2. For the remaining problems, it became prohibitively expensive to solve the large-scale problems in serial fashion.

We would like to emphasize further the simplicity of the implementation of the proposed algorithm on both the serial and parallel architectures. Indeed, at each step, the variational inequality problem was decomposed into subproblems, each of which possessed a closed form solution using a simple formula. Although our preliminary computational results clearly suggest that this algorithm's efficiencies are best realized on a parallel architecture for large-scale problems, this does not preclude researchers, students and practitioners, from using the algorithm on a serial architecture, when the size of the problem is small or medium-scale, which we expect to be the case in many applications in practice.

9.5 Summary and Conclusions

In this paper, we develop a spatial price equilibrium model with discriminatory ad valorem tariffs and apply the methodology of the theory of variational inequalities for its formulation and computation. The motivation for this work stems, in part, from the recent emphasis on tariffication as means of trade liberalization.

In particular, we introduce a spatial price equilibrium model with fixed per unit transportation costs and with supply and demand functions which may, in general, be nonlinear and asymmetric. The equilibrium conditions in the presence of discriminatory ad valorem tariffs are then formulated as a variational inequality problem.

We propose an algorithm, the modified projection method, which is guaranteed to converge for our model under reasonable assumptions on the supply and demand price functions and boundedness of the tariff rates. Research to date on algorithms for spatial price equilibrium problems based on variational inequality theory has focused primarily on models where the more restrictive strong monotonicity condition is required. The notable feature of the algorithm is that, in the context of our problem at each iteration, the commodity shipments may be computed using a simple, closed-form expression. Moreover, all the shipments may be updated simultaneously, leading one to be able to exploit novel massively parallel computer architectures.

The algorithm is then implemented on two distinct architectures, the IBM ES/9000 and the Thinking Machines CM-2. Two sets of examples, medium and large-scale (with and without tariffs, as a baseline) are then solved on the former architecture in a serial fashion and on the latter architecture in a massively parallel fashion. The numerical results suggest that for problems with fewer variables than the number of parallel processors in the 'smallest' CM-2 configuration, the serial architecture is appropriate (and more efficient). For larger problems, the parallel architecture clearly outperforms the serial one.

Our method of solving spatial price equilibrium models with discriminatory ad valorem tariffs is applicable to modelling trade agreements such as NAFTA and GATT in which tariffication plays an important role. However, other trade and domestic policies and differences in production technologies are also determinants of international trade patterns. Thus, incorporating these other trade and domestic policies into a spatial price equilibrium framework including ad valorem tariffs is a logical extension of the methods presented here.

In particular, the theory of variational inequalities could be applied to solve spatial price equilibrium models that simultaneously include ad valorem tariffs and other policy instruments such as price controls (see Nagurney and Zhao 1991; Thore 1991 and the references therein), input subsidies, export subsidies and quotas. Specific representation of these policies in a spatial price equilibrium framework often necessitates more market-level disaggregation, explicit representation of production technologies and product transformation and spatial arbitrage (transshipment) possibilities. Bishop et al. (1993) and Anania and McCalla (1991) have noted that explicit representation of other policy instruments and arbitrage possibilities is helpful to policy makers wishing to devise appropriate policies and to other economic agents who must respond to policies.

Acknowledgements

The research at the University of Massachusetts was supported by the National Science Foundation under grant DMS 9024071 under the Faculty Awards for Women Program.

This research was conducted at the National Center for Supercomputing Applications at the University of Illinois at Urbana-Champaign and at the Cornell Theory Center at Cornell University in Ithaca, New York. The use of these facilities is very much appreciated.

The authors acknowledge the comments of James E. Pratt and Andrew M. Novakovic on an earlier draft of this paper.

The authors would also like to express their appreciation to the referee for helpful comments and suggestions.

References

Anania G, McCalla (1991) Does Arbitraging Matter? Spatial Trade Models and Discriminatory Trade Policies. American Journal of Agricultural Economics 73:103-117

Bishop PM, Pratt JE, Novakovic AM (1993) Analyzing the Impacts of the Proposed North American Free Trade Agreement on European-North American Dairy Trade Using a Joint-Input, Multi-Product Approach. Department of Agricultural, Resource and Managerial Economics. Cornell University, Ithaca. New York. Staff Paper 17

Chang CC, McCarl BA, Mjelde JW, Richardson JW (1991) Sectoral Implications of Farm Program Modifications. American Journal of Agricultural Economics 74:38-49

Cramer GL, Wailes EJ, Goroski JM, Phillips SS (1991) The Impact of Liberalizing Trade on the World Rice Market. A Spatial Model Including Rice Quality. Arkansas Agricultural Experiment Station, Fayetteville, Arkansas

Dafermos S (1980) Traffic Equilibrium and Variational Inequalities. Transportation Science 14:42-54

Florian M, Los M (1982) A New Look at Static Spatial Price Equilibrium Models. Regional Science and Urban Economics 12:579-597

Friesz TL, Harker PT, Tobin RL (1984) Alternative Algorithms for the General Network Spatial Price Equilibrium Problem. Journal of Regional Science 24:475-507

Hertel TW (1990) Agricultural Trade Liberalization and the Developing Countries. A survey of the Models. In: Agricultural Trade Liberralization. Implications for Developing Countries. Goldin I, Knudsen O (Eds) Organization for Economic Cooperation and Development and the World Bank, Paris and Washington DC

Holland FD, Sharples JA (1984) World Wheat Trade. Implications for U.S. Exports. Department of Agricultural Economics, Purdue University, West Lafayette, Indiana. Staff Paper 84:20

Institute for International Legal Information (1992) The 'Dunkel Draft' from the GATT Secretariat. Multilateral Trade Negotiations, The Uruguay Round, Trade Negotiations Committee. Draft Final Act Embodying the Results of the Uruguay Round of Multilateral Trade Negotiations 20 December 1991. William S. Hein and Company Inc, Buffalo, New York

Judge GG, Takayama T (Eds) (1973) Studies in Economic Planning Over Space and Time. North-Holland, Amsterdam, The Netherlands

Korpelevich GM (1977) The Extragradient Method for Finding Saddle Points and Other Problems. Matekon 13:35-49

Labys WC (1989) Spatial and Temporal Price and Allocation Models of Mineral and Energy Markets. In: Quantitative Methods for Market-Oriented Economic Analysis Over Space and Time. Labys WC, Takayama T Uri ND (Eds). Avebury, Aldershot, England

Labys WC, Lesourd JB, Uri ND, Gövenen O (1991) New Horizons in

International Commodity Market Modelling. In: International Commodity Market Models. Advances in Methodology and Applications. Gövenen O, Labys W, Lesourd JB (Eds) International Studies in Economic Modelling. Chapman and Hall, London, England

Nagurney A (1987) Computational Comparisons of Spatial Price Equilibrium Methods. Journal of Regional Science 27:55-76

Nagurney A (1993) Network Economics. A Variational Inequality Approach. Klumer Academic Publishers, Boston, Massachusetts

Nagurney A, Takayama T, Zhang D (1995) Massively Parallel Computation of Spatial Price Equilibria as Dynamical Systems. Journal of Economic Dynamics and Control 19:3-37

Nagurney A, Zhao L (1991) A Network Equilibrium Formulation of Market Disequilibrium and Variational Inequalities. Networks 21:109-132

North American Free Trade Agreement (NAFTA) (1992) Chapter Seven, Annex 704.2, Section I, Appendix A, September 6

Samuelson P (1952) Spatial Price Equilibrium and Linear Programming. American Economic Review 42:283-303

Takayama T, Judge GG (1971) Spatial and Temporal Price and Allocation Models. Contributions to Economic Analysis 73. North-Holland, Amsterdam, The Netherlands

Thompson RL (1989) Spatial and Temporal Equilibrium Agricultural Models. In: Quantitative Methods for Market-Oriented Economic Analysis Over Space and Time. Labys WC, Takayama T, Uri ND (Eds) Avebury, Aldershot, England

Thore S (1991) Economic Logistics. Quorum Books, New York, New York

10 A Spatial Price Equilibrium Model for Environmental Policy Analysis of Mobile and Immobile Sources of Pollution

Erik T. Verhoef and Jeroen C.J.M. van den Bergh
Vrije Universiteit

10.1 Introduction

The issue of environmental and other externalities related to spatial interactions, and particularly transport, has received quite some attention recently (see, e.g. Verhoef 1994). However, more comprehensive approaches to environmental policy analysis, integrating issues of transport and spatial economic structure, are scarce (see the literature on environmental economics (Kneese and Sweeney 1985), regional and urban economics (see Nijkamp 1986), and transport economics (see, e.g. Button 1993). This is somewhat surprising, since a policy aimed at mitigating, for instance, transport externalities will certainly affect the economic structure both in a sectoral and in a spatial sense.

Spatially disaggregated, economy wide equilibrium models can be used to fill up this gap. One can consider then two possible approaches. The first is to use general equilibrium type of models with a multiregional disaggregation. Such modelling approaches have received quite some attention recently, both in the context of multiregional and network transport modelling, and in the context of international trade modelling (see Chapter 2). An alternative approach is adopted in this chapter, and employs an extension of the Spatial Price Equilibrium (SPE) model. From the perspective of the SPE framework, the proposed model is novel in the sense that it combines financial instruments for environmental and transport policy with behaviour rules on a micro level in the context of the SPE quasi-social welfare optimization. On a conceptual level, the novelty can be expressed in terms of a structure that allows for the investigation of an environmentally sustainable spatio-sectoral organisation of an economy.

SPE models have been used to analyze spatial interactions in terms of commodity (freight) flows, with flexible prices clearing spatial excess demands and supplies for given transport cost structures and given local demand and supply structures. A more general interpretation of SPE may include flows of production factors and intermediates, and even passenger transport (see, for many variations on SPE models, Takayama and Judge 1971; Takayama and Labys 1986; and various chapters in this book). The analysis of regulation in SPE models is rare. Still, the quasi-welfare approach of SPE can be combined quite naturally with a spatially and sectorally disaggregated analysis of economy wide impacts of environmental externalities as well as environmental policies.

The structure of this paper is as follows. Section 10.2 introduces a conceptual multi-layer framework for the analysis of environmental externalities in a spatial economy. In Section 10.3, this framework is formalized in terms of a spatial price equilibrium model, which combines a quasi-social welfare approach with market behaviour of producers, households and transporters. In Section 10.4, the optimal rules for environmental, transport and infrastructural policies are shown to be interdependent, which means that in order to have an optimal outcome the choice of one of these policies narrows the spectrum of the remaining policies. Section 10.5 is devoted to the case where, because of an open country situation or an independent departmental policy-making, the environmental policy is fixed and only second-best transport policies are possible. Concluding remarks are made in Section 10.6.

10.2 Environment in the Spatial Economy

The nowadays much debated interpretation of environmental sustainability is even less straightforward for a region or a sector, than it is for a national economy (see, for instance, Van den Bergh 1991; Van den Bergh and Nijkamp 1994). A main reason is that the goal of sustainability reflects an orientation towards the overall behaviour of a well defined, closed system. This implies that, unless external parameters are assumed to be fixed, indeterminacies result in case one deals with an open system such as a region or a sector. Obviously, sustainability constraints applied to an open system have only limited meaning, as their relevance critically hinges on the adequateness of the assumed fixity of the external parameters. Therefore, environmental sustainability needs to be interpreted as a characteristic of a whole system. From the perspective of this paper, sustainability should then refer to a sustainable level of overall economic activity (including transportation), a sustainable spatial organization of these activities, and a supply of infrastructure which is able to serve the transportation flows resulting from this organization. Thereby account should be taken of all interactions that may occur between these phenomena, and the way each separate element influences the sustainability indicator. Closely related to these questions is of course the question of how to realize this ideal sustainable state, given the policy instruments available, and taking serious account of most probably counteracting individual optimizing behaviour.

In Figure 10.1 four interacting layers are distinguished, which jointly represent the spatio-economic structure together with its impact sphere. Notice that the parentheses in the title of the second layer, which represents the spatial organization of economic activities, indicate that we use a broad definition of the term 'economic', including all possible kinds of productive and consumptive activities. Due to specialization of these nodes, fed by comparative advantages, scale economies or agglomeration economics, the nodes are not self-sufficient. As a result, at prevailing local market prices, the bundle of goods and factors supplied within a node is not the same as the bundle demanded. Therefore inter-nodal trade takes place, which may represent inter-regional or inter-country trade. This trade is made possible by means of the presence of infrastructure (the third layer) and gives rise to all sorts of transport activities (the fourth layer). Finally, the upper layer represents the area of environmental effects.

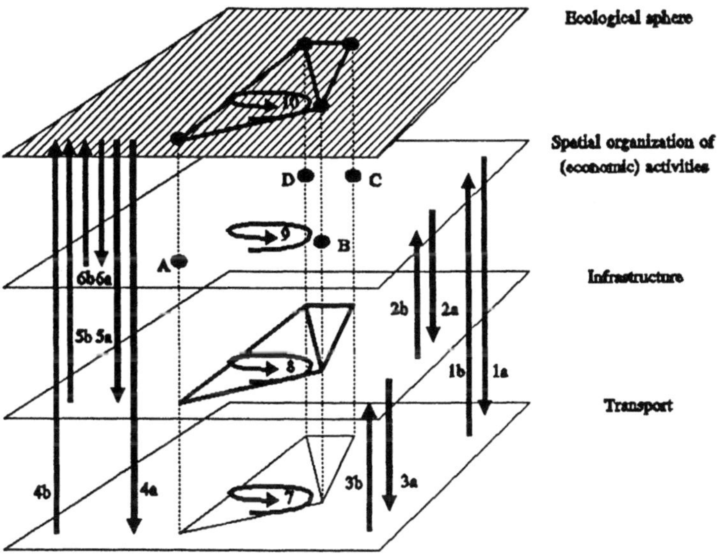

Fig 10.1. Spatial sustainability in a multi-layer approach

The arrows indicate various interactions that may occur in the system considered. The arrows on the right hand side describe issues which are traditionally in the heart of regional- and transport economics, such as: the derived demand for transport; the effect of transportation costs on the spatial distribution of activities; the spatial construction of infrastructure depending on the spatial distribution of economic activities as well as transport flows; the effect of the spatial supply of infrastructure on the spatial organisation of the economic system and on transport flows. The group of arrows on the left hand side represent the additional interactions involved when considering the question of ecological sustainability. The three ascending arrows on the left indicate the environmental degradation resulting from each of the other layers. These effects will to some extent be localized, which is represented by the reprint of the spatial structure of the other layers in the ecological sphere. Other environmental externalities will be non-localized, which is represented by the shading of the ecological layer. The three descending arrows indicate that environmental degradation may affect both the productivity and the utility in the other layers.

The figure illustrates clearly that a serious consideration of the issue of sustainable transport results in the adoption of a quite complex system of multilateral spatio-dynamic interactions. Furthermore, it may be noted that traditional economic approaches towards the relation between transport and environment usually concentrate on the outer left arrow (10.4b) only, and in addition on its static, non-localized component only.

The framework of Figure 10.1 can be cast into a formal model. In the following section, a model based on the Spatial Price Equilibrium (SPE) approach (see Takayama and Judge 1971) is used to accomplish this. From an economic perspective, a main advantage of such a model is its close relation to traditional welfare economic modelling, thus lending itself quite naturally to formulations in terms of welfare maximization, and derivations of associated optimal policies. All four layers distinguished in Figure 1 can be included, enabling consideration of the full policy complications of all possible interactions discussed above.

10.3 Environmental and Transport Policies in a Spatial Price Equilibrium Model

In this section a general type of SPE model is considered, which describes a system of $J \cdot I$ spatially differentiated markets. For J goods and I nodes, $J \cdot I$ local demand relations and $J \cdot I$ local supply relations are assumed to exist. It should be noted here that this assumption does not necessarily imply positive demand and supply quantities on each market in equilibrium. In formulating the model, we follow the *net quasi welfare* approach. For the present model, this means that the economy wide algebraic sum of Marshallian consumers' surpluses, minus all production costs, minus all transport costs (including costs of infrastructure supply), and in this model also minus all environmental costs, is assumed to give the appropriate welfare measure for the government (or regulator) to base their policies on. A discussion of the associated implicit assumptions with regard to the utility function (that is: the integrability problem) is left aside here; details can be found in Takayama and Judge (1971, chapter 6) and Varian (1992, section 8.5). The social optimization problem of maximizing this net quasi welfare, subject to a sustainability constraint, subject to individual maximizing behaviour, and subject to some relevant non-negativity constraints, can be represented by the following Kuhn-Tucker specification:

$$
\begin{aligned}
\Phi = {} & \sum_{i=1}^{I}\sum_{j=1}^{J}\int_0^{Y_i^j} b_i^j(\psi_i^j)\,d\psi_i^j - \sum_{i=1}^{I}\sum_{j=1}^{J}\int_0^{X_i^j} c_i^j\big(e_i^h(X_i^h)\,\forall h,\chi_i^j\big)\,d\chi_i^j - \sum_{i=1}^{I}\sum_{j=1}^{J}\varepsilon_i^j\cdot e_i^j(X_i^j) \\
& - \sum_{o=1}^{I}\sum_{\substack{d=1\\ d\neq o}}^{I}\sum_{j=1}^{J}\sum_{m=1}^{M} Z_{od}^{jm}\cdot t_{od}^{jm}(Z_{od}^{hm}\,\forall h, I_{od}^{m}) - \sum_{o=1}^{I}\sum_{\substack{d=1\\ d\neq o}}^{I}\sum_{m=1}^{M} C_{od}^{m}(I_{od}^{m}) \\
& - \sum_{o=1}^{I}\sum_{\substack{d=1\\ d\neq o}}^{I}\sum_{m=1}^{M} \varepsilon_{od}^{Zm}\cdot e_{od}^{Zm}(Z_{od}^{jm}\,\forall j) - \sum_{o=1}^{I}\sum_{\substack{d=1\\ d\neq o}}^{I}\sum_{m=1}^{M} \varepsilon_{od}^{Im}\cdot e_{od}^{Im}(I_{od}^{m})
\end{aligned}
\qquad (10.1a)
$$

$$
+ \lambda_E\cdot\Big[E^* - E\big(e_i^j(X_i^j), e_{od}^{Zm}(Z_{od}^{hm}\,\forall h), e_{od}^{Im}(I_{od}^{m})\big)\ \forall i\ \forall j\ \forall o\ \forall d\ \wedge o\neq d\ \forall m\Big] \qquad (10.1b)
$$

$$
+ \sum_{i=1}^{I}\sum_{j=1}^{J}\lambda_{M_i^j}\cdot\Big[c_i^j\big(e_i^h(X_i^h)\,\forall h, X_i^j\big) + \pi_i^j + \gamma_i^j - b_i^j(Y_i^j)\Big] \qquad (10.1c)
$$

$$
+ \sum_{o=1}^{I}\sum_{\substack{d=1\\ d\neq o}}^{I}\sum_{j=1}^{J}\sum_{m=1}^{M}\lambda_{N_{od}^{jm}}\cdot\Big[c_o^j\big(e_o^h(X_o^h)\,\forall h, X_o^j\big) + \pi_o^j + t_{od}^{jm}(Z_{od}^{hm}\,\forall h, I_{od}^{m}) + \tau_{od}^{jm} + \gamma_d^j - b_d^j(Y_d^j)\Big] \qquad (10.1d)
$$

$$
\begin{aligned}
\text{s.t.:}\quad & Z_{ii}^j \geq 0 \quad \forall i\ \forall j && (10.1e)\\
& Z_{od}^{jm} \geq 0 \quad \forall o\ \forall d\ \wedge o\neq d\ \forall j\ \forall m && (10.1f)\\
& I_{od}^{m} \geq 0 \quad \forall o\ \forall d\ \wedge o\neq d\ \forall m && (10.1g)\\
\text{with:}\quad & Y_i^j = \sum_{\substack{o=1\\ o\neq i}}^{I}\sum_{m=1}^{M} Z_{oi}^{jm} + Z_{ii}^j \quad \forall i\ \forall j && (10.1h)\\
& X_i^j = \sum_{\substack{d=1\\ d\neq i}}^{I}\sum_{m=1}^{M} Z_{id}^{jm} + Z_{ii}^j \quad \forall i\ \forall j && (10.1i)\\
& p_i^j = b_i^j - \gamma_i^j = c_i^j + \pi_i^j \quad \forall\ Z_{ii}^j > 0 && (10.1j)\\
& p_d^j = b_d^j\ \gamma_d^j = c_o^j + \pi_o^j + t_{od}^{jm} + \tau_{od}^{jm} = p_o^j + t_{od}^{jm} + \tau_{od}^{jm} \quad \forall\ Z_{od}^{jm} > 0 && (10.1k)\\
\text{and:}\quad & i,o,d \in \{1,2,\ldots,I\} \wedge o\neq d;\ j,h \subset \{1,2,\ldots,J\};\ m \in \{1,2,\ldots,M\} && (10.1l)
\end{aligned}
$$

The variables have the following meaning: Y_i^j is the consumption of good j ($j=1,2,\ldots,J$) in node i ($i=1,2,\ldots,I$), and X_i^j is the production of good j in node i. The variables b_i^j and c_i^j give the marginal benefits of consumption and the marginal production costs for good j in node i, respectively. Later on, we will use B_i^j and C_i^j to indicate total benefits and consumption, so that $b_i^j=\partial B_i^j/\partial Y_i^j$ and $c_i^j=\partial C_i^j/\partial X_i^j$. We assume that $B_i^j(0)=C_i^j(0)=0$: total benefits from zero consumption and total costs of zero production are both equal to zero. The physical parameters e_i^j give the total local environmental emissions associated with (and non-decreasing in) the production of good j in node i. We assume that for all sectors in all nodes, marginal production costs c_i^j rise with total production X_i^j, and are non-decreasing in any local emissions e_i^h (where $h=1,2,\ldots,J$), reflecting possible adverse effects of local environmental degradation on productive activities. Consumers' marginal benefits from consumption b_i^j are assumed to be diminishing in nodal consumption Y_i^j. Additionally, the extent to which overall nodal indirect money metric welfare in node i is directly adversely affected by local emissions is given by $\Sigma_j \varepsilon_i^j e_i^j$, where ε_i^j gives the nodal shadow price for sector j's emissions in node i associated with environmental values in addition to the direct effects on production costs (for instance, ε_i^j includes consumers' valuation of environmental goods).

Z_{ii}^j gives total intra-nodal deliveries of good j within node i, which are assumed not to require any transportation (that is, nodes are assumed to be spaceless points). Z_{od}^{jm} represents total deliveries of good j from node o (origin; $o=1,2,\ldots,I$) to node d (destination; $d=1,2,\ldots,I$), using transport mode m ($m=1,2,\ldots,M$). The average transport costs of shipping good j from node o to node d using transport mode m are given by t_{od}^{jm}. I_{od}^m gives the capacity of infrastructure of type m between nodes o and d (in that particular direction), and C_{od}^m gives the associated total costs of infrastructure provisions (including depreciation and maintenance costs). Average transport costs t_{od}^{jm} are assumed to be negatively related to the relevant infrastructure capacity I_{od}^m available, and positively related to all deliveries on that particular link, Z_{od}^{hm} (for $h=1,2,\ldots,J$), representing congestion effects. The total (non-decreasing) physical environmental effects of the usage and existence of infrastructure of type m between nodes o and d are given by e_{od}^{Zm} and e_{od}^{Im}, respectively. As is the case for production, $\varepsilon_{od}^{Im}\cdot e_{od}^{Im}$ and $\varepsilon_{od}^{Zm}\cdot e_{od}^{Zm}$ give the associated extent to which instantaneous indirect money metric welfare is adversely affected by these effects along link *o-d*. Strictly speaking, as we do not consider any economic activities to take place along these links (and hence no actors are assumed to be located along these links), these effects could best be interpreted in terms of instantaneous non-use values (option values or intrinsic existence values) of the environment along such links.

Apart from the instantaneous localized environmental effects, all physical parameters e have their respective (non-decreasing) effects on a global environmental sustainability parameter E associated with the commitment to act consistently with future needs. The restriction on the present generation's choice space, resulting from commitment to sustainable development, is represented by the exogenous parameter E^*. E^* thus represents the *environmental utilisation space* (Opschoor 1992). It may be worth underlining that we merely consider two types of environmental effects: instantaneous localized effects, and effects on the global sustainability parameter E.

Next, considering the markets, p_i^j is the local price for good j in node i. The government is assumed to be able to regulate by means of specific producers' taxes π_i^j on the production of good j in node i; consumers' taxes γ_i^j on the consumption of good j in node i; and mode specific transport taxes τ_{od}^{jm} on deliveries of good j from region o to region d. Finally, the λ's are Lagrangian multipliers. Whereas Lagrangian multipliers in more traditional SPE formulations with quantity constraints reflect shadow prices of the respective goods, the present formulation with constraints in terms of individual optimizing behaviour yields multipliers λ_M and λ_N indicating *shadow prices of market and/or policy failures*, as expressed in deviations from optimal pricing. Due to the linear nature of these restrictions, these multipliers also translate marginal changes of the respective regulatory taxes into the additional net social benefits they yield.[1] The multiplier λ_E ($=\partial\Phi/\partial E^*$) can readily be interpreted as the shadow price for meeting the sustainability constraint.

The model may now be described as follows. Expression (10.1a) gives the objective of maximizing the net quasi welfare, and (10.1b) gives the sustainability constraint. The additional constraints (10.1c) and (10.1d) bring maximizing behaviour of all actors on all markets involved into the model: (10.1c), together with (10.1j), summarize the joint effect of maximizing behaviour on intra-nodal markets of consumers on the one hand: purchasing each good in that quantity where marginal benefits are just equal to the sum of the local price and the consumers' tax, or not purchasing the particular good when this sum exceeds marginal benefits at zero consumption; and productive sectors on the other: producing that quantity for which the sum of marginal costs and producers' taxes are equal to local prices, or not producing when this sum exceeds producers' prices at zero production. Likewise, (10.1d), together with (10.1k), describes the spatial arbitrage resulting from transportation activities. Clearly, equations (10.1j) and (10.1k) ensure spatial consistency of the pattern of market prices. The distribution of these flows among the various transport modes will be such that, in equilibrium, average transport costs including regulatory taxes are equal for all modes, and are in addition equal to inter-nodal price differentials. Clearly, cross-hauling (the simultaneous shipment of a certain good in both directions along a certain link) cannot be explained from a private maximizing (or: market) point of view when transport costs (including transportation taxes) are larger than zero. Should such cross-hauling actually occur in reality, the obvious conclusion is that the purchasers apparently regard the goods as, although perhaps close substitutes, still definitely heterogeneous.

Next, (10.1e)-(10.1g) give some non-negativity conditions, and (10.1h) and (10.1i) some logical quantity relations (note in particular that we abstain from inventory-building, as is typical for static SPE-approaches). Finally, (10.1l) summarizes the domain of the various indices mentioned. It will be clear by now, that we assume atomistic markets, with price-taking behaviour throughout the economic system. Indeed, that is also why intra sectoral producers' externalities are included in the model.

With reference to Figure 10.1, it may be noted that the above model covers most of the interactions distinguished, except for those represented by arrows (5a) and

---

[1]Actually, in the present setting, where it will be assumed that the regulator has complete flexibility in the specification of the tax structure, the constraints (10.1c) and (10.1d) need not be included in the Lagrangian expression, but could also be specified as separate conditions. However, when the scope for perfect regulatory taxation is limited, formulations such as the one considered here will turn out to be very useful, as will become apparent in Section 10.5. As the multipliers λ_M and λ_N will vanish in the first-best optimum considered now, we will not engage in a discussion of their interpretation here. It should be noted already though, that at positive quantities traded, the respective constraints should be satisfied strictly. Hence, in such cases, 'traditional' Kuhn-Tucker non-negativity conditions on λ_M and λ_N are not in order, and are therefore not included in the first-order conditions (10.6) and (10.7).

(4a). Finally, although we merely distinguish single environmental parameters e_i^j, e_{od}^{Im} and e_{od}^{Zm} per source of environmental degradation, the model obtains sufficient flexibility in the specification of each source's emission composition by allowing for source specific factors ε_i^j, ε_{od}^{Im} and ε_{od}^{Zm}; as well as source specific marginal effects on the productivities in all local industries (for productive sources); and source specific effects on the sustainability parameter E.

10.4 First-best Environmental and Transport Policies

We now proceed with the solution of the SPE-model presented in the foregoing section in a first-best world, where the regulator has perfect control over all regulatory taxes (or subsidies, for negative taxes) distinguished. Problem 1 leads to the following necessary first-order conditions for the maximization of net quasi welfare:

$$\frac{\partial \Phi}{\partial Z_{ii}^j} \leq 0; \quad Z_{ii}^j \geq 0 \quad \text{and} \quad Z_{ii}^j \cdot \frac{\partial \Phi}{\partial Z_{ii}^j} = 0; \quad \forall i \ \forall j \tag{10.2}$$

$$\frac{\partial \Phi}{\partial Z_{od}^{jm}} \leq 0; \quad Z_{od}^{jm} \geq 0 \quad \text{and} \quad Z_{od}^{jm} \cdot \frac{\partial \Phi}{\partial Z_{od}^{jm}} = 0; \quad \forall o \ \forall d \wedge o \neq d \ \forall j \ \forall m \tag{10.3}$$

$$\frac{\partial \Phi}{\partial I_{od}^{m}} \leq 0; \quad I_{od}^{m} \geq 0 \quad \text{and} \quad I_{od}^{m} \cdot \frac{\partial \Phi}{\partial I_{od}^{m}} = 0; \quad \forall o \ \forall d \wedge o \neq d \ \forall m \tag{10.4}$$

$$\frac{\partial \Phi}{\partial \lambda_E} = E^{*} - E \geq 0; \quad \lambda_E \geq 0 \quad \text{and} \quad \lambda_E \cdot \frac{\partial \Phi}{\partial \lambda_E} = 0 \tag{10.5}$$

$$\frac{\partial \Phi}{\partial \lambda_{M_i^j}} = c_i^j + \pi_i^j + \gamma_i^j - b_i^j \geq 0 \quad \text{and} \quad \lambda_{M_i^j} \cdot \frac{\partial \Phi}{\partial \lambda_{M_i^j}} = 0; \quad \forall i \ \forall j \tag{10.6}$$

$$\frac{\partial \Phi}{\partial \lambda_{N_{od}^{jm}}} = c_o^j + \pi_o^j + t_{od}^{jm} + \tau_{od}^{jm} + \gamma_d^j - b_d^j \geq 0 \quad \text{and} \quad \lambda_{N_{od}^{jm}} \cdot \frac{\partial \Phi}{\partial \lambda_{N_{od}^{jm}}} = 0; \quad \forall o \ \forall d \wedge o \neq d \ \forall j \ \forall m \tag{10.7}$$

$$\frac{\partial \Phi}{\partial \pi_i^j} = \lambda_{M_i^j} + \sum_{\substack{d=1 \\ d \neq i}}^{I} \sum_{m=1}^{M} \lambda_{N_{id}^{jm}} = 0 \quad \forall i \ \forall j \tag{10.8}$$

$$\frac{\partial \Phi}{\partial \gamma_i^j} = \lambda_{M_i^j} + \sum_{\substack{o=1 \\ o \neq i}}^{I} \sum_{m=1}^{M} \lambda_{N_{oi}^{jm}} = 0 \quad \forall i \ \forall j \tag{10.9}$$

$$\frac{\partial \Phi}{\partial \tau_{od}^{jm}} = \lambda_{N_{od}^{jm}} = 0 \quad \forall o \ \forall d \wedge o \neq d \ \forall j \ \forall m \tag{10.10}$$

First of all, equations (10.8)-(10.10) imply that the regulatory taxes should be set their optimal levels: all multipliers λ_M and λ_N are equal to zero. Complex conditions resulting from (10.2)-(10.4) (not given above) may then be simplified accordingly as the terms associated with these multipliers are eliminated. Next, as our ultimate goal is the derivation of optimal policies given the workings of market mechanisms, we will solve the model in terms of the optimal taxation structure of producers' taxes π_i^j, consumers' taxes γ_i^j, and transportation taxes τ_{od}^{jm}, as well as optimal infrastructural investment policies (which we assume to be the government's responsibility).

For any optimal $Z_{ii}^j > 0$, the first expression (10.2) should hold as a strict equality. For positive intra-nodal deliveries to materialize via the market process, the first expression in (10.6) should clearly hold as a strict equality, where the nodal price will be as given by (10.1j). Therefore, optimal taxation on intra-nodal markets requires:

$$\gamma_i^j + \pi_i^j = \left[\sum_{h=1}^{J} \frac{\partial C_i^h}{\partial e_i^j} + \varepsilon_i^j + \lambda_E \cdot \frac{\partial E}{\partial e_i^j} \right] \cdot \frac{\partial e_i^j}{\partial X_i^j} \quad \forall i \ \forall j \tag{10.11}$$

Likewise, for any optimal $Z_{od}^{jm} > 0$, the first expression in (10.3) should hold as a strict equality, whereas the market mechanism will only lead to positive inter-nodal deliveries provided the first expression in (10.7) holds as a strict equality. We thus obtain:

$$\gamma_d^j + \pi_o^j + \tau_{od}^{jm} = \left[\sum_{h=1}^{J} \frac{\partial C_o^h}{\partial e_o^j} + \varepsilon_o^j + \lambda_E \cdot \frac{\partial E}{\partial e_o^j} \right] \cdot \frac{\partial e_o^j}{\partial X_o^j} + \sum_{h=1}^{J} Z_{od}^{hm} \cdot \frac{\partial t_{od}^{hm}}{\partial Z_{od}^{jm}}$$

$$+ \left[\varepsilon_{od}^{Zm} + \lambda_E \cdot \frac{\partial E}{\partial e_{od}^{Zm}} \right] \cdot \frac{\partial e_{od}^{Zm}}{\partial Z_{od}^{jm}} \quad \forall o \ \forall d \wedge o \neq d \ \forall j \ \forall m$$

(10.12)

Equations (10.11) and (10.12) define a set of $J \cdot I + J \cdot I \cdot (I-1) \cdot M$ equations. The number of unknowns (i.e. the number of regulatory taxes) is $2 \cdot J \cdot I + J \cdot I \cdot (I-1) \cdot M$. Therefore, in the present setting, there is a whole set of optimal regulatory tax schemes the regulator may choose from. Provided conditions (10.11) and (10.12) remain satisfied, $J \cdot I$ taxes may be set at an arbitrary level, and the remaining optimal taxes (or subsidies, when such taxes turn out to be negative) may then be derived. The main policy implication arising from (10.11) and (10.12), however, is that when considering regulatory taxes in a more general setting, one cannot ignore incentives associated with taxation on activities and markets related to the issues actually considered. If, for instance, consumption taxes are already in existence, (10.11) and (10.12) indicate that optimal regulatory producers' and transportation taxes are affected accordingly. This interdependence is closely related to the traditional trade-off that a government faces when taxation serves both regulatory purposes and a general goal of revenue raising (and, likewise, to the problem of optimal taxation).[2]

Clearly, the most straightforward among the solutions implied by (10.11) and (10.12) is to put consumption taxes equal to zero, and consequently to impose the following regulatory taxes:

$$\gamma_i^j = 0 \quad \forall i \ \forall j \tag{10.13}$$

$$\pi_i^j = \left[\sum_{h=1}^{J} \frac{\partial C_i^h}{\partial e_i^j} + \varepsilon_i^j + \lambda_E \cdot \frac{\partial E}{\partial e_i^j} \right] \cdot \frac{\partial e_i^j}{\partial X_i^j} \quad \forall i \ \forall j \tag{10.14}$$

$$\tau_{od}^{jm} = \sum_{h=1}^{J} Z_{od}^{hm} \cdot \frac{\partial t_{od}^{hm}}{\partial Z_{od}^{jm}} + \left[\varepsilon_{od}^{Zm} + \lambda_E \cdot \frac{\partial E}{\partial e_{od}^{Zm}} \right] \cdot \frac{\partial e_{od}^{Zm}}{\partial Z_{od}^{jm}} \quad \forall o \ \forall d \ \wedge \ o \neq d \ \forall j \ \forall m \tag{10.15}$$

Equations (10.14) and (10.15) indicate a direct application of the *polluter pays principle* -or, regulation at the source - by imposing environmental taxation on productive activities, and a composite congestion and environmental tax on transportation activities. It may be worth underlining that this choice for zero consumption taxes has no real distributive consequences, due to the market assumptions underlying the model. Indeed, the regulator is to a considerable extent free to choose the *statutory incidence* of the regulatory policies as long as (10.11) and (10.12) remain fulfilled - i.e. the regulator has some freedom in the choice of directly taxing either consumption, production, or transportation. Still, the eventual *economic incidence* - i.e. the eventual distribution of the tax burden among all actors involved, taking account of all tax shifting that will occur - only

[2]Explicit consideration of this trade-off requires a tax revenue constraint, and - in order to determine the optimal overall level of taxation - also explicit consideration of the benefits resulting from government spending. Apart from considering the social benefits of government spending on infrastructure, we abstain from such issues, and stick to the regulatory dimension.

eventual *economic incidence* - i.e. the eventual distribution of the tax burden among all actors involved, taking account of all tax shifting that will occur - only depends on the respective demand and supply elasticities over the relevant ranges, which are assumed to remain unaffected by the policies considered. In the optimum, the algebraic sum of nodal prices minus producers' taxes is the same for all optimal tax schemes, as is the sum of nodal prices and consumers' taxes (see (10.1j) and (10.1k)). Consequently, exactly the same optimal (spatial) pattern of production and consumption, and consumers' and producers' welfare will result.

Finally, optimal investment rules for infrastructure are given by (10.4), implying the following marginality rules for positive optimal investments:

$$-\sum_{j=1}^{J} Z_{od}^{jm} \cdot \frac{\partial t_{od}^{jm}}{\partial I_{od}^{m}} = \frac{\partial C_{od}^{m}}{\partial I_{od}^{m}} + \left[\varepsilon_{od}^{lm} + \lambda_E \cdot \frac{\partial E}{\partial e_{od}^{lm}} \right] \cdot \frac{\partial e_{od}^{lm}}{\partial I_{od}^{m}} \qquad \forall I_{od}^{m} > 0 \tag{10.16}$$

The left-hand side of (10.16) represents marginal benefits of infrastructural investments in terms of lower transportation cost, and the right-hand side gives the associated marginal social cost as the sum of financial outlays and environmental costs.

The interdependencies between optimal levels of economic and transportation activities, and infrastructure in a sustainability context is in the first place obviously reflected in the appearance of the multiplier λ_E in all optimality rules. Apart from the environmental taxes with regard to instantaneous effects (and congestion taxes for transport), each activity faces an additional sustainability tax, the level of which depends on the extent to which the activity gives rise to emissions, the extent to which these emissions inflict the sustainability parameter, and finally the system-wide shadow price λ_E associated with the sustainability constraint.

Apart from the interdependence via the common parameter λ_E however, more fundamental interdependences follow from the logical quantity restrictions in (10.1h) and (10.1i), together with private maximizing behaviour as implied by (10.1c) and (10.1d) together with (10.1j) and (10.1k). As the optimality rules are derived with consideration of these relations, they recognize the need for internal consistency of the spatio-economic structure in terms of absence of excess demands or supplies at equilibrium prices on all markets (that is, market clearing throughout). Finally, the associated optimal levels of transportation (and therefore also the optimal spatial pattern of production and consumption) are determined simultaneously with the optimal levels of infrastructure through the impact of both on average transport costs. In terms of Figure 10.1 therefore, the present model indeed recognizes the relevance for optimality of the interdependencies among all layers.

The 'physical characteristics' of the optimal sustainable spatio-economic structure as compared to the market based structure can, given the model's generality, only be broadly sketched here. In the first place, an overall system-wide *scale effect* follows from the commitment to sustainability. This scale effect is directly dependent on the actual level of E^*, and is reflected in the system-wide prevailing shadow price λ_E ($=\partial\Phi/\partial E^*$). Additionally, taking account of the role of optimal regulatory taxes within the model (via relations (10.11), (10.12), (10.16) and (10.1c) and (10.1d)), we find that the optimal efficient *allocation* of the limited *environmental utilisation space* E^* among the various economic activities requires a larger restriction of a certain activity, (1) the larger the activity's emissions, (2) the larger the extent to which these emissions inflict the sustainability parameter, (3) the more sensitive the quantity demanded to price changes (that is, the less the activity is valued as a necessity; either due to its very nature or due to substitution possibilities), and (4) the more sensitive the quantity supplied to price changes (that is, the less reductions in total output yield cost advantages). As a matter of fact, the extent to which the remaining parts of the regulatory taxes (as related to instantaneous environmental and congestion effects) lead to reductions in the activity depends on the activity's marginal external costs, and the sensitivities in quantities demanded and supplied to price changes, in exactly the same manner.

The inter-sectoral effects just mentioned will in addition imply a reordering of patterns of spatial specialization. In the first place, this has to do with the fact that some nodes may initially have more heavily polluting activities than others, and will therefore face larger optimal reductions in nodal productive activities than other nodes do. In the second place however, this is due to the fact that transportation is one of the activities considered. Along with the expected restrictions on transportation activities, one would foresee a process of de-specialization, and enhanced spatial self-sufficiency. This may for instance result in the revitalization of sectors which have been declining in certain nodes, due to enlarged spatial (inter-nodal) competition (along with the secular decline in transport costs). Likewise, one may expect shifts from purchases of imported goods towards locally supplied substitutes.

10.5 Second-Best Optimal Transport Policies

In this section, we have a closer look at optimal second-best transportation policies. In particular, we focus on the case where the regulator has no control over regulation in other economic activities: (s)he is simply facing given values of π_i^j and γ_i^j which are not necessarily optimal according to (10.11). Deviations from (10.11) may then be referred to as *system failures*, resulting from the simultaneous occurrence of *market failures* and, as prevailing policies do not (fully) compensate for these failures, *policy failures*. The question then becomes one of optimal second-best regulation, under the restriction of non-optimality on related markets. In terms of our SPE-model, equations (10.8) and (10.9) are no longer in order, as π_i^j and γ_i^j become given parameters rather than instrumental variables. This situation may for instance occur when different governmental bodies have their own responsibilities and interests (such as revenue raising). Also, for a small country, concerned with global environmental issues, and having to deal with considerable throughflow, this is a highly relevant problem.

Before turning to the solution, it is important to pay some closer attention to the role of the multipliers λ_M and λ_N. The associated constraints indicate the effects of market behaviour on intra-nodal and inter-nodal markets respectively, in terms of market equilibria at positive quantities traded requiring the associated marginal benefits to be equal to the sum of marginal production costs, plus relevant consumers' and producers' taxes; and, plus marginal transportation costs and taxes in case of inter-nodal deliveries. On the other hand, zero trade occurs on a market when the marginal benefits fall short of this sum at zero deliveries. As λ_M and λ_N give shadow prices of market and/or policy failures, the associated condition $\lambda_M=0$ or $\lambda_N=0$ at zero deliveries means that marginal changes in the relevant taxes then do not affect social welfare. For positive deliveries, the constraints turn into 'ordinary' Lagrangian equality constraints, which implies that in such cases the Kuhn-Tucker condition that these multipliers should be non-negative need not be satisfied. On the other hand, these multipliers can now be zero for two distinct reasons: first, if we are in the corner solution of zero deliveries, and the constraint holds as a strictly inequality; and secondly if we are in an interior solution, where the constraint is being satisfied as a strict equality, and additionally the multiplier turns out to be zero by coincidence.

With producers' taxes π_i^j and consumers' taxes γ_i^j being exogenous in the present setting, the first-order conditions (10.8) and (10.9) should be removed, and the multipliers λ_M can no longer be set equal to zero. However, (10.10) still holds, indicating that transportation taxes should be used optimally, and we may still skip the associated terms in (10.2)-(10.4). As (10.4) then turns out to yield the same outcome as given by (10.16), we pay no further attention to optimal infrastructural investments in this section.

Next, we substitute the equality from (10.7) into (10.3), and find that for socially optimal inter-nodal deliveries to materialize, transportation taxes τ_{od}^{jm} should be set as follows:

$$\tau_{od}^{jm} = \sum_{h=1}^{J} Z_{od}^{hm} \cdot \frac{\partial t_{od}^{hm}}{\partial Z_{od}^{jm}} + \left[\varepsilon_{od}^{Zm} + \lambda_E \cdot \frac{\partial E}{\partial e_{od}^{Zm}} \right] \cdot \frac{\partial e_{od}^{Zm}}{\partial Z_{od}^{jm}}$$
$$+ \left[\sum_{h=1}^{J} \left(\frac{\partial C_o^h}{\partial e_o^j} - \lambda_{M_o^h} \cdot \frac{\partial c_o^h}{\partial e_o^j} \right) + \varepsilon_o^j + \lambda_E \cdot \frac{\partial E}{\partial e_o^j} \right] \cdot \frac{\partial e_o^j}{\partial X_o^j}$$
$$- \lambda_{M_o^j} \cdot \frac{\partial c_o^j}{\partial X_o^j} + \lambda_{M_d^j} \cdot \frac{\partial b_d^j}{\partial Y_d^j} - \pi_o^j - \gamma_d^j \quad \forall o \; \forall d \wedge o \neq d \; \forall j \; \forall m \tag{10.17}$$

As the appearance of the terms λ_M in equation (10.17) readily demonstrates, optimal levels of transportation taxes now critically depend on the extent to which systems failures occur in the nodes of origin and destination in the second-best optimum. For instance, if no failures occur because intra-nodal deliveries simply do not occur on any market referred to in (10.17) in the second-best optimum, all factors λ_M are zero and (10.17) reduces to:

$$\tau_{od}^{jm} = \sum_{h=1}^{J} Z_{od}^{hm} \cdot \frac{\partial t_{od}^{hm}}{\partial Z_{od}^{jm}} + \left[\varepsilon_{od}^{Zm} + \lambda_E \cdot \frac{\partial E}{\partial e_{od}^{Zm}} \right] \cdot \frac{\partial e_{od}^{Zm}}{\partial Z_{od}^{jm}}$$

$$+ \left[\sum_{h=1}^{J} \frac{\partial C_o^h}{\partial e_o^j} + \varepsilon_o^j + \lambda_E \cdot \frac{\partial E}{\partial e_o^j} \right] \cdot \frac{\partial e_o^j}{\partial X_o^j} - \pi_o^j - \gamma_d^j$$

(10.18)

Although not identical to it, this situation is actually not far from the first-best case. In particular, if intra-nodal deliveries do not occur on any single market, equations (10.11) become irrelevant for the first-best solution, and equations (10.18) are indeed consistent with equations (10.12), giving first-best policies in that particular case. The only difference is that in the present case producers' taxes π_i^j and consumers' taxes γ_i^j are given, and therefore unique values for transportation taxes τ_{od}^{jm} will be found. Still, the first-best optimum is attainable.

As soon as (some) intra-nodal deliveries do occur, however, ending up in a second-best optimum inferior to the first-best optimum can be avoided only by sheer coincidence, as will be seen below. For transportation flows between nodes in which intra-nodal trade on the markets considered in (10.17) does not occur, equations (10.18) remain valid (although the overall first-best outcome for the whole system is no longer likely to be attainable). They simply indicate transportation taxes equal to the sum of marginal external costs of production and transport, minus the sum of contingent taxation on production in the node of origin and on consumption in the node of destination. More complex expressions for optimal second-best transportation taxes arise for transport flows if intra-nodal deliveries do take place. In such cases, the terms λ_M need certainly not be zero. Rather, it can be derived from (10.2) and (10.6) that occurrence of intra-nodal deliveries in a certain node in the second-best optimum implies that[3]:

$$\lambda_{M_i^j} = \frac{\left[\sum_{h=1}^{J} \frac{\partial C_i^h}{\partial e_i^j} - \sum_{\substack{h=1 \\ h \neq j}}^{J} \lambda_{M_i^h} \cdot \frac{\partial c_i^h}{\partial e_i^j} + \varepsilon_i^j + \lambda_E \cdot \frac{\partial E}{\partial e_i^j} \right] \cdot \frac{\partial e_i^j}{\partial X_i^j} - \left[\pi_i^j + \gamma_i^j \right]}{\frac{\partial c_i^j}{\partial X_i^j} + \frac{\partial c_i^j}{\partial e_i^j} \cdot \frac{\partial e_i^j}{\partial X_i^j} - \frac{\partial b_i^j}{\partial Y_i^j}}$$

(10.19)

[3]Unfortunately, we are not able to give a general expression for λ_M, independent of the other terms λ_M occurring on the right hand side of (10.19). The reason is that such an expression will depend on the number of sectors involved in the node considered, and therefore on the number of λ_M's on the right hand side of (10.19). For a given number of sectors, however, an explicit expression for λ_M can be found.

It follows from (10.19) that intra-nodal deliveries may now cause the first-best spatio-economic structure not being attainable, which will then be reflected in a non-zero shadow price λ_M. There are two exceptions.

First, the numerator may be zero. This is for instance the case as the other terms λ_M in the numerator are zero, and if the sum of producers' taxes π_i^j and consumers' taxes γ_i^j happen to fulfil the first-best condition (10.11). The resulting outcome that λ_M equals zero indicates that in such cases, there is no social cost associated with the regulator's incapability of affecting producers' and consumers' taxes, as there are no system failures. Assuming the terms $\lambda_{M_i^*}$ in the numerator zero, and given a finite (and non-negative) value of the denominator in (10.19), the larger the divergence between the sum of these two taxes and their first-best optimal value, the larger the shadow price of system failures as reflected in the absolute value of $\lambda_{M_i^j}$. However, the terms $\lambda_{M_i^*}$ in the numerator need of course not be zero. In such cases, the associated first term in the numerator, representing the first-order inter-sectoral external cost, is corrected for by what we may regard to be a second-order effect, in order to arrive at the overall social evaluation of this inter-sectoral externality.[4]

Secondly, if either the demand or supply (or both) on the intra-nodal market under consideration is completely inelastic in the second-best optimum, the denominator of (10.19) becomes (plus) infinity and λ_M equals zero. This reflects that if the quantity traded on the intra-nodal market is fixed due to the particular inelasticity, deviations from the first-best conditions (10.11) have a zero shadow price, as they do not lead to non-optimal responses in terms of quantity traded. Clearly, given a non-zero value of the numerator in (10.19), the more responsive the quantity traded to price differentials (that is, the flatter both the demand curve and the supply curve), the larger the absolute value of λ_M. The second term in the denominator shows that any intra-sectoral externalities (within the node) should be taken into account here in order to be sure to consider the socially relevant marginal cost curve.

Before proceeding, it may be noteworthy that the actual form of (10.19) reveals the nature of the multipliers λ_M and λ_N. Being the *shadow price of system failures*, which are expressed in *price differentials*, it should not be too surprising that these multipliers are actually expressed in quantities. More precisely (and concentrating on λ_M), (10.19) shows that, when evaluated in the second-best optimum, λ_M is a *linear approximation* of the excess quantity of second-best intra-nodal deliveries over the deliveries in the first-best optimum, evaluated for the slope of the demand and supply curves in this second-best optimum. A positive value of λ_M means that second-best deliveries exceed first-best deliveries, whereas a negative value implies the opposite. Hence, these multipliers give in fact shadow excess deliveries resulting from system failures (where these latter are expressed in non-optimal prices). It is important to stress the 'shadow' nature of these multipliers. In

---

[4]In particular, we will see below that the value of λ_M is a linear approximation of excess trade on market M. Therefore, if any λ_M on the right hand side is larger (smaller) than zero, the social value of the inter-sectoral externality (ceteris paribus leading to higher production costs and therefore a smaller quantity traded on the affected market) is smaller (larger) than represented by the first-order effect only, due to its restrictive (stimulating) effect on the quantity traded.

general, the multipliers λ_M will not give the exact excess deliveries, as they are merely valid at the margin. Only if the nodal demand and supply relations were linear throughout (or when λ_M equals zero at positive deliveries)[5], λ_M would give an exact measure of the excess deliveries in the second-best optimum over deliveries in the first-best optimum. In general, they give a linear approximation of these excess deliveries for the reason just mentioned.

Figure 10.2 provides an illustration of λ_M according to (10.19) (that is: under optimal second-best regulation) for positive values (Figure 10.2a) and negative values (Figure 10.2b). The horizontal axis depicts volumes delivered on the intra-nodal market considered. Q_0 gives the non-intervention market outcome as the intersection of the demand (or marginal benefit) curve $D=b$, and the supply (or marginal cost) curve $S=c$. For graphical convenience, we assume the intra-sectoral externality to be zero, so that the second term in the denominator of (10.19) drops out. The greek letter η is a shorthand for the marginal external costs as given by the composite term in the numerator of (10.19). Therefore, Q_1 is the first-best optimal level of intra-nodal deliveries. However, suppose Q_2 is the level of intra-nodal deliveries in the second-best optimum, where nodal producers' taxes π and consumers' taxes γ prevail. In the first case (Figure 10.2a), this second-best trade volume Q_2 exceeds the first-best volume Q_1, whereas the opposite holds in the second case in (Figure 10.2b). In both cases, λ_M can be seen to be a linear approximation of the difference between the first-best and the second-best quantity traded. It is found by shifting the marginal cost curve up by the value of η in the second-best optimum (indicated η_2) to the curve S', and determining the intersection of the tangent lines to S' and D. It is easy to verify that this is the value of λ_M in the second-best optimum as defined by (10.19).

One might now proceed by substituting the values of λ_M into (10.17), in order to find their impact on the optimal second-best transportation taxes. However, due to the mutual dependencies of the terms λ_M, an explicit expression cannot be given (see footnote 4). These interdependencies however, disappear if we ignore the intra-sectoral and inter-sectoral externalities altogether, and merely consider the effects related to the terms e_i^j and to the sustainability constraint (as we did for the transportation sector throughout). If we then take the 'worst' case, where intra-nodal deliveries occur on both markets connected through the inter-nodal delivery, we find that the following optimal second-best regulatory transportation taxes apply for such cases:

---

[5]When λ_M equals zero at zero deliveries in the second-best optimum, we cannot be sure whether it gives an exact measure of (negative) excess deliveries, as λ_M may then be zero because the associated constraint then holds as a strict inequality, whereas first-best deliveries might still have been positive.

$$\tau_{od}^{jm} = \sum_{h=1}^{J} Z_{od}^{hm} \cdot \frac{\partial t_{od}^{hm}}{\partial Z_{od}^{jm}} + \left[\varepsilon_{od}^{Zm} + \lambda_E \cdot \frac{\partial E}{\partial e_{od}^{Zm}} \right] \cdot \frac{\partial e_{od}^{Zm}}{\partial Z_{od}^{jm}}$$

$$+ \frac{\left[\varepsilon_o^j + \lambda_E \cdot \frac{\partial E}{\partial e_o^j} \right] \cdot \frac{\partial e_o^j}{\partial X_o^j} - \pi_o^j}{1 - \frac{\partial c_o^j}{\partial X_o^j} \Big/ \frac{\partial b_o^j}{\partial Y_o^j}} + \frac{\gamma_o^j}{1 - \frac{\partial b_o^j}{\partial Y_o^j} \Big/ \frac{\partial c_o^j}{\partial X_o^j}}$$

$$- \frac{\left[\varepsilon_d^j + \lambda_E \cdot \frac{\partial E}{\partial e_d^j} \right] \cdot \frac{\partial e_d^j}{\partial X_d^j} - \pi_d^j}{1 - \frac{\partial c_d^j}{\partial X_d^j} \Big/ \frac{\partial b_d^j}{\partial Y_d^j}} - \frac{\gamma_d^j}{1 - \frac{\partial b_d^j}{\partial Y_d^j} \Big/ \frac{\partial c_d^j}{\partial X_d^j}} \qquad (10.20)$$

Equation (10.20) falls apart in three distinct parts. The first part, expressed in two terms, is equal to the marginal external (congestion and environmental) costs of the transportation activity as such, and therefore requires no further comments. The other two parts, both also expressed in two terms, are symmetric expressions, related to the two intra-nodal markets concerned. These parts deserve some closer attention.

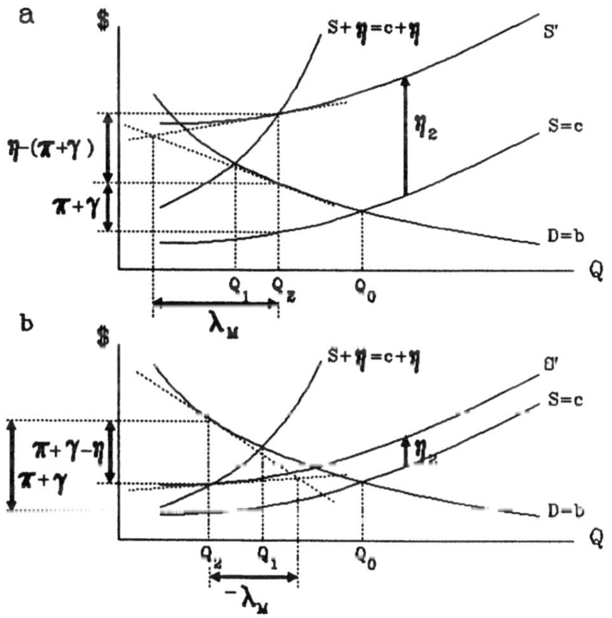

Fig. 10.2 Graphical representation of the shadow price of system failures λ_M for positive values (a) and negative values (b)

First, given the slopes of the demand and supply curves, it turns out that the optimal second-best transportation tax should be higher, the larger the extent to which marginal external costs of production exceed producers' taxes in the node of origin, and the smaller the extent to which marginal external costs of production exceed producers' taxes in the node of destination.[6] The first outcome reflects that transportation taxes may serve as a (limited) substitute for lacking producers' taxes in the node of origin. The second effect is typical for second-best situations. It can best be interpreted as follows: if the transport flow considered concerns goods which serve in the node of destination as a substitute for locally produced goods, the less this local production is environmentally regulated, the less it pays off to restrict transportation of that particular good to that particular destination. The reason is that a restriction of such transport will, due to the workings of the market, lead to a certain expansion of local (polluting) production in the node of destination. The larger the extent to which the marginal external costs exceed producers' taxation in the node of destination, the more harmful to social welfare this adverse substitution effect will be, and hence the smaller the second-best transportation tax should be. In other words: there is a trade-off between pollution resulting from transportation, and pollution resulting from local production that would arise if transportation were to be restricted.

Secondly, still given the slopes of the demand and supply curves, the optimal second-best transportation tax should be higher, the smaller the consumers' tax in the node of destination, and the higher the consumers' tax in the node of origin. The first effect is again in line with our findings for the first-best case, indicating that consumers' taxation in the node of destination and transportation taxes may to a certain extent serve as substitutes in the optimization of inter-nodal deliveries. The second effect reflects that higher transportation taxes will in general lead to lower exports, and therefore lower local production, implying lower production costs and hence lower nodal prices in the node of origin. The more nodal consumption in the node of origin is already restricted below optimal levels because of consumers' taxes, the more it becomes worthwhile to increase transportation taxes in order to keep nodal consumers' prices down as much as possible.

The denominators of the four terms just discussed may vary between 1 and (plus) infinity, where the first polar case implies that the effect considered in the nominator of the respective term should be fully accounted for in the determination of the optimal second-best regulatory transportation tax, whereas the latter polar case implies that this effect vanishes completely. Indeed, these denominators can be seen as some sort of weights, trading off the severity of system failures on the side of the market considered against failures on the other side of that market, and against failures on the other two markets present in the equation (that is, the other intra-nodal market and the inter-nodal market). Hence, the weights are such that the inevitable failures resulting from the use of the second-best taxes are distributed as efficiently as possible among the three markets in order to keep the overall welfare loss as small as possible.

---

[6] These conditions can of course readily be extended to cases where producers' taxes exceed marginal external costs, rather than falling short of marginal external costs.

The actual values of these weights depend on the slopes of both the relevant supply curve and the relevant demand curve. For the two terms related to divergences between marginal environmental costs of local production and local producers' taxation, it turns out that these weights become smaller, the less responsive local supply to such divergences is. The correction for the steepness of the demand curve indicates that, in addition, the weights become smaller the less the extent to which local consumers' surplus is affected by such supply responses. Likewise, for the two terms related to local consumers' taxes, it turns out that the effect in the numerator becomes less important, the less responsive local demand for price variations. Here too, the weights become smaller, the less local producers' surplus is affected by such demand responses. In this way, the four 'second-best effects' present in (10.20) are weighted against the 'first-best' direct impact of the regulatory tax on transport externalities, in order to arrive at the optimal second-best fee. One of the striking features of this optimal second-best fee is that it may actually take on negative values; in particular if, in the node of destination, local production is relatively polluting, the producers' regulation is very modest, while consumers' taxes are relatively high.

10.6 Conclusion

The present paper stressed the need for integration of elements concerning economic structure, infrastructure, transportation and environmental quality in one analytical framework when considering questions on sustainability in a spatial economy. The adapted SPE model presented indeed seems to open a wide range of possibilities for investigating the differences between market based versus sustainable spatio-economic structures. In particular the possibility of explicitly recognizing the simultaneous effects and inevitable interdependencies of individual optimizing behaviour and associated market processes on the one hand, and regulatory policies on the other, as two fundamental forces in the development of spatio-economic structures, seems to indicate a large potential of such type of modelling in investigating the issues at hand. In this approach, the concept of *system failures* as resulting from simultaneous occurrence of *market failures* and *policy failures* turned out to be a useful concept.

The first-best and second-best analyses performed in this paper demonstrate that a common physical sustainability constraint should not be distributed among the various activities according to some naive rule. Rather, the optimal distribution of the 'burden' resulting from the commitment to sustainability should be such that the constraint is met in the most efficient way, implying that the internal consistency of the system should not be violated. As a result, the spatio-economic system has to undergo structural changes, both in a sectoral and in a spatial sense, which requires the use of internally consistent optimal regulatory tax structures.

One can think of many possible extensions to the analyis presented here. To name just five options: (i) dynamic extensions, in which the processes of transition towards sustainable spatio-economic structures are investigated; (ii) inclusion of spatial factor markets (e.g., for labour) and spatial markets for intermediates in order to investigate sectoral interdependencies and the process of income and purchasing power generation; (iii) investigation of other second-best aspects in policy formulation; (iv) empirical operationalization of the type of models presented. Such issues, unfortunately, have to be deferred to subsequent papers.

References

Bergh JCJM van den (1991) Dynamic Models for Sustainable Development. Thesis. Publishers, Amsterdam

Bergh JCJM van den, Nijkamp P (Eds) (1994) Sustainability, Resources and Region. Special Issue of the Annals of Regional Science 28

Button KJ (1993) Transport Economics. 2nd ed. Edward Elgar Publ, Aldershot

Kneese AV, Sweeney JL (Eds) (1985) Handbook of Natural Resource and Energy Economics. 3 Volumes. North-Holland, Amsterdam

Nijkamp P (Ed) (1986) Handbook of Regional and Urban Economics 1. North Holland, Amsterdam

Opschoor JB (1992) Sustainable Development, the Economic Process and Economic Analysis. In: Opschoor JB (Ed) (1992) Environment, Economy and Sustainable Development. Wolters-Noordhoff, Groningen, pp 25-52.

Takayama T, Judge GG (1971) Spatial and Temporal Price and Allocation Models. Contributions to Economic Analysis 73. North-Holland, Amsterdam

Takayama T, Labys WC (1986) Spatial Equilibrium Analysis. In: Nijkamp P (Ed) (1986) Handbook of Regional and Urban Economics 1. Elsevier Science Publishers, Amsterdam

Varian HR (1992) Microeconomic Analysis. Third Edition. Norton, New York

Verhoef ET (1994) External Effects of Road Transport. Some Theory and a Survey of Empirical Results. Transportation Research A 28:273-287

11 Congested Roads and General Equilibrium Pigouvian Tax Solutions[1]

Inge Mayeres and Stef Proost
Katholieke Universiteit Leuven

11.1 Introduction

In this paper we use Pigouvian taxation theory to solve network congestion problems in a general equilibrium framework. In *network congestion problems*, the cost of using the network increases as a function of the total flow on the network. The present problems on our roads and airports are the best illustrations of such disequilibria between the demand for transportation services and the available capacity. Congestion problems also occur in other public facilities like telephone networks, swimming pools, etc. where the quality of services decreases when the total level of demand is in disproportion to the available capacity. We will however concentrate on the road transport example.

Solutions to network congestion problems are different in *the short, medium and long term.* In the short term, the available capacity is fixed and the use of the network is rationed when demand is too high. We are used to implicit rationing solutions where delay times decrease the quality of service and discourage users to some extent. Explicit rationing may use other methods such as arbitrary number plate rationing. However, in general, economists prefer congestion pricing. In the medium term, the capacity of the network can be extended so as to ease the disequilibria. In the long term, demand for transportation services itself can drastically change through relocation of demand for transportation services. We will focus on the short to medium term dimension and disregard the relocation possibilities. Medium and long term instruments are of course complementary. The use of existing infrastructure has to be optimized via prices and these prices will guide opportunities to invest in capacity expansion and will also guide relocation decisions of the agents demanding transportation services. We will mainly concentrate on optimal short run pricing. Coordination is much more

[1]The research reported in this paper has been conducted under contract TR/C8/019 (promoter S. Proost) of the National Impulse Programme Transport and Mobility initiated by the Belgian State - Prime Minister's Service - Science Policy. The scientific responsibility is assumed by its authors. S. Proost is a Research Fellow of the Belgian National Fund for Scientific Research.

difficult between investments in road capacity and relocation because of the irreversibility or high sunk costs of investment and location decisions.

In the literature one can distinguish *three partial equilibrium approaches* to congestion pricing: the network analysis in spatial and regional economics, the transport economics approach and the environmental economics approach. In the spatial economics literature, the solution of spatial interaction models with costs dependent on the flows is a common problem since Beckman et al. (1956)[2]. The demand for trips is taken as a decreasing function of the total transport cost which contains monetary costs and time costs. The total travel cost is an increasing function of the flows on the route used. Minimizing total travel costs yields equilibrium flows on the network. The emphasis is on the explanation of the behaviour of the agents in the network: which routes do they take and why.

In the transport economics literature, there is a long tradition of optimal congestion pricing and its second best derivatives such as parking fees or priority lanes. Important contributions are Walters (1961) and Vickrey (1969), focusing on one mode and Glaister and Lewis (1978) and Small (1983), who include several modes. Demand for transport is considered a function of generalized cost (money plus time costs) and the emphasis is on the measurement of the social marginal cost (equal to the sum of marginal resource costs and marginal congestion costs), on the translation of this notion into pricing structures and on the new equilibrium which can be obtained in this way. It mostly concerns partial equilibrium approaches where all users of transport, be it private cars or trucks, are treated equally and where the use of revenues from the congestion fees is not relevant because they are mere transfers and do not affect the total consumers' and producers' surplus.

More recently the transport problem is approached from the environmental economics tradition. There are many efforts continuing to try to apply the Pigouvian idea of making all activities pay their marginal social costs, to the road transportation sector (see Newbery 1990; Button 1990). The social costs cover congestion costs, air pollution, accident costs, noise and other amenity values. Basically this effort can be seen as an extension of the optimal transport pricing tradition. Congestion usually comes out as the dominant externality cost (Mayeres 1993,1994) and one finds that in peak periods the present prices for urban transport are far below the social marginal costs (De Borger et al 1994). There is an increased interest to treat the optimal congestion pricing problem in a *general equilibrium context*. There are two reasons for this. The use of the Pigouvian tax idea could lead to an important extension of the tax base and of the tax revenues. In an economy with already large existing tax rates this can offer opportunities to lower the most inefficient tax rates (discussed by Bovenberg and Van der Ploeg 1994). Moreover, as the introduction of Pigouvian or congestion taxes will affect the welfare positions of all agents considerably, there is a need to analyze the overall welfare effect of this policy change for each group of individuals involved. Hence the need to consider explicitly the use of the Pigouvian tax revenues to reduce other taxes or to increase other public expenditures.

---

[2]This area is surveyed by Batten & Boyce (1986).

11.2 The Optimal Tax Approach

In this paper we address the problem of the optimal design of a taxation system for externalities that affect the demand for other goods: there are feedback effects. A typical example of this is congestion on roads: the use of the private car at certain moments creates congestion and this congestion influences itself the use of the private car and the demand for other goods: possibly a higher demand for public transportation but also for other goods, e.g. when excessive congestion decreases the overall transport volume and the income is spent on non-transport goods. The same framework could be used for other public facilities.

Optimal congestion taxes and congestion tax reform rules are derived for a traditional second best taxation model based on Diamond and Mirrlees (1971), Sandmo (1975), Atkinson and Stiglitz (1976) and Dixit (1985). It is a general equilibrium model for an open economy with as tax instruments a head tax, commodity taxes on all goods consumed and an excise on congestion-generating activities in the production sector. One of the commodity taxes is a tax on the consumption of transport services in peak periods. This set of tax instruments is chosen because it approaches reasonably well the existing tax possibilities in most industrialized economies except for the tax on transport in peak periods. Since we analyze the problem by means of a general equilibrium approach, we focus on the simultaneous determination of all tax rates. In addition, it is assumed that the congestion externality can also be reduced by public abatement. In the particular case of congestion problems this can be seen e.g. as government investments in infrastructure in order to alleviate the externality.

The optimal congestion tax problem addressed in this paper is highly simplified by the following two assumptions. The first assumption is the measurability and taxability of the externality generating private goods. This means that transport use can be taxed at the time and place of congestion. The second assumption is that the externality is of the 'atmosphere' type: only the sum of the individual uses of the externality creating commodity matters and not the distribution over individuals. Both assumptions are important to keep the problem structure simple: taxation of the use of externality generating goods is feasible and the Pigouvian prescription (tax equal to marginal externality damage) is not ruled out from the beginning. Indeed, when the individual consumptions all have a different effect on the level of the externality, the Pigouvian tax per unit of consumption has to be individualised which is infeasible in most tax systems. Diamond (1973), and more recently Balcer (1980) and Wijkander (1985) show in a simplified model how the optimal Pigouvian tax structure is affected when it is impossible to tax peak and off-peak transport use differently. In the present paper however, the emphasis lies on the derivation of optimal taxes in a second best framework where the second best refers to the absence of lump sum redistribution instruments rather than to the absence of discrimination possibilities between the peak and the off-peak period.

In the first part of the paper we concentrate on the determination of the optimal government policy. Typical questions are: which goods should be taxed (i.e. only the transport good or also complementary goods), how should the tax rate be computed, should we tax the transport use by producers as well as by consumers, how to use the proceeds of the congestion taxes and what is the optimal level of public infrastructure.

First the basic model is presented, next we discuss the structure of optimal consumption taxes, the structure of taxes on producers and on import as well as the optimal level of government investment in public infrastructure. The analysis includes congestion as well as non-congestion taxes. There is a strong interdependence between both as all taxes have to fulfil common goals like income distribution and the raising of sufficient public revenue. This simultaneous analysis of all taxes allows for consideration of double benefit effects (deadweight loss savings through externality taxes) and of income distribution effects of congestion taxes, dimensions which are absent in the partial equilibrium models used in the transport economics literature.

In the second part of the paper we turn to the less ambitious problem of characterizing the optimal marginal tax reform starting from an arbitrary equilibrium of an economy with congestion.

11.2.1 The Structure of the Economy

We consider an open economy in which there are M goods and factors, denoted $k=1,\ldots,M$ and one externality whose level is denoted by Z. The level of the externality will be the central variable in this paper. The level of Z (e.g. the level of congestion) affects the welfare of consumers negatively in a direct and in an indirect way. The direct effect consists of the influence of Z on the utility of the consumers (discomfort, time losses). The indirect effect will take place through the increased production costs caused by extra congestion (need for extra trucks, time losses...). The level of Z is a function of the use of the externality generating good M (transport in peak periods) by consumers and by the production sector. It can be decreased by public investments in infrastructure.

There are N nonidentical consumers (indexed $i=1,\ldots,N$) who differ in their earning capacity e^i. The consumption vector of consumer i is $x^i=(x_1^i,\ldots,x_M^i)$, where x_k^i denotes his consumption of good k $(x_k^i>0)$. x_1^i is assumed to represent his consumption of leisure. Leisure is taken to be the numéraire. With I denoting total time available to each consumer, $(I-x_1^i)$ gives the labour supply of consumer i. Since the productivity per hour worked by consumer i is given by e^i, the wage earned can be represented by $e^i(I-x_1^i)$. In addition each consumer receives a head subsidy T and thus faces the following budget constraint:

$$\sum_{k=2}^{M} q_k x_k^i \leq T + e^i (I - x_1^i) \qquad \forall i \tag{11.1}$$

The utility function of consumer i is given by

$$u^i = u^i (x_1^i, \ldots, x_M^i, Z) \qquad \forall i$$

u^i is concave in x_k^i and twice continuously differentiable. It is assumed that the externality Z (congestion) is an external diseconomy, or:

$$\frac{\delta u^i}{\delta Z} < 0 \qquad \forall i$$

Each individual i chooses his/her consumption bundle x^i in such a way as to maximize his/her utility function subject to his/her budget constraint.

$$\begin{aligned} MAX \quad & u^i (x_1^i, \ldots , x_M^i, Z) \\ s.t. \quad & \sum_{k=2}^{M} q_k x_k^i \leq T + e^i (I - x_1^i) \end{aligned} \tag{11.2}$$

It is assumed that in this maximization problem the consumers ignore their own impact on the externality. They consider themselves to be infinitely small compared to the total traffic flow.

Assuming an interior solution exists, the first order conditions are given by

$$\frac{\delta u^i}{\delta x_1^i} = \alpha^i e^i \qquad \forall i$$

$$\frac{\delta u^i}{\delta x_k^i} = \alpha^i q_k \qquad \forall i \qquad k=2,\ldots,M$$

where α^i stands for the marginal private utility of income.

This implies demand functions of the form

$$x_k^i = x_k^i (q, T, Z)$$

The maximum utility individual i can achieve when facing prices q, lump sum subsidy T and externality Z is given by the indirect utility function:

$$V^i (q, T, Z) = u^i (x_1^i (q, T, Z),\ldots, x_M^i (q, T, Z), Z) \tag{11.3}$$

There are J firms (or sectors), indexed $j=1,\ldots,J$, that use a net quantity of inputs y_k^j, outputs are measured positively and inputs negatively. The production possibilities are described by the implicit production functions which contain the level of the externality Z as argument.

$$F^j (y_1^j, \ldots , y_M^j, Z) \leq 0 \qquad \forall j \tag{11.4}$$

For a given level of the externality Z national production activities are characterized by constant returns to scale and diminishing marginal rates of transformation.

The externality Z is an external diseconomy:

$$\frac{\delta F^j}{\delta Z} > 0 \qquad \forall j$$

The level of the externality Z is determined by the use of good M, the externality generating good (e.g. car use at peak hours). Good M produces the externality in a quantity γ per unit of use in private consumption and at a rate δ in production processes. The government can reduce the externality level by investments R (e.g. enlarging roads). The level of Z is thus given by

$$Z = Z\left(\gamma \sum_{i=1}^{N} x_M^i - \delta \sum_{j=1}^{J} y_M^j \,,\, R\right) = Z\left(P \,,\, R\right) \tag{11.5}$$

where γ and δ stand for the differential impact of private cars and trucks on congestion.

To save on notation we assume that neither good M nor R are produced in the country but that they are imported. This avoids an explicit treatment of all interindustry deliveries and uses of these goods.

Net imports from the rest of the world are denoted m_k and the import possibilities are given by the offer curve G. This surface defines all possible net trades which are feasible, including the import of the externality reducing good :

$$G\left(m_1, \ldots\ldots, m_k, \ldots\ldots, m_M\right) + E\left(R\right) \leq 0 \tag{11.6}$$

The government requires for its functioning given quantities g_k of all goods except good M and has to pay for the externality reducing investments R. The expenditures are financed by indirect taxes t_k on the consumption of all goods and by a head tax T. The possibility of levying indirect taxes on all consumption goods, the absence of other income than labour income and the presence of a head tax can also be interpreted as a linear income tax combined with a full set of indirect taxes. The government maximizes an individualistic social welfare function $W(V^1, \ldots, V^N)$.

To gain insight in the taxation problem we assume as a start that the government has complete control over production y_k^j and over net imports m_k. Its problem is then to decide on the taxes on consumption t_k, and on the head tax T and on the ultimate aggregate externality level via the level of R, production variables being controlled directly. This is the familiar Diamond and Mirrlees (1971) framework to discuss optimal tax problems. Later on we will discuss the decentralization of production decisions and their impact on the optimal tax structure.

The government's problem consists of maximizing social welfare with respect to t_k, T, y_k^j, and R, respecting the production possibility constraints and the material balance constraints. The latter are given by

$$-\sum_{i=1}^{N} e^{i}\,(\,I - x_1^i\,) \leq \sum_{j=1}^{J} y_1^j - g_1$$

$$\sum_{i=1}^{N} x_k^i + z_k \leq \sum_{j=1}^{J} y_k^j \qquad k=2,\ldots,M$$

(11.7)

The government's problem is given by

$$MAX \quad W\,(\,V^1\,(\,q,\,T,\,Z(\,P,\,R\,)),\;.\;.\;.\;,\;V^N\,(\,q,\,T,\,Z(\,P,\,R\,))$$

$s.t.$

$$F^j\,(\,y_1^j,\;.\;.\;.\;,\;y_M^j,\;Z\,) \leq 0 \qquad \forall j \qquad (\,\sigma_j\,)$$

$$G\,(\,z_1,\;.\;.\;.\;,\;z_M\,,R) \leq 0 \qquad (\,\lambda\,)$$

$$-\sum_{i=1}^{N} e^{i}\,(\,I - x_1^i\,) \leq \sum_{j=1}^{J} y_1^j - g_1 \qquad (\,\pi_1\,)$$

$$\sum_{i=1}^{N} x_k^i + g_k \leq \sum_{j=1}^{J} y_k^j + m_k \qquad k=2,\ldots,M \qquad (\,\pi_k\,)$$

(11.8)

Forming the Lagrangean for this maximisation problem and using multipliers π_k for the material balances, σ_j for the production constraints and λ for the public infrastructure production constraint, we can derive the following necessary first-order conditions for an interior maximum:

$$\sum_{i=1}^{N} \frac{\delta W}{\delta V^i}\frac{\delta V^i}{\delta q_k}\frac{\delta q_k}{\delta t_k} + \sum_{i=1}^{N} \frac{\delta W}{\delta V^i}\frac{\delta V^i}{\delta Z}\frac{dZ}{dq_k}\frac{\delta q_k}{\delta t_k} - \sum_{j=1}^{J} \sigma_j \frac{\delta F^j}{\delta Z}\frac{dZ}{dq_k}\frac{\delta q_k}{\delta t_k}$$
$$- \pi_1 \sum_{i} e^i \frac{dx_1^i}{dq_k}\frac{\delta q_k}{\delta t_k} - \sum_{l=2}^{M} \pi_l \sum_{i=1}^{N} \frac{dx_l^i}{dq_k}\frac{\delta q_k}{\delta t_k} = 0 \qquad k=2,\ldots,M$$

(11.9)

$$\sum_{i=1}^{N} \frac{\delta W}{\delta V^i} \frac{\delta V^i}{\delta T} + \sum_{i=1}^{N} \frac{\delta W}{\delta V^i} \frac{\delta V^i}{\delta Z} \frac{dZ}{dT} - \sum_{j=1}^{J} \sigma_j \frac{\delta F^j}{\delta Z} \frac{dZ}{dT} - \pi_1 \sum_i e^i \frac{dx_1^i}{dT} - \sum_{l=2}^{M} \pi_l \sum_{i=1}^{N} \frac{dx_l^i}{dT} = 0$$

(11.10)

$$\pi_k - \sigma_j \frac{\delta F^j}{\delta y_k^j} = 0 \qquad \forall k \neq M$$

(11.11)

$$\pi_M - \sigma_j \frac{\delta F^j}{\delta y_M^j} - \sum_{l=1}^{J} \sigma_l \frac{\delta F^l}{\delta Z} \frac{\delta Z}{\delta y_M^j} + \sum_{i=1}^{N} \frac{\delta W}{\delta V^i} \frac{\delta V^i}{\delta Z} \frac{\delta Z}{\delta y_M^j} - \pi_1 \sum_i e^i \frac{\delta x_1^i}{\delta Z} \frac{\delta Z}{\delta y_M^j} - \sum_{l=1}^{M} \pi_l \cdot \sum_{i=1}^{N} \frac{\delta x_l^i}{\delta Z} \frac{\delta Z}{\delta y_M^j} = 0$$

(11.12)

$$\pi_k - \lambda \frac{\delta G}{\delta z_k} = 0 \qquad \forall k$$

(11.13)

$$\sum_{i=1}^{N} \frac{\delta W}{\delta V^i} \frac{\delta V^i}{\delta R} - \sum_{j=1}^{J} \sigma_j \frac{\delta F^j}{\delta R} - \lambda \frac{\delta G}{\delta R} - \pi_1 \sum_i e^i \frac{\delta x_1^i}{\delta R} - \sum_{l=1}^{M} \pi_l \sum_{i=1}^{N} \frac{\delta x_l^i}{\delta R} = 0$$

(11.14)

11.2.2 Consumer Taxes

Before proceeding, several conventions are necessary.

Following Dixit (1985), we use three sets of prices in the model: shadow prices π_k, associated with the material balance equations, producer prices p_k and consumer prices q_k. The following normalizations are applied. The shadow price of the first good is put equal to one. As we assume that the government fully controls all production, we provisionally take the producer prices equal to the shadow prices π_k . As regards consumer prices, which equal the sum of the producer price and the indirect tax rate, we can without loss of generality (Diamond and Mirrlees 1971) take one good as untaxed. Here we will take the first good as untaxed.

The consumer tax results can best be illustrated, taking (optimal) producer prices as given. This gives the following conventions for consumer prices :

$$\begin{aligned} \pi_k &= p_k \\ q_k &= p_k + t_k \quad \Rightarrow \text{ for given } p_k \Rightarrow dq_k = dt_k \\ &\text{for } k = 1, \ldots, M \end{aligned} \tag{11.15}$$

$$\pi_1 = p_1 = q_1 \qquad t_1 = 0 \tag{11.16}$$

Using Roy's identity

$$\frac{\delta V^i}{\delta q_k} = - \alpha^i x_k^{\ i} \tag{11.17}$$

together with expression (11.15) and the consumer's budget constraint, (11.9) can be transformed into (11.18). Applying a similar procedure to (11.10), one obtains (11.19).

$$-\sum_{i=1}^{N} \frac{\delta W}{\delta V^i} \alpha^i x_k^{\ i} + \left(\sum_{i=1}^{N} \frac{\delta W}{\delta V^i} \frac{\delta V^i}{\delta Z} - \sum_{j=1}^{J} \sigma_j \frac{\delta F^j}{\delta Z} \right) \frac{dZ}{dq_k} + X_k + \sum_{l=2}^{M} t_l \sum_{i=1}^{N} \frac{dx_l^{\ i}}{dq_k} = 0 \tag{11.18}$$

$$\sum_{i=1}^{N} \frac{\delta W}{\delta V^i} \alpha^i + \left(\sum_{i=1}^{N} \frac{\delta W}{\delta V^i} \frac{\delta V^i}{\delta Z} - \sum_{j=1}^{J} \sigma_j \frac{\delta F^j}{\delta Z} \right) \frac{dZ}{dT} - N + \sum_{l=2}^{M} t_l \sum_{i=1}^{N} \frac{dx_l^{\ i}}{dT} = 0 \tag{11.19}$$

Compared to the case without externalities the second term in (11.18) and (11.19) is added. It represents the effect of each tax instrument on the level of congestion, which is present also for taxes on other goods than good M. In order to analyze the structure of optimal taxes we have to examine the effect on congestion more deeply.

In order to know the full effect of tax changes on congestion we first have to define the full effect of tax changes on demand. The full effect of a tax change on the demand of a good is obtained by adding to the direct effect the indirect effect of the tax change on the congestion level and the impact of this change in congestion on the demand of the good. We use the following conventions:

$$\frac{dx_t^i}{dq_k} = \frac{\delta x_t^i}{\delta q_k} + \frac{\delta x_t^i}{\delta Z}\frac{dZ}{dq_k} \qquad \forall t,k \tag{11.20}$$

The first term at the right hand side of expression (11.20) gives the direct effect of q_k on x_t^i, holding Z constant. The second term gives the indirect effect on x_t^i through the influence of q_k on Z. Adding these two terms, one obtains the total effect dx_t^i/dq_k. Similarly, one has

$$\frac{dx_t^i}{dT} = \frac{\delta x_t^i}{\delta T} + \frac{\delta x_t^i}{\delta Z}\frac{dZ}{dT} \qquad \forall t \tag{11.21}$$

Using expression (11.5), (11.20) and (11.21) one obtains the following expressions for dZ/dq_k and dZ/dT:

$$\frac{dZ}{dq_k} = \frac{\delta Z}{\delta P}\gamma\sum_{i=1}^{N}\frac{dx_M^i}{dq_k} = \xi\frac{\delta Z}{\delta P}\gamma\sum_{i=1}^{N}\frac{\delta x_M^i}{\delta q_k} \tag{11.22}$$

$$\frac{dZ}{dT} = \frac{\delta Z}{\delta P}\gamma\sum_{i=1}^{N}\frac{dx_M^i}{dT} = \xi\frac{\delta Z}{\delta P}\gamma\sum_{i=1}^{N}\frac{\delta x_M^i}{\delta T} \tag{11.23}$$

where

$$\xi = \frac{1}{1 - \frac{\delta Z}{\delta P}\gamma\sum_{i=1}^{N}\frac{\delta x_M^i}{\delta Z}} \qquad 0 \le \xi \le 1 \tag{11.24}$$

is the congestion feedback parameter. (1-ξ) gives the proportion of the initial reduction in congestion which is eaten up by the additional traffic generated by the initial improvement in traffic conditions. Conversely, the congestion feedback parameter measures the extent to which an increase in congestion with one unit is reduced by the autonomous decrease in the demand for transport in response to that initial unitary increase in congestion.

Define β^i, the direct social marginal utility of income of consumer i as:

$$\beta^i = \frac{\delta W}{\delta V^i}\alpha^i \tag{11.25}$$

β^i does not consider the impact of T on Z. Z is held constant.

The social marginal disutility of Z for consumer i is defined as:

$$\zeta^i = \frac{\delta W}{\delta V^i}\frac{\delta V^i}{\delta Z} \tag{11.26}$$

The Slutsky equation is given by (holding Z constant)

$$\frac{\delta x_l^i}{\delta q_k} = s_{lk}^i - x_k^i\frac{\delta x_l^i}{\delta T} \tag{11.27}$$

and by definition

$$\sum_{i=1}^{N} s_{lk}^i = S_{lk} \tag{11.28}$$

We now have all elements for an interpretation of the optimal tax conditions. We proceed in two steps: we first define an extended version of the social marginal utility of transferring one unit of government income to a consumer. This concept is then used to derive simpler commodity tax rules. This way of proceeding allows to contrast our results with the optimal tax literature.

Define b^i, the net social marginal utility of an additional unit of income in the hands of consumer i normalized by the marginal welfare cost of government income (here equal to 1) as (all partial derivatives other than w.r.t. Z are defined for constant Z)[3]:

[3] Since all terms of b^i are normalized by dividing through $\pi^1=p^1=q^1=1$, b^i is expressed in monetary terms. By definition, π^1, which is the multiplier of the market equilibrium constraint for good 1, measures the welfare cost of an extra unit of expenditure by the government on good 1.

$$b^i = \beta^i + \sum_{t=1}^{M} t_t \frac{\delta x_t^{\ i}}{\delta T} + \frac{\delta x_M^{\ i}}{\delta T} Y$$

(11.29)

where

$$Y = \xi \frac{\delta Z}{\delta P} \gamma \left(\sum_{i=1}^{N} \zeta^i - \sum_{j=1}^{J} \sigma_j \frac{\delta F^j}{\delta Z} + \sum_{t=1}^{M} t_t \sum_{i=1}^{N} \frac{\delta x_t^{\ i}}{\delta Z}\right)$$

(11.30)

The first term of b^i gives the value in terms of government revenue of an increase in the consumer's income. The second term measures the money gain to the government resulting from the increased taxes paid by consumer i, when additional income is spent. The sum of these two terms gives the total direct effect of a change in T. It corresponds to the definition of b^i in models which do not incorporate externalities (Atkinson and Stiglitz 1976). The inclusion of externalities leads to an additional term which reflects the indirect effect of T. T will influence $x_M^{\ i}$ and hence Z. The change in Z will have an impact on the welfare of all consumers, on production and on tax revenue. The effect of this is measured by Y, which is defined by expression (11.30).

The externality term Y defined in (11.30) consists of three elements: the marginal externality damage of an additional unit of consumption of good M for consumers and producers and the induced government revenue. The first two elements correspond to the elements put forward by Lévy-Lambert (1968). The last element is added in our analysis because of the explicit treatment of the feedback effect of Z on the consumption of all goods k.

Using the definitions for β^i, ζ^i and b^i, together with (11.20), (11.21) and the Slutsky equation, (11.18) can be transformed into (cfr. Atkinson and Stiglitz 1976)

$$\sum_{t=2}^{M} t_t S_{tk} = \sum_{i=1}^{N} b^i x_k^{\ i} - X_k - S_{Mk} Y \qquad k=2,\ldots,M$$

(11.31)

Similarly, (11.19) can be transformed into (11.32) which states that on average the net social benefit of transferring one unit of income equals 1.

$$\bar{b} = 1$$

(11.32)

Using (11.32) one can simplify (11.31) into the more common form

$$\sum_{t=2}^{M} t_t \, S_{tk} = X_k \, \phi_k - S_{Mk} \, Y \qquad k=2,\ldots,M$$

$$\text{where} \quad \phi_k = \sum_{i=1}^{N} \frac{x_k^{\,i}}{X_k} \, b^i - 1$$

(11.33)

These are the fundamental conditions for optimality. Φ_k is the normalized covariance between the consumption of good k and the net social marginal utility of income. It is equal to the distributional characteristic of good k (r_k) minus 1. The distributional characteristic r_k measures the extent to which good k is consumed by people with a high net social marginal utility of income b^i.

Finally one can solve (11.33) for the indirect tax rates. Let $\underline{S}$ be the determinant of the coefficient matrix of t_k and $\underline{S}_{mk}$ the cofactor of the element in the *mth* row, *kth* column of this matrix. The determinant has dimension M-1 because the tax rate for the first good is assumed to be zero. Using

$$\sum_{t=2}^{M} S_{Mt} \, \underline{S}_{tk} \;\; = 0 \qquad k=2,\ldots,M-1$$

$$= \underline{S} \qquad k=M$$

(11.34)

we obtain expression (11.35).

$$\text{for } k = 2, \ldots, M-1 \quad t_k = \sum_{m=2}^{M} (X_m \, \phi_m) \, \frac{\underline{S}_{mk}}{\underline{S}}$$

$$\text{for good } M \quad t_k = \sum_{m=2}^{M} (X_m \, \phi_m) \, \frac{\underline{S}_{mk}}{\underline{S}} - Y$$

(11.35)

The optimal tax structure has the so-called 'additivity property' which has been put forward by Sandmo (1975): only the expression for the indirect tax rate of the externality generating good M contains an externality term. The tax formula for the other goods is unchanged. So if transport use during the peak period can be taxed, one does not have to levy taxes on complementary goods (e.g. fuel) or give subsidies to substitutes (e.g. public transport).

However, one should not be misled in the interpretation of the additivity property. Although the tax formula for the other goods remains unchanged, the introduction of a tax on externalities necessitates a change of all taxes to satisfy the budget constraint and to re-optimize the welfare distribution. Moreover, the tax on the externality generating good M cannot be interpreted as a sum of a distributional (and revenue) element and a pure externality correcting element. The externality term Y (see expression (11.30)) includes the sum over all consumers of ζ^i which is defined by expression (11.26) and which can alternatively be written as

$$\zeta^i = \frac{\beta^i}{\alpha^i} \frac{\delta V^i}{\delta Z}$$

(11.36)

So a distributionally weighed sum of externality damages is considered in the Pigouvian part of the optimal tax. This is logical because there are no perfect distribution instruments.

In the absence of externalities and compensated cross-price effects, the tax rate equals the inverse compensated price elasticity (Ramsey rule) corrected for the distributional characteristic of the good Φ which will take higher values for goods proportionally more consumed by people with a low net social marginal utility of income, as stated by Atkinson and Stiglitz (1976).

The introduction of the head tax as an instrument diminishes the need for non-uniform indirect taxes. Under certain restrictions on the preferences (e.g. present in the LES demand system) and in the absence of externalities the optimal tax is a linear income tax (Atkinson 1977). The additivity property implies then the addition to the linear income tax of a proportional congestion tax or equivalently the addition to the linear income tax of uniform commodity taxes and an extra excise on the consumption of transport during the peak periods.

11.2.3 Taxes on the Production Sector

In the problem under study industry also contributes to the congestion level. To show how the contribution to congestion by the production sector can be handled efficiently by a price and tax policy, we use the consumer's budget constraint (11.1), the definition of ζ^i in (11.36) and expression (11.15) to transform (11.11) and (11.12) and disregard for the moment the import possibilities. Dividing by the price of good 1, one obtains for good M, which is used as an input, the optimal producer price:

$$p_M = \sigma_j \frac{\delta F^j}{\delta y_M^j} + \frac{\delta Z}{\delta P} \delta \left(\sum_{i=1}^{N} \zeta^i - \sum_{l=1}^{J} \sigma_l \frac{\delta F^l}{\delta Z} + \sum_{l=2}^{M} t_l \sum_{i=1}^{N} \frac{\delta x_l^i}{\delta Z} \right) \quad \forall j$$

(11.37)

For all other goods k $(k \neq M)$ one obtains

$$p_k = \sigma_j \frac{\delta F^j}{\delta y_k^j} \quad \forall k \neq M, \forall j$$

(11.38)

Good M in our example being an input factor, this means that the production price p_M should equal its marginal productivity minus the marginal externality cost linked to the use of this factor in the production process. Both terms are expressed in terms of the producer price of good 1.

Decentralization of the production decisions is straightforward when an excise equal to

$$\frac{\delta Z}{\delta P}\,\delta\,\Big(\sum_{i=1}^{N} \zeta^{i} - \sum_{l=1}^{J} \sigma_l \frac{\delta F^l}{\delta Z} + \sum_{\iota=1}^{M} t_\iota \frac{\delta x_\iota^{\ i}}{\delta Z}\Big)$$

(11.39)

can be used. The excise expresses the external cost of the use of transportation facilities at peak periods by the production sector. It equals the marginal external damage of an increase in the level of congestion (the term with brackets) times the congestion generating effect of an increased use of trucks. The marginal external damage of congestion is the sum of three terms: the distributionally weighted sum of direct externality damage to the consumers, the increase in production costs induced by an increase in congestion and finally a correction for the induced government revenue effects due to a marginal increase in congestion. The excise is almost the same as the 'externality term' in the optimal commodity tax on consumer purchases of good M. It differs in the parameter δ (the contribution of trucks to congestion differs from the effect of cars on congestion) and in the feedback parameter ξ. To avoid too much complexities in notation we avoided an explicit representation of all intermediate uses of goods. It is clear that an excise can influence the relative price of a good through the whole Input-Output structure even if there is no direct externality involved in the production of that good.

11.2.4 The Impact of the Rest of the World

When we want to decentralize import decisions the most simple case to consider is the assumption of relative world prices which are taken as given. In that case using (11.15), relative shadow prices equal relative producer prices and relative international trade prices:

$$G(m_1,\ldots,m_M) = r_1\ m_1\ +\ldots+\ r_M\ m_M\ = 0$$

$$\frac{\pi_k}{\pi_\iota} = \frac{\dfrac{\delta G}{\delta m_k}}{\dfrac{\delta G}{\delta m_\iota}} = \frac{r_k}{r_\iota}$$

(11.40)

So in this case no excise or other tax or subsidy instrument should be used in relation to imports and exports. This is a simple restatement of the free trade theorem in international trade (Dixit 1985). Imported goods are taxed like home-produced goods and because imported goods do not generate externalities at the level of their production (assumption of no spill-overs) there is no other need to tax them.

If the world prices are no longer considered as given it becomes interesting to use import and export taxes in order to improve the terms of trade. If import and export taxes are not allowed indirect taxes will be used as second best instruments for the same goal.

In both cases the structure of the externality excise is not affected by the trade possibilities if the actions of the trade partners are given. If there is a strategic element in the externality-policies adopted abroad, this is no longer true. This needs more study.

11.2.5 The Optimal Provision of R

In our model, the government can influence the level of congestion by infrastructure investments R. These should be jointly optimized with the externality tax. Based on King (1986) we may write the following analogue of Roy's identity for the externality Z

$$\frac{\delta V^i}{\delta Z} = - w_Z^i \frac{dV^i}{dT} \tag{11.41}$$

where the negative of w_Z^i is defined as the marginal willingness to pay of consumer i for reduction of the externality Z. Comparison with (11.17) shows that the quantity demanded of the private good is replaced here by the willingness to pay for Z. The negative sign is explained by the fact that V^i decreases in Z. dV^i/dT is given by

$$\frac{dV^i}{dT} = \frac{\delta V^i}{\delta T} + \frac{\delta V^i}{\delta Z}\frac{dZ}{dT} \tag{11.42}$$

King also derives the Slutsky equation for a public good. In our case this becomes:

$$\frac{\delta x_t^i}{\delta Z} = s_{tZ}^i - w_Z^i \frac{dx_t^i}{dT} \tag{11.43}$$

s_{tZ}^i is the compensated effect of a change in the level of Z on the demand for good t by consumer i.

Using these two expressions (11.41), (11.42) and (11.43) and combining them with (11.14) and the following traditional definition of the net direct social marginal utility of an additional unit of income accruing to consumer i
we obtain

$$b^{i*} = b^i - \frac{\delta x_M^i}{\delta T} Y = \beta^i + \sum_{l=2}^{M} t_l \frac{\delta x_l^i}{\delta T} \tag{11.44}$$

$$\sum_{i=1}^{N} b^{i*} \frac{w_Z^i}{1 + w_Z^i \frac{dZ}{dT}} \frac{\delta Z}{\delta R} = - \sum_{j=1}^{M} \sigma_j \frac{\delta F^j}{\delta Z} \frac{\delta Z}{\delta R} - \frac{\lambda}{\xi} \frac{\delta G}{\delta R} + \sum_{l=2}^{M} t_l \sum_{i=1}^{N} \frac{s_{lZ}^i}{1 + w_Z^i \frac{dZ}{dT}} \frac{\delta Z}{\delta R} \tag{11.45}$$

The term on the left hand side of expression (11.45) denotes the direct marginal benefit for consumers of a decrease in Z due to an improved infrastructure, weighed by net direct social marginal values of income b^{i*}. The first term on the right gives the indirect benefit of the reduction of Z to the production sector. The second term on the right corresponds with the net direct marginal cost of the reduction of Z. Finally, the third term on the right corresponds with what is labelled the 'Pigou term' by King. This term measures the indirect deadweight loss effects due to the substitution effects in taxed goods caused by the change in Z. For example, if peak transport is already highly taxed, an improved infrastructure might increase traffic and tax proceeds thereby decreasing the net tax revenue to be raised (in this model the social cost of raising one unit of government revenue is normalized to one).

11.3 The Tax Reform Problem

11.3.1 Generalities

Following Feldstein (1976) and Guesnerie (1977) the tax reform approach can solve two criticisms to the optimal tax approach. The first problem in the optimal tax approach is that the optimum is highly dependent on the knowledge of the individual utility functions (and demand functions) in a domain which is far from the present equilibrium and for which consequently empirical information is lacking. Much more information is available to study small changes around the initial equilibrium.

The second criticism to the optimal tax approach points at its dependence on the Social Welfare Function as a normative objective function which cannot be observed or even does not exist (cfr. Arrow impossibility theorem). The tax reform approach can alleviate this problem by analyzing only Pareto-Improving (=*P-I* in the rest of the text) directions of tax reform. However, as there may be many *P-I* tax reform directions, one cannot escape the use of a normative criterion to select one particular solution.

Guesnerie (1977) used the Diamond and Mirrlees framework without externalities for an analysis of Pareto-Improving tax changes in consumer taxes. He showed under what conditions a *P-I* change is feasible and can be obtained as an equilibrium with adapted production prices. He also had a major negative result: production efficiency is not necessarily required for an optimal marginal tax change.

We will concentrate on the simplest marginal tax reform problem addressed by Ahmad and Stern (1984): we use constant rather than variable production prices and focus on tax changes that improve a given welfare function rather than being *P-I*. In this model we simplify the congestion problem: congestion is only generated by consumers and not by the production sector.

11.3.2 Optimal Marginal Tax Reform Results

We use the model of the previous theoretical sections with the following two amendments:

(i) Production prices are fixed. As was pointed out before, this property combined with constant returns to scale implies a linear production function of which the coefficients will in equilibrium (and in an efficient solution) equal the production prices p_k. The price of the untaxed good labour is normalized to 1: $p_1 = 1$;

(ii) Congestion is only generated by the consumers. In addition it is assumed that the production sector is not affected by congestion. So in (11.5), δ equals zero and Z is absent in (11.4);

Under these assumptions we can, given an initial equilibrium $q(0)$, $x(0)$, $Z(0)$, $t(0)$, $T(0)$, $R(0)$, $W(0)$, look for marginal tax changes dt_k $(k=2,....,M)$, dT, and marginal changes in infrastructure dR such that $dW \geq 0$, the production remains feasible and the government budget constraint is satisfied.

This problem can be simplified if we restrict ourselves to efficient solutions in production[4]. In that case, using Walràs' law and keeping initial production prices $p(0)$ constant, we can look for changes in tax variables which improve welfare while keeping the government budget balanced. Following Ahmad and Stern (1984) the idea is then to compute the welfare cost and the revenue generated by a marginal change in the different policy variables. Taxes with a relatively higher welfare cost per unit of revenue should be decreased and vice-versa. This procedure guarantees an increase in social welfare and the respect of the government budget constraint.

The marginal welfare cost of an extra unit of revenue generated by increasing the tax on good k equals:

Using similar techniques as in the previous part we obtain the following

[4] Remember that Guesnerie (1977) has shown that production efficiency is not necessarily required for an optimal marginal tax change.

$$MC_k = -\frac{dW / dt_k}{dREV / dt_k} \qquad k=2,\ldots,M$$

(11.46)

expressions for the numerator and the denominator:

$$\frac{dW}{dt_k} = -\sum_i \beta_i x_k^i + \sum_i \zeta_i \frac{dZ}{dq_k} \qquad k=2,\ldots,M$$

(11.47)

$$\frac{dREV}{dt_k} = X_k + \sum_{j=2}^{M} t_j \frac{\delta X_j}{\delta q_k} + \sum_{j=2}^{M} t_j \frac{\delta X_j}{\delta Z} \frac{dZ}{dq_k} \qquad k=2,\ldots,M$$

(11.48)

In these expressions dZ/dq_k is defined as in (11.21).

We see that in order to find an optimal tax reform one needs information on the distribution of the consumption of the good in the initial state (x^i_k) and on the total (not the individual) demand derivatives. The introduction of congestion adds two elements to the information requirements: the general equilibrium effect on congestion of a marginal tax change in good k and the individual marginal willingness to pay for congestion alleviation. This type of individual valuation is hard to obtain so that the tax reform approach economizes less on information than in the case without externalities. The reason is that in the case of private goods, the relative valuation for a good is revealed by the individual quantity one buys - for public goods there is no such valuation as everybody consumes the same quantity of congestion. Contrary to the optimal tax rules, a marginal reform in any tax rate always has to consider the indirect welfare (see expression (11.47)) and revenue (see expression (11.48)) effects on congestion levels.

Similarly, we can define the welfare cost of changing the head tax and compare it to the welfare cost of changes in indirect taxes:

$$MC_T = -\frac{dW / dT}{dREV / dT}$$

(11.49)

$$\frac{dW}{dT} = -\sum_i \beta_i + \sum_i \zeta_i \frac{dZ}{dT}$$

(11.50)

$$\frac{dREV}{dT} = \sum_{j=2}^{M} t_j \frac{\delta X_j}{\delta T} + \sum_{j=2}^{M} t_j \frac{\delta X_j}{\delta Z}\frac{dZ}{dT} - N$$

(11.51)

Finally, nothing prevents us from considering marginal changes in the expenditure on infrastructure as another control variable of the government which takes resources from the government budget and influences welfare of the different individuals:

$$MC_R = - \frac{dW / dR}{dREV / dR}$$

(11.52)

$$\frac{dW}{dR} = \left(\sum_i \zeta_i \right) \frac{dZ}{dR}$$

(11.53)

$$\frac{dREV}{dR} = \sum_{j=2}^{M} t_j \frac{\delta X_j}{\delta Z}\frac{dZ}{dR} - 1$$

(11.54)

The marginal welfare cost of changes in the government infrastructure can then be compared with the welfare cost of marginal tax changes.

11.4 Conclusions

The theoretical general equilibrium model presented in this paper, is used to derive optimal tax and marginal tax reform rules in an economy in which both consumers and producers cause congestion externalities. The government is assumed to have the dual objective of correcting for these externalities and of raising enough revenue in an efficient and equitable way. The analysis takes into account the so-called feedback effect associated with congestion externalities. In contrast to partial equilibrium models for the control of traffic externalities, the use of a second best general equilibrium tax model allows us to view instruments for the control of those externalities as an element of a broader package of government instruments which all have to be optimized simultaneously. In the analysis a number of simplifying assumptions are made. Relaxing these assumptions may generate interesting insights. It is assumed that the activity generating the externality is taxable. In the case of road traffic congestion this amounts to assuming that it is feasible to introduce perfect road pricing which takes into account the different contribution to congestion by cars and trucks at different traffic levels. In reality, this may be infeasible, due to informational, practical, political or other restrictions. An extension of the model may therefore consist of studying the consequences of restrictions on the pricing instruments available. We also assumed that apart from road infrastructure, pricing instruments are the only instruments available to the government. The model could be extended further by introducing the possibility of imposing quantity controls. Finally, the marginal tax reform analysis assumes fixed production prices and the absence of producer externalities or externalities affecting producers. Dropping these assumptions would mean endogenizing the production prices and introducing into the analysis the welfare and revenue effects of a marginal change in the excise on externality creating activities in the production sector.

References

Ahmad E, Stern N (1984) The Theory of Tax Reform and Indian Indirect Taxes. Journal of Public Economics 25:259-298

Atkinson AB, Stiglitz JE (1976) The Design of Tax Structure. Direct Versus Indirect Taxation. Journal of Public Economics 6:55-75

Atkinson AB (1977) Optimal Taxation and the Direct Versus Indirect Tax Controversy. Canadian Journal of Economics 10:590-606

Balcer Y (1980) Taxation of Externalities. Direct Versus Indirect. Journal of Public Economics 13:121-129

Batten DF, Boyce DE (1986) Spatial Interaction, Transportation, and Interregional Commodity Flow Models. In: Handbook of Urban and Regional Economics I. Nijkamp P (Ed). Elsevier Science Publishers (North-Holland). pp 357-406

Beckmann M, McGuire CB, Winsten CB (1956) Studies in the Economics of Transportation. Yale University Press, New Haven

Bovenberg AL, Ploeg F van der (1994) Environmental Policy, Public Finance and the Labour Market in a Second-Best World. Journal of Public Economics 55:349-390

Button K (1990) Environmental Externalities and Transport Policy. Oxford Review of Economic Policy 6:61-75

Borger B de, Mayeres I, Proost S, Wouters S (1994) Social Cost Pricing of Urban Passenger Transport - With an Illustration for Belgium. SESO Discussion Paper.

Diamond PA, Mirrlees JA (1971) Optimal Taxation and Public Production I. Production Efficiency and II. Tax Rules. American Economic Review 61: 8-27 and 261-278

Diamond PA (1973) Consumption Externalities and Imperfect Corrective Pricing. Bell Journal of Economics and Management Science 4:526-538

Dixit A (1985) Tax Policy in Open Economies. In: Handbook of Public Economics. Auerbach AJ, Feldstein M (Eds) Elsevier Science Publishers, North Holland pp 313-374

Feldstein M (1976) On the Theory of Tax Reform. Journal of Public Economics 6:77-104

Glaister S, Lewis D (1978) An Integrated Farés Policy for Transport in London. Journal of Public Economics 9:341-355

Guesnerie R (1977) On the Direction of Tax Reform. Journal of Public Economics 7:179-202.

King MA (1986) A Pigovian Rule for the Optimal Provision of Public Goods Journal of Public Economics 30:273-291

Levy-Lambert H (1968) Tarification des Services à Qualité Variable - Application aux péages de circulation. Econometrica 36:564-574

Mayeres I (1993) The Marginal External Costs of Car Use - With an Application to Belgium. Tijdschrift voor Economie en Management 38:225-258

Mayeres I (1994) The Marginal External Costs of Trucks. Tijdschrift Vervoerwetenschap 30:121-138

Newbery D (1990) Pricing and Congestion. Economic Principles Relevant to Pricing Roads. Oxford Review of Economic Policy 6:22-38

Pigou AC (1920) The Economics of Welfare, London

Sandmo A (1975) Optimal Taxation in the Presence of Externalities. Swedish Journal of Economics 77:86-98

Small K (1983) The Incidence of Congestion Tolls on Urban Highways. Journal of Urban Economics 13:90-111

Vickrey WS (1969) Congestion Theory and Transport Investment. American Economic Review 59:251-260

Walters AA (1961) The Theory and Measurement of Private and Social Cost of Highway Congestion. Econometrica 29:676-699

Wijkander H (1985) Correcting Externalities Through Taxes on/Subsidies to Related Goods. Journal of Public Economics 28:111-125

IV

Space in General Equilibrium Models

12 Linking CGE Models: Modelling the Transport Sector and Spatially Homogeneous Goods

Chris Elbers
Vrije Universiteit

12.1 Introduction

Researchers in the field of Computable General Equilibrium (CGE) modelling have recently come up with a number of models specifically designed to study the interaction of countries and *regions*. These models address strategic questions like the formation of trade blocks (e.g. Kehoe and Kehoe 1994, discuss a number of CGE-models dedicated to the NAFTA trade block), the incidence of environmental taxes across nations (e.g. Whalley and Wigle 1991), or more analytical questions like whether a country like Canada can be treated as a single location (e.g. Wigle 1992).

These developments have widened the field of CGE-modelling which was traditionally devoted to single-country models in a two-country framework: national economy and Rest-of-the-World. And, quite naturally, because of these developments the question of the *costs* of regional interaction is revisited.

The costs of interregional trade and their consequences are a well-established theme in economics. Takayama and Judge's book (1971) can be considered as a more or less definitive treatment of the subject. It is usually classified with regional and transport economics.[1] Thus, a great deal of attention has been paid to transport flows, congestion, the structure of the market for transport services etc. These are not necessarily the first concerns of CGE-modellers, who emphasize national or regional models and are looking for a convenient framework to link these models into a global setting.

The national CGE-models almost invariably employ the so-called Armington specification (or a close relative) for modelling foreign trade. According to this approach a sector in the national economy produces goods which are imperfect substitutes of goods produced abroad by a sector of the same classification. Any activity (production or consumption) using goods from the sector must do so in a combination of foreign and national varieties. In general the particular combination depends on prices and is interpreted as the result of a cost minimization, 'usable' goods being 'produced' with national and foreign varieties as inputs. In most cases

[1]The hunt for efficient algorithms is perhaps best classified as Mathematical Programming.

CES- or nested CES-functions are used as production, or aggregator functions. This idea has been generalized in a straightforward fashion in the multiregional or multinational CGE-models.

The Armington specification (and its cousins) is based on the heterogeneity of goods from different origins, and rejects the original approach followed in Spatial Equilibrium Theory, which assumes homogeneity across regions. The Armington specification results in a kind of gravity model of trade; one which has a clear economic foundation. As mentioned in Armington's original paper (Armington, 1969) heterogeneity of goods across origins of production is consistent with the phenomenon of cross-hauling: the statistical fact of intra-sector trade in both directions.

A technical advantage of the Armington specification is that there is never the question of choosing optimal origin-destination pairs, a question which is the very substance in, for instance, the Transportation Problem of Linear Programming. With the Armington specification (in the form of nested CES-functions) market-clearing for any good can ultimately be reduced to a single region (the producing region); the set of 'basic equations' to solve in equilibrium is always the same, 'corner solutions' do not occur, and solving for general equilibrium amounts to solving a fixed set of equations.

With CES aggregator functions, demand for foreign varieties is proportional to $(p_{foreign}/p_{home})^{-\sigma}$, where $p_{foreign}$ and p_{home} denote the prices of foreign and home goods and σ is the CES-function's substitution elasticity. Thus, trade elasticities compiled from empirical studies in international economics can be used for numerical specification of the model.[2]

Because of all these advantages it is not surprising that using some version of the Armington specification is the *de facto* standard for modelling international trade in CGE-models.

The linkage of various national (or regional) models introduces a global element if the international exchange of goods requires inputs: transportation costs. The treatment of this point in models of interregional freight flows is not completely satisfactory from the perspective of CGE-analysis. These partial equilibrium models focus on the dependence of transport costs on traffic flows in the transport network, and on the market organization of the transport sector. However, they ignore the fact that transport services are produced from inputs which are produced in other sectors of the global economy. As I will show in section 12.2, accounting for the production of transport services from model inputs considerably complicates the computation of (global) general equilibrium, unless one is willing to make some drastic assumptions.

Despite the success of the Armington specification in multiregional CGE-models it is of course very well conceivable that a good is produced at more than one location in the model and is *not* distinguished according to its origin. Primary commodities are an obvious example. It is usually argued that the Armington specification can handle this case by fixing substitution elasticities at very high

---

[2]There is a flip side to this convenience, however. Since elasticities must be finite, each country in a multi-regional CGE has monopoly power over the varieties it alone produces. The standard 'small country' assumption of international economics can therefore not be implemented with the Armington specification.

values. However, numerical solution of the model becomes increasingly difficult with higher substitution elasticities. Hence it would seem that the approach of Spatial Equilibrium Theory with spatially homogeneous goods is a more appropriate tool for this situation, both from a theoretical and a practical point of view. Accordingly, in section 12.3, I propose a hybrid model for linking CGE-models, incorporating the Armington specification for some sectors and Spatial Equilibrium Theory for others.

Computable General Equilibrium models claim superiority over other multi-sectoral models because a proper emphasis is laid on budget equations of agents and balance equations of goods and services. From this perspective the eager adoption of the Armington specification is a bit surprising. The Armington specification effectively amounts to replacing a balance equation by two or more sub-equations. Thus the impact of exogenous shocks on the system is greatly reduced.

The rest of this paper is organized as follows. Section 12.2 discusses the linking of CGE-models using the Armington specification. The numerical solution of such models is loosely discussed and attention is drawn to the problem introduced by modelling the production of transport services endogenously. Section 12.3 discusses the introduction of goods which are not distinguished by origin of production. Section 12.4 concludes and mentions a number of unsolved issues. An Appendix lists the ingredients of the hybrid model with spatially homogeneous and heterogeneous goods.

12.2 Linking CGE Models and the Armington Specification

Suppose that goods $i = 1,\ldots,n$ are produced in a number of locations, $r = 1,\ldots,R$. I will call these locations *regions*. To emphasize the origin of production of goods I use a double subscript (i,r). According to the Armington specification produced goods cannot be used directly in consumption or production. Usable goods are produced in a second stage, using the regional 'raw' goods as inputs:

$$x_i - f_i(x_{(i,1)},\ldots,x_{(i,R)}).$$

The precise choice of regional goods $x_{(i,r)}$ is determined by a cost minimization: subject to

$$\min \sum_r p_{(i,r)} x_{(i,r)}$$

subject to

$$f_i(x_{(i,1)},\ldots,x_{(i,R)}) \geq \bar{x}_i$$

where $p_{(i,r)}$ denotes the user price of $x_{(i,r)}$. If $f()$ is constant returns to scale this results in a unit price π_i of the composite good, and a unit-'cost' function

$$\pi_i = \pi_i(p_{(i,1)},\ldots, p_{(i,R)}).$$

The second-stage production function $f()$ may differ according to good and eventual use of the good. For simplicity I use the same production function f_i for all uses within a region r. Hence f_i and π_i are further labelled as f_i^r and π_i^r, where r is the region of destination for goods $(i,1),\ldots,(i,R)$.

In Armington's original paper $f_i^r()$ is a CES production function which has the convenient property that

$$\lim_{x_{(i,r)} \downarrow 0} \frac{\partial f_i^r}{\partial x_{(i,r)}} = +\infty,$$

so that in a cost-minimizing combination $(x_{(i,1)},\ldots,x_{(i,R)})$ all $x_{(i,r)}$ are strictly positive.

12.2.1 Counting and Solving Equations

It is not without interest to count the equations of a multi-regional CGE-model. Assume for the moment that trade between regions is costless (as, throughout this article, will always be the case for trade *within* regions), so that the $p_{(i,r)}$ are the same for all regions: write $p^{r'}_{(i,r)}$ for the price of good (i,r) in region r' then $p^{r'}_{(i,r)} = p^r_{(i,r)}$ for all r'. Suppose further that each region has a number m^r of non-traded goods with prices q^r, including, among others, immobile factors of production. Denote excess supply of non-traded goods as $F^r(p,q^r)$ and excess supply of traded goods as $G^r(p,q^r)$. Assume for the moment that these are point-valued functions. The functions depend on local prices p^r, q^r of the region itself, but, because of the Armington specification, also on prices $p^1,\ldots,p^R$ in other regions. Finally, assume that full competition prevails on all markets. The equilibrium equations are

$$\begin{aligned} F^r(p,q^r) &= 0; \; (r = 1,\ldots,R) \\ G^r_{(i,r)}(p,q^r) &= -\sum_{r \neq r'} G^{r'}_{(i,r)}(p,q^{r'}); \; (i = 1,\ldots,n; \; r = 1,\ldots,R). \end{aligned}$$

In total there are $\sum_{r=1}^{R} m^r + n \times R$ equations to determine as many unknowns: the prices of non-traded goods q^r and the prices $p_{(i,r)}$ of traded goods. As usual, if no money illusion has krept into economic behaviour underlying the equations, prices are only determined up to a scalar factor and one of the equations can be dropped (Walras identity).

More often than not production functions in these models are constant-returns-to-scale, so that supply (and hence excess supply) is not a point-valued function. In this case prices are determined by unit-cost functions and quantities by demand. As an example, suppose this is the case for all traded goods, while for non-traded goods excess supply is a point-valued function of prices. Write S_F and S_G for supply functions of non-traded and traded goods, then a system of equations like the following results:

$$p^r_{(i,r)} = c^r_{(i,r)}(p,q^r) \tag{12.1}$$

$$S^r_G = \sum_{r'=1}^{R} A^{rr'}S^{r'}_G + \sum_{r'} D^{r'}_G(p,q^{r'}) \tag{12.2}$$

$$S^r_F(p,q^r) = D^r_F(p,q^r) + \sum_{j=1}^{n} \frac{\partial c^r_{(j,r)}}{\partial q^r} S^r_{G(j,r)};\ \forall r, \tag{12.3}$$

where $S^r_G = (S^r_{G(1,r)},\ldots,S^r_{G(n,r)})$ (and similar for $D^{r'}_G$) and

$A^{rr'}_{(i,j)} = \partial c^{r'}_{(j,r')}/\partial p^{r'}_{(i,r)}$. $D^{r'}_F$ denotes (point-valued) demand for the non-traded goods sectors, both final and intermediate.

The solution of system (12.1 - 12.3) hinges around finding the equilibrium values of prices for non-traded goods q^r. Given q^r, equation (12.1) determines prices p^r of traded goods. Then, the supplied quantities of constant-returns-to-scale goods, S^r_G, adjusts to intermediate demand[3] generated in its own production plus other (point-valued) demand $\Sigma_{r'} D^{r'}_G$: equation (12.2). Finally, the balance in equation (12.3) checks whether prices of non-traded goods q^r were at their equilibrium values.

This system has $n \times R$ more equations than the previous one to determine the extra unknowns S^r_G. A possible solution scheme for this set of equations is given in Table 12.1

It is rather obvious to try to implement such a solution scheme as a sequence of regional equilibrium computations: fix prices and export quantities from other regions and solve for regional equilibrium. Proceed to the next region and repeat. With any luck this sequence will converge to global equilibrium. Thus existing tailor-made software for equilibrium computations in each separate region can in principle be assembled into a global computational scheme with minimal effort.

Table 12.1. Solving for multi-regional general equilibrium

0.	Choose a proper normalization, say $q^1_1 = 1$.
1.	Start with values for the other q^r_j.
2.	Determine the p^r from equation (12.1).
3.	Determine S^r_G from equation (12.2).
4.	Check balance in equation (12.3).
5.	If necessary, adjust q^r and go back to step 2.

[3] This is basically the input-output equation $x = Ax + f$.

12.2.2 Costly Inter-regional Trade

To incorporate costly trade assume first that transport services are imported from some artificial region at constant costs: $\tau^{rr'}_{(i,r)}$ to transport a unit of good *(i,r)* from r to r'. Now prices $p_{(i,r)}$ are no longer equal for all regions. A new block of $n \times R^2$ equations and unknowns is added:

$$p^{r'}_{(i,r)} = p^{r}_{(i,r)} + \tau^{rr'}_{(i,r)}, \text{ for all } r,r',i.$$

To avoid unnecessary complications, put $\tau^{rr} = 0$.

The price equations (12.1) must now be rewritten to reflect that cost minimization is based on *local* prices:

$$p^{r}_{(i,r)} = c^{r}_{(i,r)}(\bar{p}^{r},q^{r}), \tag{12.4}$$

where

$$\bar{p}^{r} = (p^{r}_{(1,1)},\ldots,p^{r}_{(n,1)};\ldots;p^{r}_{(1,R)},\ldots,p^{r}_{(n,R)})$$

is the vecotr of local prices of traded goods in region r. These changes only marginally affect the computational scheme of Table 12.1.

The next step is that transport services are being produced within the system. A simple approach is to specify the $\tau^{rr'}_{(i,r)}$ as unit-cost functions with prices of traded goods as arguments (and perhaps some non-traded goods[4]). The use of a good *(i,r)* in trade per unit of good transported can be determined from Shephard's lemma and the trade flows, and can then be added as an extra demand term D^{r}_{trade} in equation (12.2):[5]

$$D^{r}_{trade} = \sum_{s} \sum_{s' \neq s} [\frac{\partial \tau^{ss'}}{\partial p^{r}}]^{T} A^{ss'} S^{s'}_{G} + \sum_{r' \neq r} [\frac{\partial \tau^{rr'}}{\partial p^{r}}]^{T} D^{r'}_{G}$$

Alternatively, the use of transport inputs could be subsumed in the cost functions $c^{r}()$. In both cases the computational scheme sketched above can still be applied.

This way of endogenizing trading costs is not entirely satisfactory from the point of view of *transport economics*. In moving a good from one place to another, the transport service delivered is purely complementary. Any other service achieving the same result is a perfect substitute. If different routings or different modes of transportation have different cost functions, it is not *a priori* clear which cost function is relevant in equilibrium. This has consequences for the computation of equilibrium since the relevant cost function will have to be determined during the

---

[4]In a sense these would then become tradable.

[5]Here it is assumed that deliveries to the interregional-trade sector require no transport inputs themselves.

computation. Even worse, if there are inputs to transportation (directly or indirectly) with less-than-perfect elasticity of supply (in particular fixed endowments and production factors) market clearing for these goods may result in equalization of transport costs between two nodes, for different routings or modes of transportation. Thus an equilibrium computation involves assigning the *volumes* of transportation according to each of the equally costly and perfectly substitutable transport services. The usual equilibrium searching in the space of prices is quite incapable of handling such complications.

The reason for these difficulties lies in the special character of the transport industry. Most CGE-models tacitly assume that a production sector's output can be represented by a single production function and consists of only a single homogeneous good. There is an implicit identification of (representative) firm, sector and good. This approach is untenable in the case of transportation: the transport industry is both 'multi-firm' (there is more than one way of moving a good from *A* to *B*) and 'multi-product' (a truck can move more than one kind of good, and service more than one origin-destination pair).[6]

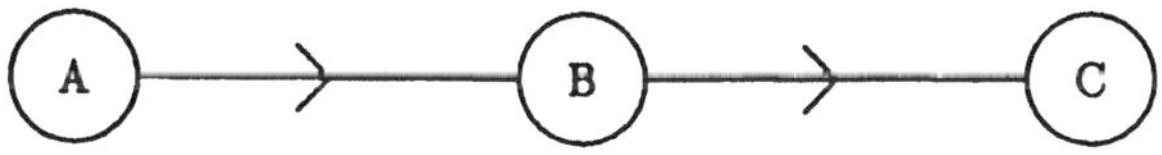

Fig. 12.1. A simple transport network

As an example let me illustrate the complication introduced by the 'multi-firm' character of transportation. Suppose there are two locations, *A* and *B*, and suppose that there is a market for 'transport services' between the two. One unit of transport services in either direction costs one unit of factor *i* or factor *j* as input. In the tradition of transport economics these different 'modes of transportation' are depicted as separate links between *A* and *B*. See figure (12.1). Normally, market clearing for factors *i* and *j* would be expressed as (using the notation adopted above and assuming that functional forms prevent prices from dropping to zero)

$$S_i(p,q) = 0$$
$$S_j(p,q) = 0.$$

Suppose that, apart from their use in *A-B* transport, factors *i* and *j* are in excess supply. General equilibrium prices are denoted as $(\hat{p},\hat{q})$:

$$\bar{S}_i(\hat{p},\hat{q}) > 0$$
$$\bar{S}_j(\hat{p},\hat{q}) > 0,$$

where the bar denotes excess supply, not accounting for the use in *A-B* transport.

[6]An assumption like the Armington specification solves the multi-firm (countries!) problem by splitting sectors into single-country subsectors.

This would give $q_i = 0$ and $q_j = 0$, if there were no transport demand for factors i and j. In fact, the presence of *A-B* transport changes the two equations above to the *five* complementarity conditions:[7]

$$\begin{aligned} \bar{S}_i(p,q) - S_i^{AB} &\geq 0 \perp q_i \geq 0 \\ \bar{S}_j(p,q) - S_j^{AB} &\geq 0 \perp q_j \geq 0 \\ S_i^{AB} + S_j^{AB} - D_{trans}^{AB} &\geq 0 \perp \tau^{AB} \geq 0 \\ \tau^{AB} &\leq q_i \perp S_i^{AB} \geq 0 \\ \tau^{AB} &\leq q_j \perp S_j^{AB} \geq 0. \end{aligned}$$

The first two conditions express market clearing for factors i and j. S_i^{AB} denotes the use of factor i as input to *A-B* transport and similarly for S_j^{AB}. The third condition expresses that the sum of both must be sufficient as inputs for *A-B* transport. τ^{AB} denotes unit transportation cost. The last two conditions show that transportation costs are minimal.

For the particular equilibrium $(\hat{p},\hat{q})$ these conditions reduce to the conditions

$$\begin{aligned} \hat{q}_i = \hat{q}_j &= \tau^{AB} \\ \bar{S}_i + \bar{S}_j - D_{trans}^{AB} &\geq 0 \perp \tau^{AB} \geq 0. \end{aligned}$$

Thus, for all purposes, factors i and j have become perfect substitutes in equilibrium! By the fact that i and j can be used interchangeably in *A-B* transport, it is no longer clear whether markets for i and j clear 'separately' ($S_i = 0$, $S_j = 0$), or in combination ($S_i + S_j = 0$, $q_i = q_j$). The equilibrium finding algorithm will have to find the particular regime which applies in equilibrium.

Another way of expressing the same complication is as follows. The network models of transport economics usually solve for an equilibrium set of unit transportation costs or charges between origin-destination pairs.[8] From the point of view of CGE-analysis unit cost - a scalar - must be mapped into inputs per unit of transportation - a vector. If this mapping is point-valued (and continuous) then ordinary equation solving techniques can be used in the computation of equilibrium. If the mapping is set-valued (as it is in the example above) then more advanced methods must be used.

I can offer no satisfactory set of assumptions that will result in a multiregional CGE-model with endogenous production of transport services, which is still easy to solve. A popular assumption is to let the importing region produce the necessary transportation services. This (partially) solves the 'multi-firm' problem.

---

[7]The notation $a \geq b \perp c \geq d$ summarizes the three conditions $a \geq b$, $c \geq d$ and $(a - b)(c - d) = 0$.

[8]These costs may or may not be specific for the goods transported.

I consider the problem above a challenging subject in the interface between CGE-analysis and transport economics. But the gap between these two disciplines is still considerable: while CGE-modellers tend to relate costs to use of inputs, transport economists tend to relate costs to transport flows and congestion.

I conclude this section in a positive mood by briefly sketching a very simple model for the production of transport services which leads to only minimal modification of the solution scheme in Table (12.1).

Suppose all regions in the multiregional CGE-model are nodes in a network. All traded goods (i,r) can be transported over all links in the network. Transportation services are produced by a single factor M, 'motion', which is used for transport purposes only. The total supply $\underline{M}$ of motion is fixed and inelastic. Moving a unit of good (i,r) over link ss' requires $\theta(i,r)\, a^{ss'}$ units of M, where $\theta(i,r) \geq 0$ depends only on the good-index (i,r). Transport costs may depend on the direction of trade, so that in general $a^{ss'} \neq a^{s's}$. Assume that transport costs over a path are equal to the sum of transport costs over the links in the path. Now, for each origin-destination pair (r,r') compute a minimum-cost path. Thus,

$$b^{rr'} = \min\{ \sum_{i=0}^{n} a^{o_i o_{i+1}} \mid o_0 = r, o_n = r' \text{ for some } n \}.$$

Then, in the notation adopted earlier:

$$\tau^{rr'}_{(i,r)}(p,q) = \theta_{(i,r)} b^{rr'} q_M.$$

The total use of M can be calculated as

$$M^d = - \sum_{(i,r)} \sum_{r' \neq r} \theta_{(i,r)} b^{rr'} G^{r'}_{(i,r)}.$$

Table 12.1's solution scheme can now be modified by adding an extra loop to it:[9]

Table 12.2. Solving a multiregional CGE-model with endogenous production of transport services

0. Choose a proper normalization, say $q^1_1 = 1$.[10]
1. Choose an initial value q_M (e.g. $q_M = 0$).
2. Solve equilibrium, given transport prices τ as in Table 12.1.
3. Check balance $\underline{M} - M^d$.
4. If necessary, adjust q_M and go back to step 2.

[9] I do not claim that this is an efficient procedure.

[10] It is tempting to mornalize on q_M, but q_M may happen to equal zero in equilibrium.

Note that some agents in the model actually *own* M. Hence some interesting income effects may be introduced by endogenizing the production of transport services. Elbers (1992 p 105-107) gives an example which shows that such income effects may actually lead to multiple equilibria.

12.3 Linking CGE Models and Spatial Equilibrium Theory

Several reasons were mentioned in the Introduction why one would want to include spatially homogeneous tradable goods in a model. Primary commodities, at a sufficient level of disaggregation, provide an obvious case. Or one might want to link regional models into a global model. If there are no data on inter-regional trade, the obvious way to proceed is to treat traded goods as spatially homogeneous. Even in the absence of trade data, global equilibrium requires that total excess supplies are zero (or equal to external exports) and, hence, equilibrium puts a constraint on the values of regional excess supplies: $\Sigma_r G^r = 0$. Moreover, the global system can be given more structure by endowing it with an infra-structure for trade as well as a technology for producing transport services.

In what follows I will always assume point-valued regional excess supply functions G^r for tradable goods that are not distinguished by region of origin. For such goods the double index (i,r) is not necessary: a simple index i suffices.

12.3.1 The Equations of Spatial Equilibrium Theory

Spatial Equilibrium Theory and Armington specification lead to different sets of equations. For the moment assume constant and given transport costs $\tau^{rr'}$ between all pairs of regions r and r'. The Armington specification leads to the equations:

$$G^r_{(i,r)}(q^r,p^r) = -\sum_{r'\neq r} G^{r'}_{(i,r)}(q^{r'},p^{r'})$$

$$p^{r'}_{(i,r)} = p^{r}_{(i,r)} + \tau^{rr'}_{(i,r)}.$$

The conditions of Spatial Equilibrium are

$$G^r_i(p^r,q^r) + \sum_{r'=1}^{R} (x_i^{r'r} - x_i^{rr'}) = 0 \text{ (or } \geq 0 \perp p_i^r \geq 0) \qquad (12.5)$$

$$p_i \leq p_i^r + \tau_i^{rr'} \perp x_i^{rr'} \geq 0, \qquad (12.6)$$

where $x_i^{rr'}$ denotes a flow of good i from region r to region r'. Equation (12.5) states that excess supply of a region equals net exports.[11]

The equations of the Armington specification were explained above: in the typical case each region exports its traded goods to all other regions in the model; *only* the producing region r exports the good (i,r), all other regions import it:

$$G^{r'}_{(i,r)} < 0 \text{ for all } r' \neq r$$

There are many more configurations in Spatial Equilibrium. Of course, summation of the regional balance equations leads to $\Sigma_r G^r_i = 0$, just like the Armington case, but now there may be more than one exporting region. Moreover, the set of regions can be split into several subsets with market clearing within subsets of regions, and no trade between them. E.g. $G^1 + G^2 = 0$, $G^3 + G^4 + G^5 = 0$. In other words, while there is exactly one market clearing equation for each Armington good (i,r), there may be up to R equations in Spatial Equilibrium. This is partly governed by the transport costs τ: if trading good i is costless, there is only one market clearing equation, $\Sigma^r_i G^r_i = 0$. If transport costs are prohibitively high, all regions are in autarky for good i and there are R equations active in equilibrium: $G^r_i = 0$ for all r. In general, both transport costs τ and the excess supply functions G^r_i determine which clusters of regions engage in mutual trade and, hence, which equations are active in equilibrium. The number of such 'regimes' grows very rapidly with the number of regions.[12]

In the Spatial Equilibrium model, contrary to the Armington case, the concept of bi-lateral trade flows is meaningless. Upon superficial inspection one might be tempted to interpret the variables $x^{rr'}_i$ as region r's exports of good i to regions r′. However, many other patterns $\{x^{rr'}_i\}$ may be compatible with a certain price solution of the Spatial Equilibrium model as a glance at figure 14.2 reveals: is region C importing from A or from B?

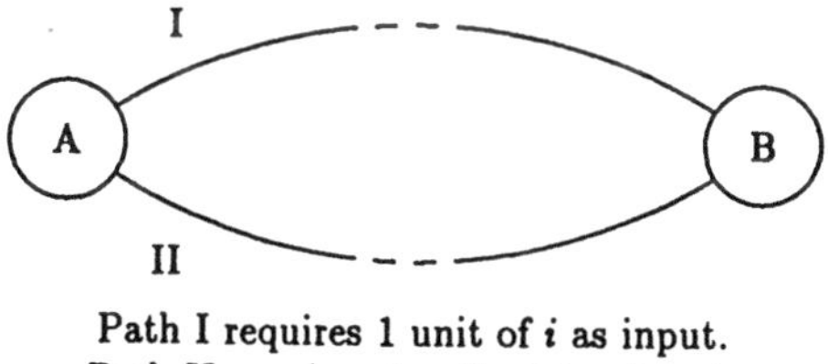

Fig. 12.2. The origin of C's imports is not determined

What *is* determined in equilibrium are transport costs. For all sets of equilibrium trade flows $\{x^{rr'}_i\}$ total transport costs $\Sigma_{r,r'}\tau^{rr'}_i x^{rr'}_i$ are the same: minimal for the given set of excess demands.

A stronger result can be established than uniqueness of total transport costs. To

[11]For a discussion of the Spatial Equilibrium Problem see e.g. Pang (1984).

[12]For instance, if $R = 10$ and all bilateral transport costs are independent parameters, there are 115,975 regimes possible - for each good! For $R = 20$ the number is 51,724,158,235,372.

state the result a distinction must be made between *links* and *paths*. Think of links as physical connections between neighbouring regions. Paths are chains of links. Referring to figure 14.2, links exist between regions A and B, and B and C, but not between A and C. These last two regions are connected by the path consisting of links $A \rightarrow B$ and $B \rightarrow C$. If cost-minimizing paths between all pairs of regions are unique (which they are generically if the transport costs are independent parameters) then - as figure 14.2 again suggests - the *flows* of goods over *links* are uniquely determined in Spatial Equilibrium (see Elbers 1992 p 63-65).

12.3.2 Computing Equilibrium

The problem of computing Spatial Price Equilibrium can be said to be practically solved. Many good algorithms are available from the literature (e.g. Nagurney 1993 section 3.4). How can these algorithms be employed within a solution strategy for multi-regional CGE models? A spatial equilibrium is necessarily global, involving many regions. In the computational procedure described in Table 14.2 the global element (market clearing for transport services) was handled in an outer loop, the inner loop consisting of a sequence of regional equilibrium computations, taking transport prices as given. Likewise, solving for spatial price equilibrium of spatially homogeneous goods is done in an outer loop. For the inner loop computations, the resulting regional prices p_i^r are kept constant, as if these goods are costslessly importable and exportable at prices p_i^r.

Two main problems are connected to including spatially homogeneous goods in a multi-regional CGE. First, during each main loop a spatial price equilibrium is calculated for all these goods. What if there are more such equilibria? The approach to general equilibrium would be frustrated if the outer loop switches from one spatial equilibrium to another. The second problem concerns the handling of inputs into transportation. This problem was already mentioned in section 12.2.

Regarding the first problem, most uniqueness results for Spatial Price Equilibrium put conditions on the functions G^r (see equation (12.5)) which are not likely to be encountered in CGE models. In particular, the G^r are often assumed to be positive definite or monotonic. Elbers (1992, theorem 13 and proposition 20) shows that a strong kind of diagonal dominance is sufficient for uniqueness of spatial equilibrium (i.e. uniqueness of prices p_i^r, not of path flows $x_i^{rr'}$). Diagonal dominance, at least locally, is often encountered in empirical excess supply functions.

The second problem can be solved if link flows are known: with production or cost functions specifying the technology for producing transport services over links, transport flows generate demand for inputs (which must subsequently be supplied from regions according to some rule - e.g. cheapest region of supply).

But *are* link flows known from the conditions of Spatial Equilibrium? As mentioned earlier, if transport costs τ are independent parameters, link flows are generically uniquely determined in spatial price equilibrium. In CGE models however, transport costs are endogenous and may adjust such that the uniqueness of link flows is jeopardized, see the example of figure 12.1. If this is the case the allocation of flows over links must be derived from factors other than the Spatial

Equilibrium conditions. I have not investigated this line of research. A short (but brutal) cut is made by assuming that *if a path minimizes transport costs from region r to region r′, then it minimizes the use of every input into transportation between the regions as well.* If this assumption holds, link flows, and from them input demand by the transport sector, follow (generically) uniquely from a spatial equilibrium. Then, computation of equilibrium in a multi-regional CGE can be attempted by carrying out the steps listed in Table 12.3.[13]

Table 14.3. Solution stragegy for a multi-regional CGE with spatially homogeneous good and endogenous production of transport services.

0. Choose a proper Normalization, say $p^{1}_{(1,1)} = 1$ (it is best to use a single-origin good, used in all regions, as the mumeraire good).
1. Choose initial values for transport costs and demand for transport inputs (e.g. by setting them to 0). Allocate transport inputs to regions according to some fixed rule. Choose initial values for prices of any transport-specific inputs (like the *M*-good of Table 12.2).
2. Choose initial values for spatially homogeneous goods.
3. Solve for regional equilibria as in Table 12.1, taking transport costs, prices of spatially homogeneous goods and demand by the transport sector as given.
4. Solve spatial price equilibria for spatially homogeneous goods; this gives new prices p^r_i. This step and the previous give new demand for inputs by the transport sector. Check convergence and, if necessary, set prices p^r_i and input demands for next main interation.
5. Check balance and adjust prices for transport-specific inputs.
6. Repeat from step 3 until convergence is reached.

The computational approach of Table 12.3 was successfully employed in a multi-regional CGE model for Nepal (Elbers 1992 chapters 4-6). In this model agricultural goods are assumed to be homogeneous across regions. All other traded goods follow the Armington specification. Convergence to equilibrium is very much helped by the fact that one of the foreign regions in the model, India, is assumed to be a perfectly elastic source and sink of agricultural goods. Typically, convergence is reached after 12-13 main iterations. The Appendix gives the main ingredients for a multi-regional involving traded goods of both the Armington and spatially homogeneous kind. This list of ingredients is derived from the Nepal model.

12.4 Conclusions

Two problems have been addressed in this article, arising in connection with the construction of multi-regional CGE models. The first concerns the production of transport services. Production of transport services must generate demand for other goods and factors in the model. From a computational point of view one would like this demand to be point-valued. If it is, multi-regional equilibrium can

[13]Note that this assumption excludes link-specific capacity variables.

essentially be computed by iterating a sequence of regional equilibrium computations. Unfortunately, this calls for assumptions excluding the possibility of substitution between modes of transportation or switches in cost-minimizing paths between regions.

The second problem concerns the inclusion of spatially homogeneous goods in multi-regional CGE models. For such goods the usual Armington specification cannot be employed. Instead, the older concept of Spatial Price Equilibrium must be used. Thus a hybrid model results, involving both Armington and non-Armington traded goods. The article sketches a procedure for solving such a hybrid model.

Fischer et al (1988) present a multi-national CGE-model also containing spatially homogeneous traded goods. The difference between their model and the approach taken in this article is that the underlying transport network is star-shaped (international market clearing takes place in an artificial central node called 'clearing level' which has links to all nations) and that there is no global transport sector: transport services are produced by the nations importing from or exporting to the clearing level.

A number of important difficulties remain. First, from the point of view of Transport Economics the model of the sector producing transport services is rather primitive. Assumptions to the effect that cost-minimizing paths are unique and do not change during computations, are too restrictive. Such assumptions preempt the study of mode changes and congestion. Also, the assumption of competitive market clearing in the transportation sector is overly simple, where more sophisticated models of the transport market experiment with oligopolistic market solutions (e.g. Harker 1987). The article makes clear that modelling the production of transport services in a more satisfactory way, will greatly complicate model solution. A second difficulty concerns model calibration. Spatial equilibrium theory predicts that the same good does not flow in both directions over a link. But what if data do show bi-directional flows? Even if a modeller is still convinced that spatial homogeneity describes a good better than the Armington specification, simply taking the net flow between regions will not do: it would lead to underestimating the demand for transport services. Perhaps the best solution here is to assume that some *intra*-regional trade is connected to production and consumption of goods.

Summarizing: by employing existing algorithms for solving the Spatial Equilibrium Problem it becomes possible to include spatially homogeneous goods in a multi-regional CGE model, if the sub-model of the transport sector is kept relatively simple. Including a more realistic sub-model of the transport sector in such models appears to be a greater challenge.

References

Armington, PS (1969) A Theory of Demand for Products Distinguished by Place of Production. IMF Staff Papers 16:159-176

Elbers C (1992) Spatial Disaggregation in General Equilibrium Models, with an Application to the Nepalese Economy. PhD thesis, Vrije Universiteit, Amsterdam

Elbers C (1993) Trading Networks and Endogenous Trade Regimes in Applied General Equilibrium Modelling. Discussion Paper 93-129, Tinbergen Institute, Amsterdam pp 93-129

Fischer G, Frohberg K, Keyzer MA, Parikh KS (1988) Linked National Models. A Tool for International Food Policy Analysis. Kluwer, Dordrecht

Harker PT (1987) Predicting Intercity Freight Flows. VNU Science Press, Utrecht

Kehoe PJ, Kehoe TJ (1994) Capturing NAFTA's Impact with Applied General Equilibrium Models. Federal Reserve Bank of Minneapolis Quarterly Review 18:17-34

Nagurney A (1993) Network Economics. A Variational Inequality Approachy. Kluwer, Dordrecht

Pang J-S (1984) Solution of the General Multicommodity Spatial Equilibrium Problem by Variational and Complementarity Methods. Journal of Regional Science 24:403-414

Takayama T, Judge GG (1971) Spatial and Temporal Price and Allocation Models. Contributions to Economic Analysis. North Holland, Amsterdam

Whalley J, Wigle R (1991) Cutting CO_2 Emissions. The Effects of Alternative Policy Approaches. Energy Journal 12:109-124

Wigle R (1992) Transportation Costs in Regional Models of Foreign Trade. An Application to Canada-US Trade. Journal of Regional Science 32:185-207

Appendix 12.1

Ingredients for a Multi-Regional CGE.

In this appendix a list of ingredients for a 'hybrid' multi-regional CGE. Traded goods can be of the Armington or spatially homogeneous type. The list is based on the Nepal model in Elbers (1992, chapters 4-6). An abstract equation listing is given in Elbers (1993).

regions/locations: Determines the spatial disaggregation in the model. Locations include domestic and foreign regions.

transport network: Defines the links between neighbouring regions. It is assumed that the resulting graph is complete: i.e., every pair of regions is connected by a sequence of links.

agents, local: Specifies agents and their behaviour located in specific regions: household, firms, local government, export demand, etc.

agents, global: national government, transport firms.

goods and factors, non-traded: The goods and factors which are specific to each region. Along with these goods endowments and/or production functions must be specified.

goods and factors, spatially heterogeneous: The Armington goods. All traded goods produced under constant returns to scale are in this category.

goods and factors, spatially homogeneous: Only goods in fixed supply or produced under decreasing returns to scale can be in this category.

transport technology: Gives the use of inputs per unit of transport service over links. The Nepal model uses a Leontief technology. The solution strategy of Table 12.3 requires that transport costs over paths are equal to the sum of transport costs over the in the path. Moreover, the cost-minimizing path must be unique and may not change during computations.

allocation of inputs into transportation: The inputs must be supplied according to some rule (e.g. all demand to cheapest region). Moreover, the Nepal model assumed that supplies to the transport sector do not need transport services themselves.

market clearing: The Nepal model assumes competitive market clearing.

13 The Macroeconomic Impact of Traffic Congestion: A CGE Analysis

Roberto Roson
University of Venice

13.1 Introduction

This paper illustrates the general structure and a simulation exercise carried out with a model for Italy, integrating a Computable General Equilibrium (CGE) model, a Dispersed Spatial Price Equilibrium (DSPE) model and a Freight Network Equilibrium (FNE) model.

The model, called MITER, was developed with the aim of providing a tool for the analysis of interaction processes between the economic and the transport system. By using this model it is possible to highlight how demand for freight transport is generated by trade relationships among regions, how transport costs vary to match demand and supply of transport infrastructure and how transport costs - via market prices - influence economic competitiveness of regions and industries.

The general equilibrium approach adopted in the model is especially suited for the analysis of the market of transport services, since transport demand is a derived demand generated by inter-industrial linkages and spatial trade patterns. For this reason, it is not possible to study this market in isolation but market interdependencies must be fully taken into account.

Previous research (e.g. for Italy, Leontief and Costa 1987) has pointed out that demand for transport services in most industrialized countries is likely to continue its growth at a rate that far exceeds the rate of increase in transport infrastructure supply. However, little has been said on the way the spatial economic structure would adjust to relative scarcity of transport infrastructure as well as on the global impact of traffic congestion on regional and national economic performance.

In this respect, a model like MITER, based on the mathematical complementarity linking spatial economic and transport network models (Roson 1993) may help to get some useful insights.

The complex analytical structure of MITER as well as several issues associated with the development of the data base and the estimation of the model's parameters are fully described in Roson (1994). For the sake of simplicity we shall only outline the logical structure of the model, in the second section of this paper, in order to better understand the results obtained by a simulation exercise, which are presented in the third section.

The simulation was carried out by interfacing the MITER model with a regional econometric model (MORE)[1], used to obtain final demand forecasts for the year 2000. The MITER model was therefore run to derive estimated congestion levels in the transport system, equilibrium values for relative prices, regional and industrial production levels consistent with the given exogenous scenario, assuming fixed and invariant the characteristics of the transport network at their 1988 values, which is the calibration year of the model. These results are compared with those obtained under the hypothesis of constant transport costs.

Clearly, the assumption of invariance in the stock of transport infrastructures makes our experiment unsuitable for drawing conclusions about the future evolution of the Italian transport system. Why we introduced such an assumption is apparent in our willingness to test the MITER model against other approaches, such as the Leontief-Costa model (Leontief and Costa, op cit), where supply constraints in the transport system are not explicitly taken into account.

13.2 The Model Structure

MITER is based on three submodels linked by an iterative procedure for the determination of the simultaneous equilibrium state: the non-linear price system, the input-output quantity system, and the non-linear network system (Figure 13.1).

As in many other applied equilibrium models, MITER has been 'calibrated' such as to reproduce an initial equilibrium state which is assumed to occur in a base year (1988). Simulation exercises can be performed by changing exogenous variables and looking for 'counterfactual' equilibrium states. From a mathematical point of view, this amounts to determine the fixed point of a complex non-linear system expressing the equilibrium conditions for both the economic and the transport system.

In the case at hand, final demand exogenous variables are changed and assumed to be equal to some forecasts for the year 2000 produced by the MORE econometric model. These are used as an input in the quantity system.

---

[1]MORE has been independently developed by a team from GRETA Econometrics srl, led by Professor Giorgio Brunello of the University of Venice

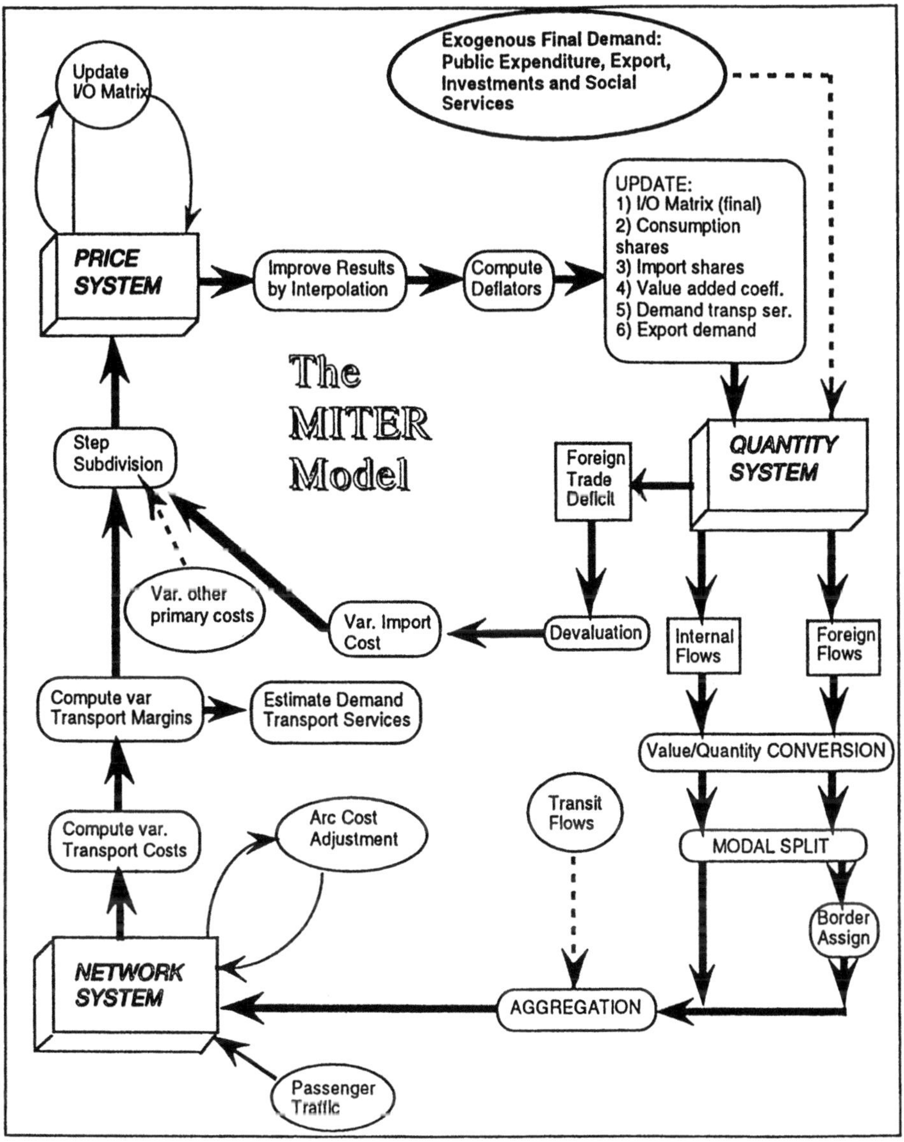

Fig. 13.1. General structure of MITER

The quantity system is a multiregional extended input-output model, estimating trade flows and production levels for 17 industries in 20 regions. Since constant returns to scale are assumed and relative prices are - at this stage - given, the system is a linear one. Its formulation is based on the macroeconomic identity between supply and demand:

$$X = AX + H + G + I + (E - M) \qquad (13.1)$$

Production levels for each industry in each region (X) must match intermediate demand, domestic absorption and net foreign trade. Intermediate demand is determined by multiplying production levels by a multiregional input-output matrix (A)[2]. Household consumption (H) is partially endogenous. It has been divided into two parts: the first one is made dependent on regional value added and, since prices are exogenously given, on regional production volumes; the second is fixed in nominal terms and is introduced to simulate the effect of pensions and transfers by the State. Public expenditure (G) is given in nominal terms and is based on the assumption of constant expenditure shares[3]. Real investments are exogenous. Both exports (E) and imports (M) are sensitive to relative domestic prices, but foreign prices are fixed. In addition, imports also depend on internal activity levels.

Since industries and final demand are specified by region, every transaction involving physical goods implies a demand for transport services. Aggregating by industry, trade flows may be computed among regions and to/from foreign countries. Both of these flows may also be expressed in quantity units of measure and split by transport modes, producing a series of origin-destination matrices.

These matrices express the potential demand for transport services generated by the economic system. In the first iteration round, when prices are unaltered, these flows may also be considered as expressing the demand generated when constraints in the supply of transport infrastructure are not taken into account.

In the current version of the model, modal split is carried out using fixed shares and, as congestion effects are considered only for road transport, only road traffic flows are actually considered when transport costs are changed. Since all flows are expressed in tons they are aggregated in a single 20x20 o-d matrix, comprehensive of traffic to and from national borders (import, export, transit). Passenger traffic flows, obtained by applying specific growth rates to the base traffic volumes, are also added once freight flows have been translated into vehicle equivalent units.

Assignment of total traffic lows to the interregional network is performed by the network system, where transport costs are updated so as to balance supply and demand in each link of the network. This is represented as an aggregated virtual graph connecting regional centroids, whose parameters have nonetheless been derived from data referring to real road networks. Network structure is internally described by means of a rectangular incidence matrix Λ, whose generic element λ_{ij} is equal to one if arc i belongs to route j, and zero otherwise. Vectors of path flows may then be translated into arc flows simply by multiplying these vectors by the incidence matrix.

---

[2]This is spatially disaggregated. It has been obtained by combining a matrix of trade coefficients and a block diagonal matrix or regional (traditional) I/O matrices, following the so-called Chenery-Moses formulation.

[3]This amounts to assume a Cobb-Douglas expenditure function.

For each o/d pair, it is assumed that there exist one or more alternative paths (sequences of arcs connecting origin to destination), and that these paths are known in advance. Route choice is based on decentralized cost minimization (Wardropian user equilibrium), where the cost to be minimized comprises all direct and indirect costs associated with moving goods (Beckmann, McGuire and Winsten 1956; Gartner 1980). This 'generalized' cost is not unique, but depends on the characteristics of the transported items, so that optimal choices may differ among agents.

Origin-destination flows, re-arranged in a column vector ϕ, may be easily translated into path flows by multiplying them by a rectangular shares matrix, whose values depend on path costs. On the other hand, path costs are simply the sum of arc costs for all arcs comprised in the paths.

The network equilibrium problem may then be stated as a single matrix equation:

$$F = \Omega(\Lambda' c(\Lambda F)\Phi \tag{13.2}$$

where F is the vector of path flows and c is the vector of arc costs computed by the arc congestion functions. In our case, the functions are strictly convex, differentiable ones (Cascetta 1990):

$$c_a = \bar{c}_a \left\{ =1 + \beta \left(\frac{f_a}{\bar{f}_a} \right)^{\lambda} \right\} \tag{13.3}$$

where c_a and f_a are arc cost and flow, and all remaining symbols are parameters.

Transport costs in the various regions can be computed as averages of path costs, with the relative incidence of road transport and the existence of fixed costs also being taken into consideration.

The non-linear price system may then be used to determine equilibrium market prices. These are computed by equating prices with industrial cost functions:

$$\begin{cases} p_1 = p_1(p_1, \cdots, p_n, va_1) \\ \vdots \\ p_n = p_n(p_1, \cdots, p_n, va_n) \end{cases} \tag{13.4}$$

where n is the number of industries, va is the value added and p are cost functions derived from the basic cost minimization problem:

$$\begin{aligned} \underset{x_{ij}}{Min}\ & p_j = \sum_i x_{ij} p_i \\ s.t.\ & x_j = f_j(x_{1j}, \cdots, x_{nj}) \end{aligned} \tag{13.5}$$

where f is a production function representing available technology and x are input and output flows.

In a spatial open economy the above system must be modified in order to distinguish domestic and import prices, and to introduce transport cost margins as additional costs on producer prices. Although the price system could in many cases be directly formulated as a set of non-linear equations, we decided to follow the so-called Johansen approach (from Johansen 1960) which involves linearizing the system expressed in logarithmic differentials.

This modelling strategy proved to be especially suited to this type of model, because it eliminates the need to know the value of all exogenous variables (only the percentage change of variables whose values are changed from the base reference equilibrium state are actually required) because the cost functions do not have to be expressed in closed form; furthermore, the model can easily be solved by software designed to handle large linear systems (e.g. GAUSS, a feature which is particularly convenient in disaggregated multiregional models).

The model is solved by stepwise linearization, also making use of interpolation formulas to reduce approximation errors (Dixon, Parmenter, Powell and Wilcoxen 1992). This approach allows to trade off estimation accuracy and computational costs. At each step the system may be formulated in matrix form as[4]:

$$\begin{aligned} P_d &= S_1\, p_d = S_2\, p_m + W\, c \\ &= (I - S_1)^{-1}(S_2\, p_m + W\, c) \end{aligned} \tag{13.6}$$

where (n = number of industries, m = number of regions):

- p_d, p_m = (nmx1) vectors of percentage change in domestic and import prices;
- c = (nmmx1) vector of percentage change in transport margins;
- S_1, S_2 = (nmxnm) value shares matrices;
- W = (nmxnmm) rectangular matrix obtained from the S matrices.

Value shares matrices are updated, at each round of the linearization process, according to the rules derived from some adopted production functions. In this case we assumed a two-stage substitution mechanism: a first level involving domestic and imported goods within the same industry (modelled by a CES function) and a second level involving all inputs of each industrial production process (modelled by a CRESH function[5]).

The solution of the price system provides a new set of equilibrium prices that are used to re-compute optimal demand patterns in the quantity system. Before repeating the cycle, several parameters are therefore updated and some new variables are computed. These include a new multiregional input/output matrix, made consistent with the new set of relative prices, consumer price indices, consumption and import shares, exports, and value added coefficients.

---

[4]In this expression, possible changes in value added components are not taken into account.

[5]This is a sort of non-linear generalization of multifactor CES function, having different elasticity values for each input factor.

In addition, demand for transport services and input/output coefficients of the transport industries are also modified in order to take variations in costs and trade patterns into account. This is due to the fact that most transport services are domestically produced and increasing traffic congestion implies higher activity levels for transport industries.

Foreign trade deficit, due to higher import and lower export flows, can also be easily computed at this state. Trade deficits or surpluses (at the national level) are not consistent with the equilibrium paradigm adopted in the model, so that a real devaluation, necessary to achieve the trade balance, is simulated by augmenting the domestic price of imports. Hence, after the first iteration round, new equilibrium prices are computed taking into account variations both in transport and import costs.

Convergence of the entire model to the global equilibrium state, where all choices are mutually consistent, is assured by the mathematical structure of sub-problems. A reasonable approximation for the global equilibrium point was found in our numerical experiments only after a few iteration rounds, with the model approaching the new equilibrium state in a (apparently) monothonical way.

13.3 Results

Figure 13.2 graphically displays forecasted national growth in GDP and its components between 1988 and 2000, produced by the MORE econometric model. In this period, real GDP is expected to rise by about 30%, averaging rather different growth rates in final demand elements.

The highest growth rates are assumed to occur in import and export flows, testifying to the increasing internationalization of the Italian economy; on the other hand, consumption by households and private institutions, the most important component of GDP, is expected to slow down the growth process observed in the 80's.

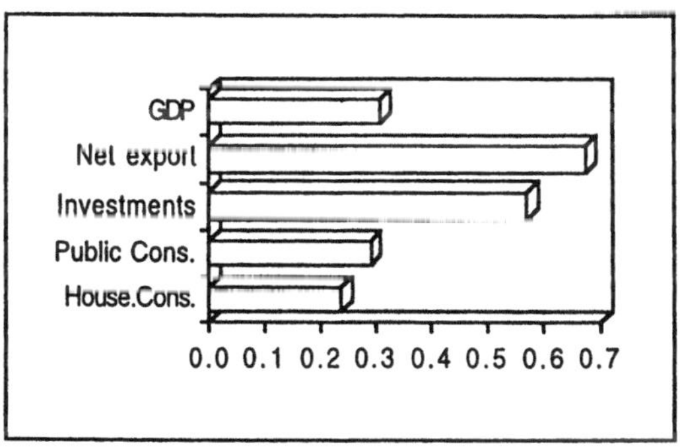

Fig. 13.2. Relative growth in the aggregated GDP components

Total increase in GDP amounts to an average yearly growth rate of 2.26%. Higher or lower growth rates are assigned to the regional economies, depending on the composition of regional industrial structures.

Since in our simulation exercise the supply of transport infrastructure is held fixed at 1988 levels, the higher demand for transport services generate traffic congestion and increasing transport costs in some links of the interregional road network.

An increase in traffic congestion can be considered as a negative technological shock for the national economic system, since overall efficiency is reduced and, with a given amount of resources, less goods and services can be produced. Private and public consumption, as well as investments (future consumption), are therefore expected to decrease, reducing welfare levels.

Alternatively, more primary resources (E.G. capital, labour) could be employed if their supply is sufficiently elastic. In our model, the price of primary resources is exogenously given and supply constraints in these markets are not taken into consideration. Imperfect competition, persistent involuntary unemployment and sticky prices may justify this assumption, implying horizontal supply curves.

As has already been mentioned, in the short run the system may also borrow from abroad by increasing imports and running a trade deficit[6]. In the long run equilibrium state, however, trade deficits are not allowed and increased real imports must be matched by increased real exports, so that domestic consumption must necessarily be further reduced.

The simulation results illustrated in this section depict these effects in a more precise and detailed way, by comparing the two cases of absence of traffic congestion[7] and congestion generated by a constant stock of transport infrastructure. In this simulation exercise, however, the foreign trade balance constraint has not been imposed, and both exports and public expenditure have been assumed price inelastic. The results may therefore interpreted as depicting the system performance in the short-medium term.

Table 13.1 summarizes the percentage variations observed in the macroeconomic accounting identities between supply and demand for the year 2000, on the basis of the two alternative assumptions.

On the left side, forecasted values are shown for total internal production and imports, and for all regions and sectors. As already mentioned, domestic production declines and, despite a substantial increase in imports from abroad, there is also a small reduction in total distributed resources. On the right side, gross demand is subdivided into intermediate, households and other, which in turn comprises exports, investments and public expenditure. It should be

---

[6]This alternative makes sense only if imports become relatively cheaper than domestic goods. This is often the case because imports are more intensively carried by uncongested transport modes, like sea transport and rail.

[7]More precisely, congestion costs are assumed to be constant and equal to their 1988 values. This amounts to disregarding the feedback effect generated by the transport system or, similarly, to assuming that supply of transport infrastructures is always updated to match demand.

recalled that these three variables were assumed exogenous in this exercise. The most important effect on the demand side is the reduction of private consumption, which actually proves to be larger if consumption of transport services is not considered.

Table 13.1. Identity of macroeconomic resources

MACROECONOMIC BALANCE Supply		Demand	
2000 without transport congestion			
Domestic prod.	2,615,977,718	Intermediate	1,220,626,478
Imports	350,521,195	Households	836,286,453
		Other	909,585,983
Total	2,966,498,914	Total	2,966,498,914
2000 with transport congestion			
Domestic prod.	2,599,697,924	Intermediate	1,222,892,057
Imports	357,753,803	Households	824,973,688
		Other	909,585,983
Total	2,957,451,727	Total	2,957,451,727
Percentage variations			
Domestic prod.	-0.62%	Intermediate	0.19%
Imports	2.06%	Households	-1.35%
		Other	0.00%
Total	-0.30%	Total	-0.30%

Intermediate demand is expected to have a small increase, as a result of substitution processes of different sign. Since production levels decline, as it was seen on the supply side, the increase in intermediate demand must be attributed to an augmented unitary demand of input factors. This is generated by the combination of two effects: a 'price effect', induced by higher relative prices of intermediate factors, bringing about substitution of intermediate with primary factors, and an 'inefficiency effect', which simply takes into account the increased demand of transport services generated by congestion. This second effect appears to prevail because the demand for transport is inelastic, and because of import substitution, which lowers the price increase for intermediate factors.

The impact of rising congestion costs may also be ascertained in terms of Gross Domestic Product, amounting to a significant 1.33% reduction in national GDP. This can, of course, be explained from both the demand and the supply side: on the demand side, GDP may be expressed as net final demand, the reduction of which stems from decreased private consumption and net exports; on the supply side, GDP is given by total (real) value added, showing a slight increase in nominal terms offset by a more than proportional price inflation.

Figure 13.3 compares the variation of production levels and GDP at a regional scale. It should be observed that, because of modification in regional competitiveness, regional production volumes are actually positive in a few cases; namely in Calabria and Basilicata, where a situation of excess supply of transport infrastructure is likely to endure, and in Lazio, whose positive sign is probably due to the rather low economic growth assumed in the exogenous scenario.

On the other hand, all regional GDPs are reduced, with significant differences among regions. The strongest reduction is found in the three regions in the Po Valley, Italy's most industrialized area, with a maximum of more than 2% for Piedmont (Piemonte). It is also interesting to note that Sardinia (Sardegna), which, being an island, is not very close to congested areas, is nonetheless significantly affected by congestion costs because of its high economic interdependence with northern regions.

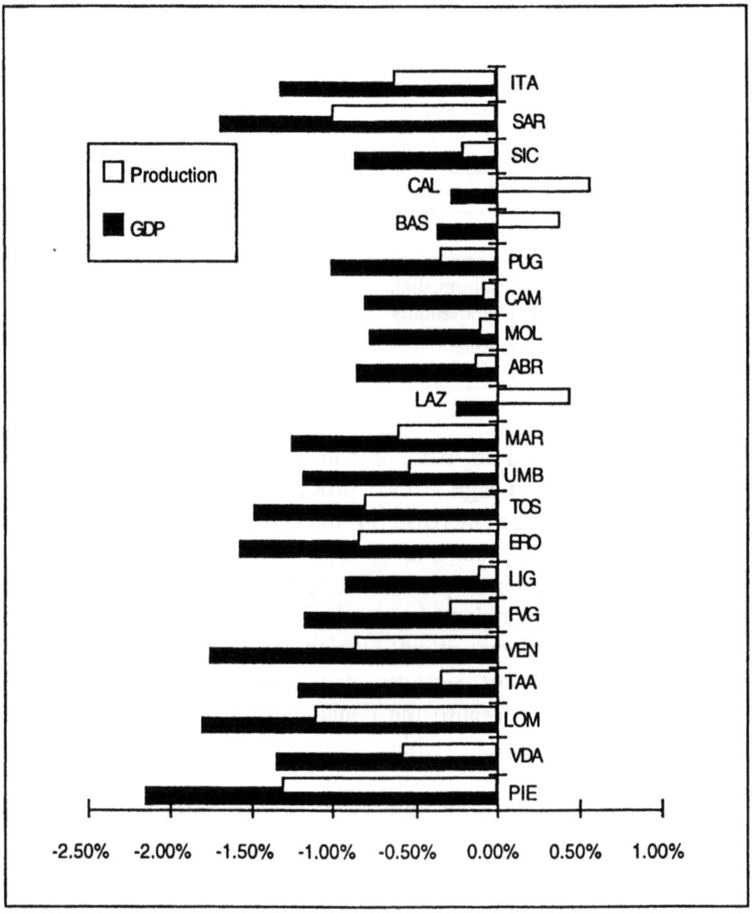

Fig. 13.3. Aggregate regional variations in production and GDP

Figure 13.5 illustrates variations in value added and production levels aggregated by industry.

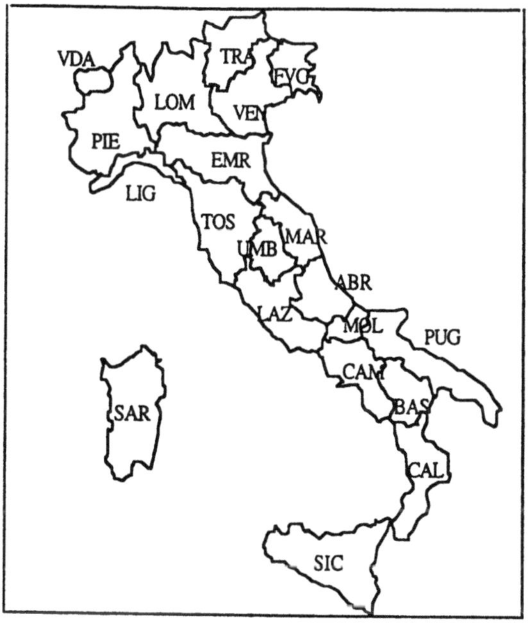

Fig. 13.4. A map of Italy, showing the twenty regions and their acronyms

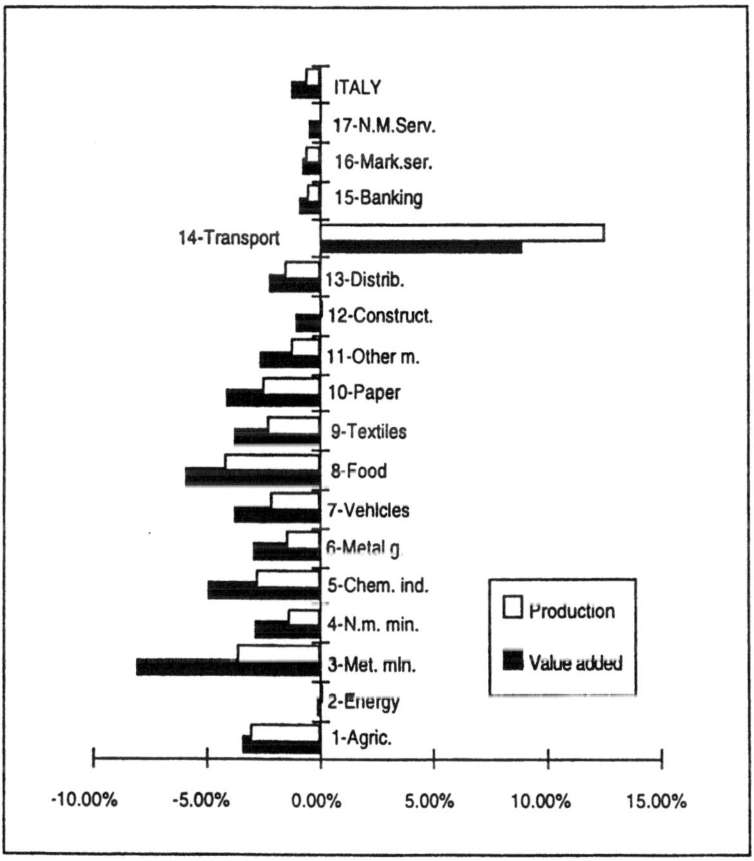

Fig. 13.5. Aggregate industrial variations in production and value added

The big jump in transport industry activity forecasted by the model is immediately apparent. This is the result of linking variations in transport margins to production levels in the transport industry and of modelling the increase in real demand for passenger transport.

Variations in value added and production levels are given by the combination of two overlapping phenomena: modified relative competitiveness among sectors and additional demand for transport services. The first effect is expected to benefit service industries, producing non-physical output and therefore only indirectly affected by rising transport costs. The second effect produces a 'demand pull' for industries that are technically linked to the transport sector. At the same time, industries requiring passenger transport services are also affected by increased production costs.

Figure 13.5 suggests that the two effects approximately compensate each other in services. The construction industry retains a very small but positive variation in production volumes.

Apart from the transport industry, the energy sector also shows a small positive variation in production levels, clearly induced by the additional demand coming from transport activities. Furthermore, this is also due to the fact that the energy industry uses several transport modes, like shipping transport, which are not affected by congestion.

Figure 13.6 illustrates inflation effects generated in the regions by rising transport costs. Two different sets of regional indices are compared: Consumer Price Index (CPI) and Producer Price Index (PPI), evaluated by weighting price variations with the 1988 industrial distribution of production and consumption. Consumer prices include transport costs associated with market distribution and the price of imports devoted to household consumption (which here are actually fixed).

At national level, the two indices are both approximately equal to 0.9%, pointing to a rather significant impact on prices, particularly since these are measures of real prices.

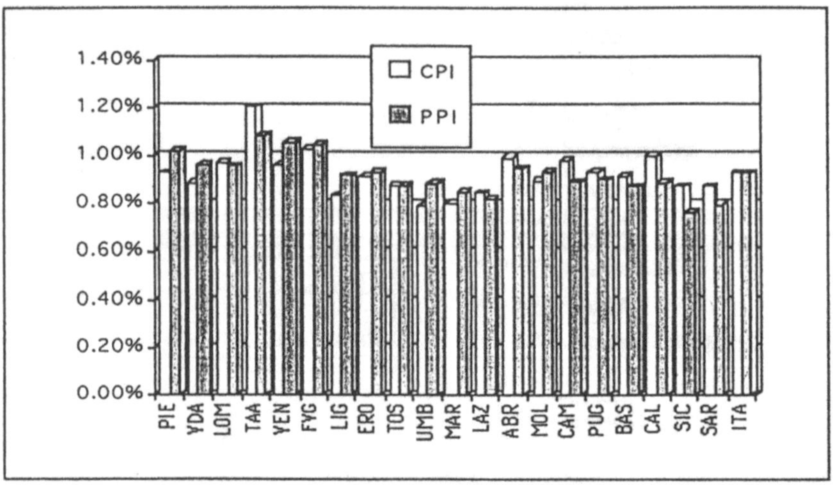

Fig. 13.6. Variations in price indexes, by region

Since the two indices are approximately equal at the national level, differences in regional indices seem to be attributed mainly to the spatial redistribution of demand rather than to differences in the industrial mix. In this case it must be remembered that, on the one hand, congestion costs are mainly felt in the norther regions and, on the other, almost all northern regions are net exporters in interregional trade (souther regions are, of course, net importers). Consequently, the spatial structure of demand contributes to the redistribution, throughout all regions, of the negative effects of the negative effects of scarcity in transport infrastructures.

Another set of numerical results produced by the model concerns the transport system and the distribution of freight flows, where the state of the system may be described in terms of origin-destination flows, path flows or arc flows. Table 13.2 display the effect of congestion costs on physical trade flows, involving all transport modes, among five macro-zones aggregating the twenty administrative regions and the rest of the world.

Table 13.2. O-D freight flows (percentage variations due to congestion costs

	North-West	North-East	Central	South	Islands	Abroad
North-West	-1.61%	-1.57%	-0.65%	-1.19%	-1.12%	0.00%
North-East	-1.80%	-1.18%	-1.56%	-1.29%	-1.09%	0.00%
Central	-1.61%	-1.45%	-1.90%	-1.03%	-1.35%	0.00%
South	-2.30%	-1.99%	-1.64%	-1.05%	-1.64%	0.00%
Islands	-2.19%	-0.97%	-0.19%	-0.05%	-1.06%	0.00%
Abroad	0.84%	0.99%	1.18%	0.51%		

Changes in O-D flows can be explained as the result of two substitution mechanisms: interregional trade is substituted with intraregional production, since intraregional transport costs are assumed unchanged, and all domestic production is substituted with imports.

Table 13.2 shows that the second process reduces all internal flows, although flows accounted for in the main diagonal also include interregional flows within the macrozones.

Another factor influencing the model's results is given by relative intensity in the use of road transport, which, in the simulation, is the only mode affected by congestion. Since the share of road transport in total carried flows appears to be inversely related to distance, the impact of higher road transport costs turns out to be more significant for short distances.

In the following three figures, flow diagrams are used to summarize results obtained by assigning physical flows to the arcs of the transport network.

Figure 13.7 shows the distribution of flows at the year 2000, relative to the (fixed) carrying capacity. This picture is similar to the analogous one obtained for the calibration year (not shown here). This outcome may easily be attributed to the growth in passenger traffic flows, which were assumed to be price independent.

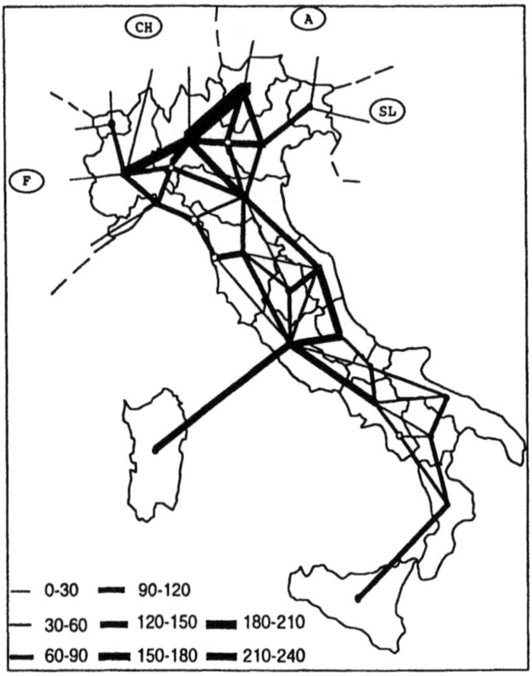

Fig. 13.7. Relative arc flows (% of carrying capacity) - year 2000

Some spatial redistribution of traffic lows is, on the other hand, generated by the reallocation of freight traffic, as is apparent in figure 13.8, which displays the percentage changes in freight flows.

This picture clearly demonstrates a tendency towards a more intensive use of alternative routes in highly congested areas, as in the case of lows between Lombardia and Piemonte and for other North-South flows, creating a significant increase in traffic on some routes crossing the Apennines. Because of competition with passenger transport, one link (Trentino-Lombardia) actually shows a decline in freight traffic.

By combining all these effects, the forecasted assignment of freight traffic to the network may be drawn in a chart (Figure 13.9) which summarizes the basic information produced by the transport system submodel.

A similar kind of result could have been obtained for all other transport modes by inserting their network structures in the model and estimating congestion functions for each arc and/or node in the graph.

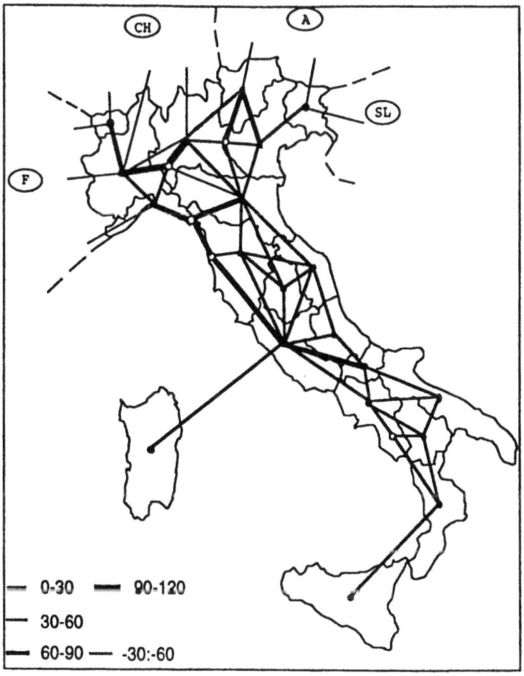

Fig. 13.8. Percentage variations in freight traffic flows 1988-2000

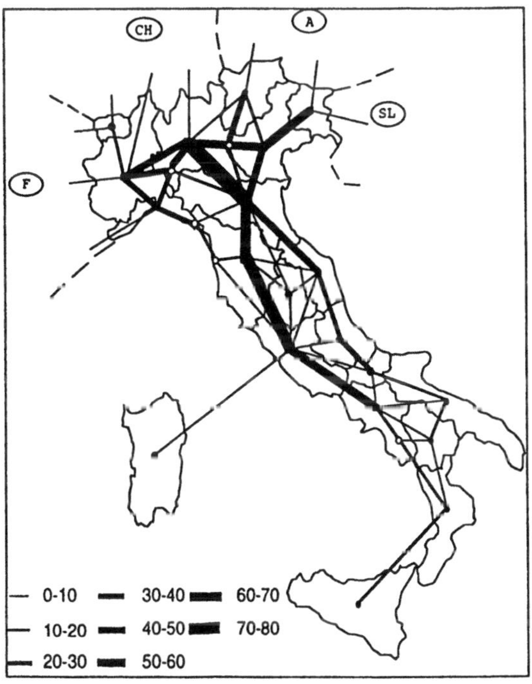

Fig. 13.9. Total forecasted freight traffic - Year 2000 (millions of tons)

13.4 Concluding Remarks

The quality of transport infrastructures has always been a key factor in national and regional economic development, although during the last century the rapid diffusion of new, efficient modes of transportation and communication has reduced the importance of physical distance on the spatial distribution of economic activities.

It is becoming increasingly clear, however, that the period of falling transport costs is now coming to an end in many countries. Congestion in transport infrastructures, as well as a growing concern about the environmental impact of transport activities, calls for a new approach in the management of transport systems.

Models like MITER can provide useful insights on the interaction process between demand and supply of transport infrastructures. Better knowledge of spatial economic systems, in turn, would improve the effectiveness of many economic policies.

The simulation results show that constraints in the supply of transport infrastructures may have a significant impact on national and regional economic performances, as well as on social welfare, and that the transport system influences economic activities and their spatial locations in many different ways: we considered the effect of changing transport margins on market prices, the input factors associated with people's mobility in production processes and the consumption of transport services. Other effects, such as those associated with land use, could also have been included.

Acknowledgements

Technical assistance by Piero Vianelli and Stefano Pomaro and financial support from the Italian National Research Council (CNR - Progetto Finalizzato Trasporti 2) are gratefully acknowledged.

References

Beckmann M, McGuire CB, Winsten CB (1956) Studies in the Economics of Transportation. Yale University Press, New Haven

Cascetta E (1990) Metodi quantitativi per l'analisi di sistemi de trasporto. CEDAM, Padova

Dixon PB, Parmenter BR, Powell AA, Wilcoxen PJ (1992) Notes and Problems in Applied General Equilibrium Economics. North Holland, Amsterdam

Gartner NH (1980) Optimal Traffic Assignment with Elastic Demands. A Review (Part I. Analysis Framework. Part II. Algorithmic Approaches). Transportation Science 14:174-208

Leontief W, Costa P (1987) Il trasporto merci e l'economia italiana. Scenari di interazione al 2000 e 2015. Sistemi Operativi, Venezia

Roson R (1993) Spatial Computable Economic Equilibria and Freight Network Models. International Journal of Transportation Economics 20:51-66

Roson R (1994) Transport Networks and the Spatial Economy. A General Equilibrium Analysis. Umeå Economic Studies, University of Umeå

14 Trade, Sensitivity and Feedbacks: Interregional Impacts of the US-Canada Free Trade Agreement

Ricardo Gazel, Geoffrey J.D. Hewings and Michael Sonis
University of Illinois

14.1 Introduction

Most analyses of the impacts of existing and proposed free trade agreements have focused on the impacts measured at the national level. Given the significant differences in the composition of industry across states within nations, concern has been raised about the possibility of there being differences in impacts of free trade at the sub-national level. However, the states' economies of the United States do not function in isolation; there is substantial inter-state trading in goods and services and a complex pattern of flows of funds of various kinds (investments, taxes, transfers, pensions, dividends and so forth). As a result, international trading relationships between one state and a foreign country will have the capacity to create a complex set of *ripple or multiplier effects* on the rest of the United States. The basic nature of the interregional relationships are presented graphically for a four-region version of the interregional computable general equilibrium model in an attempt to reveal the subtle nature of the linkages and feedback effects.

Initial attention has been focused on the potential impacts of the Free Trade Agreement between Canada and the United States. National-level assessments suggested that the impacts for the United States would be very small (about 0.1% of GNP); this result would appear to be surprising in view of the fact the US-Canada trade is by far the largest bi-lateral volume in the world. However, over 70% of items traded between the two countries are currently free of duty, thus the muted impact of additional reduction in tariffs. At the sub-national level, attention was focused on a four-region division of the US, with Canada as a fifth North American 'region'.

Analysis reported in this paper focused on the degree to which the interregional model is sensitive to parameter estimates. A limited number of experiments were performed, with attention to the elasticity estimates and the interregional flows. The variations produced, in comparison to the base case, reflect considerable uncertainty about the reliability of some of the derived elasticities and interregional trade flows. The sensitivity analysis is complemented with an evaluation of some of the more important feedback loops in the trading relationships among the regions. In the next section, a brief introduction to the US-Canada Free Trade Agreement will be provided, together with an overview of the nature of trade between the two economies. Section 14.3 will describe the major features of the model used in the analysis; following this review, the concept of feedback loops will be introdcued to assist in the interpretation of the sensitivity analysis performed in section 14.5. The paper concludes with some brief remarks.

14.2 The US-Canada Free Trade Agreement

The Free Trade Agreement (FTA) between the US and Canada came into force on January 1, 1989. The three most important points of the FTA are the elimination of all tariffs and nontariff barriers over a period of ten years, the promotion of a better environment for investment and for the trade of services between the two countries, and to create specific channels to solve trade disputes under the FTA. The first of these objectives, the elimination of tariffs on all goods traded between the US and Canada, was scheduled in three parts: some goods such as computers, fur, leather, whiskey, animal feeds, ferro alloys, satellites, telephone sets, and motorcycles would be free of duty immediately; other products such as machinery, chemicals, furniture, printed matter, paper and paperboard, plastics, photographic films, and telecommunication equipment would face reduced tariffs over a five-year period (20% reduction on tariffs per year); goods such as agriculture products, medical equipment, auto parts, auto engine parts, steel, textiles, tyres, and appliances would be free of duty at the end of a ten-year period. Besides the tariff elimination schedule, the agreement moves forward toward elimination of nontariff barriers such as reduction of quantitative restrictions, export subsidies, import licenses, etc. This proposed schedule of tariff and nontariff barrier reduction came in response to different levels of protectionism and consequent efficiency in different industries in both countries. The transition period allows industries to become more efficient before facing higher foreign competition under the agreement. This schedule was more important to Canada than to the US given the smaller size of the Canadian economy and its higher level of protectionism.

All goods with 50 per cent or more of their cost incurred in either country are eligible under the FTA. This *rule of origin* protects both countries against foreign competitors that could use one country as the back door to access the other country's markets. This clause would benefit the US more than Canada because of the relative market sizes; third countries would be more interested in using Canada as a back door to enter the American markets than the other way around.[1] The second point, trade in services and investment, deals basically with the problem of national treatment for services and investment. Both countries agreed

[1]The recent dispute (February, 1992) concerning Honda is an illustration of this issue. The US claimed that the percentage of components in the car assembled in Canada were less than the 50% minimum required.

to grant national status to the other country's service providers and investors in the future. They also agreed to reduce barriers on federal government procurement. The third point, institutional provisions, create mechanisms to settle trade disputes under the agreement. A Canadian-United States Trade Commission would supervise the implementation of the FTA and solve trade disputes. Furthermore, a binational panel would be formed at the request of either country to solve disputes related to antidumping and countervailing cases as well as to cases involving the interpretation of the agreement itself. The FTA allows each country to apply their antidumping and countervailing laws at their discretion and the binational panel could review the cases instead of other institutions such as the Court of International Trade.

14.2.1 The National-Level Results

Most studies dealing with the FTA between the US and Canada use computable general equilibrium (CGE) models as the methodological tool to measure the economic impact resulting from the agreement (see Hewings and Gazel, 1993 and Brown and Stern, 1988 for a detailed evaluation of the various approaches). The major findings suggest that (i) the impacts are projected to be much larger in Canada (up to 8.9% increase in real income) than in the US; but (ii) there is considerable range for the projected impacts in Canada and (iii) the findings for the US are close to zero or mildly negative in the case of a study by Hamilton and Whalley (1985) that also reflected results of prior studies revealing a pattern of negative impacts of free trade for the US. Given the diversity in the structure of the US sub-national (i.e. state) economies, the question obviously arises as to whether these national-level results for the US are reasonable predictors of the impact at the regional level. Prior to addressing this issue, section 14.2 will provide a brief overview of US-Canada trade.

14.2.2 US-Canada Trade: A Brief Overview

Canada and the US are the largest trading partners in the world. Canadian exports to the US increased over time from 63 per cent of total Canadian exports in 1980 to 78 per cent in 1985. They decreased to 73 per cent in 1988 and increased again to 75 per cent in 1990. The United States is the major supplier of Canadian imports. Imports from the US represented 71 per cent of total Canadian imports in 1980, decreasing over the last years to 64 per cent in 1990.

Although these are impressive numbers, bilateral trade among the US and Canada was already quite liberated before the FTA. More than 70 per cent of American exports entered Canadian markets free of duty and the average tariff rates on dutiable imports were very low when compared to Canada's other trade partners. On average, the tariff rate was less than 4 per cent on imports from the US and more than 7 per cent on imports from other countries. On the American side, the average tariff imposed on imports from Canada was even lower, around 1 per cent, while other countries faced a higher average tariff, more than 4 per cent.

Nonagricultural products dominate the trade between the US and Canada. Transportation equipment and machinery represent more than 40 per cent of total trade in both directions. The US shows a surplus in trading machinery while transportation equipment accounts for the largest trade deficit with Canada. Imports of paper, paperboard, iron and steel, and nonferrous metals are substantial, accounting for 14 per cent of total imports from Canada. Canada is an important supplier of mineral fuels and related materials (9 per cent of total US imports from Canada in 1989) and represents a safe source of energy for the US

Given these very low levels of protectionism between the two countries, a relatively small economic impact of the FTA in both economies should be expected. However, trade between the US and Canada is extremely concentrated in the northern part of the US (see Table 14.1). The Great Lakes region (Illinois, Indiana, Ohio, Michigan, and Wisconsin), New England, Pennsylvania, New York, and New Jersey accounted for more than 70 per cent of total imports from Canada. Michigan alone was responsible for 30 per cent of total imports from Canada. This represented more than 5 per cent of the Canadian GNP. These imports were not exclusively consumed or used as intermediate goods in Michigan but were also redistributed to the rest of the country, demonstrating how the regional pattern of international trade is important when the level of trade is highly concentrated.

Table 14.1. Canadian Exports and Imports to/from the US States (Can$billion)

	Exports			Imports		
	88	89	90	88	89	90
ME/NH/VT	3.11	3.30	2.38	1.33	1.72	1.83
CT/MA/RI	4.86	4.41	5.01	3.53	3.72	3.73
NJ/NY/PA	20.68	19.92	17.91	11.71	11.44	12.53
MI/OH	29.37	29.10	36.30	23.54	24.14	23.96
IL/IN/WI	9.97	9.65	8.54	12.35	11.92	10.95
RUS	29.86	32.16	35.13	33.96	35.04	34.87
TOTAL	97.84	98.55	105.20	86.44	88.01	87.89

Source: Statistics Canada

Note: Maine (ME), New Hampshire (NH), Vermont (VT), Connecticut (CT), Massachusetts (MA), Rhode Island (RI), New Jersey (NJ), New York (NY), Pennsylvania (PA), Michigan (MI), Ohio (OH), Illinois (IL), Indiana (IN), Wisconsin (WI), Rest of the US (RUS)

On the export side (see Table 14.1), the sum of the exports of these three regions accounted for 60 per cent of the total US exports to Canada in 1990. The strong concentration of exports in the northern part of the US suggests that this area would accrue most of the benefits from the FTA. This is true when, as a result of lower prices of American products in Canada, the direct effects of a stronger export demand are the only effects measured by the model. This is not necessarily true if indirect interregional effects are strong enough to compensate the rest of the country for the lack of larger direct trade with Canada.

US-Canadian trade is not only geographically concentrated but also sectorally concentrated. Each region in the US presents a different structure of trade with Canada. Michigan, the largest importer from Canada, shows most of its trade concentrated in transportation equipment while other states, such as Illinois and New York, import substantial amounts of intermediate goods such as crude materials, wood, paper, non-ferrous metals, and other fabricated materials. These different import schedules suggest different levels of direct and indirect effects from the FTA in each state within the US. The low level of textile imports from Canada reflects the barriers the US imposes on these imports in order to protect the American textile industries against the Canadian competitors which are intensive users of low priced foreign material.

14.3 The Interregional Computable General Equilibrium Model

Many models have been developed and implemented in the US and in Canada to evaluate the economic impact of the US-Canada Free Trade Agreement (FTA). The main difference between the computational general equilibrium model discussed here and previous models is this CGE's ability to measure the effect of the FTA at a regional level. We started with a very aggregated model and increased the level of disaggregation in two directions, the number of trading blocks in the model and the number of sectors in each block. The analysis began with the model structure developed by Roland-Holst and Tokarick (1990) for the International Trade Commission, hereafter, called the RHT model.

14.3.1 The RHT Model

The prototype RHT model is a single-country, two-factor, three-sector model specified in a flexible way permitting individual investigators to adapt it to their specific demands. The generic version of the model divides the economy into three basic sectors: agriculture, manufacturing, and services; two factors of production: labour and capital; and two institutions: households and government. The model presents twelve blocks of equations specifying the relationships of the different agents in the economy and the conditions for a *Walrasian equilibrium*. In such an equilibrium, there exists prices for all commodities and factors such that all markets clear, implying that there are no excess demands for any good or factor in the economy. These modeling blocks are as follows: Consumer Behavior, Production Technology, Factor Supplies, Factor Demands, Domestic Demand and Allocation of Traded Goods, Domestic Prices, Domestic Market Equilibrium, Income and Government Revenue, Foreign Balance, Foreign Commodity Prices, Foreign Commodity Supply and Demand Functions, and Numeraire.

The RHT model does not differ from other models in the literature in its theoretical specification; the most innovative feature of the model is the flexible way it was constructed as a *computational* model with the capability to change specifications, institutions, or any other feature in the model, such as the changes made to present an interregional perspective.

14.3.2 The Interregional CGE Model (IRCGE)

Several versions of the IRCGE model were developed (these are described in Hewings and Gazel, 1993); in the analysis to be described in this paper, the following regional specification was used - three regions inside the US, Canada, and ROW, with each of these five trading blocks divided into eight sectors.

This operational IRCGE model is a standard CGE model which is specified as follows.

The Demand Side. The total demand for goods and services is the sum of private consumption demand, government demand, investment demand, and exports. Government and investment demands are specified exogenously.

Consumer Behaviour: In the first block of equations, consumer behavior is specified as one in which the private demand for goods is divided into those goods produced at home and imported goods. Consumer demand for any good is a function of disposable income in the economy and the prices (domestic and foreign) of goods. These demand functions are the result of maximization behavior of consumers in each region. A typical consumer maximizes a Stone-Geary (xxx) type utility function subject to his/her budget constraint. In this formulation, a minimum consumption of each good (avoiding the zero consumption cases) is specified. An individual's utility is assumed to be a function of consumption in excess of these minimum consumption levels. The resulting demand functions allow for income, own and cross price elasticities that depart from zero, implying the consumption of one good is not independent of consumption of other goods. Consumers are assumed to respond to changes in relatives prices of goods and to changes in their own income levels.

Production Side. The second block contains the specifications of the production side of the economy resulting in the supply of domestic goods as well as the demands for capital, labour, and intermediate goods.

A *CET* (constant elasticity of transformation) function was used to specify the production of each good. This function allows substitution among a composite primary input (a mix of capital and labour) and other intermediate goods. The demand for capital and labour is a function of the price of capital (rental rate) and price of labour (wage rate) as well as of the elasticity of substitution between the two inputs (exogenous parameter). The other inputs entered the production function as fixed coefficients in a classical Leontief-type formulation. The data for this block of equations came from previously built input-output matrices.

The Allocation of Traded Goods. The domestic consumer can choose between domestically produced goods and imported goods. The goods are differentiated by source of production. In this way, extreme specialization is avoided. The demand for goods is specified in two steps. First, the consumer maximizes utility, constrained by budget as the choice is made between domestic and imported goods. In the second step, the consumer minimizes expenditure on imported goods by choosing from among the imported goods produced by different sources. An Armington-type of constant elasticity of substitution was used (see Armington, 1965). The calibration of the model using the data base gives the values of the share and shift parameters for these equations, while the first-order conditions for the maximization of utility allow for the efficient allocation of income by the consumer. The elasticities of substitution between domestic and imported goods and between imported goods from different sources are exogenously specified.

Home producers allocate their production between domestic and foreign markets in the same way, with prices acting as a strong signaling mechanism. They maximize revenue by allocating their production between domestic sales and exports and in the second step they allocate exports between the different regions. Again the calibration of the model results in the values of the structural parameters and the first order conditions allow for the efficient allocation of production. Trade is balanced between regions and countries through foreign savings which is equal to the surplus or deficit in the trade balance.

Prices, Income and the Government. The third block specifies domestic and foreign prices. For perfectly competitive sectors, price is equal to marginal cost, which is a function of quantities and prices of the respective inputs. In the case of other market structures, prices are related to the size and number of firms in each market. For the time being, all sectors have been modeled under perfect competition. The domestic prices for imported goods are subject to import tariffs. As was stated earlier, the tariffs faced by the ROW in the US and Canada are higher than those faced by the producers in their respective countries. There are no tariffs between regions within the same country although transportation costs allow different prices for the same good in different regions. Another important component of the domestic price of foreign goods is the exchange rate of the domestic country. Price is the mechanism of clearing the markets and achieving the general equilibrium for the final solution of the model.

The price of composite goods (imported good, sale price of a given good, etc.) is specified in accordance with the maximization or minimization problem for consumers and producers. Therefore, the price of an imported good is a function of both the prices of different goods of the same type produced by different sources and their respective shares of the market.

Total income is the sum of labour and capital income, and the model is solved by maximizing the total income in the economy, equal here to the gross domestic product. Government expenditure is equal to the government revenue plus or minus government savings, allowing for a balanced government budget. The government revenues come from direct and indirect taxes, tariff revenues, and subsidies.

Market Equilibrium. The fourth block of equations specifies the equilibrium for each market previously specified (home market for goods, international markets, labour and capital markets, etc.). For each market, demand must be equal to supply. For the foreign balance, a surplus or deficit in the trade balance must be equal to positive or negative foreign savings in the home country. In the regional model, a price index is chosen as a numeraire.

This highly nonlinear model was calibrated to a set of estimated data (derived from primary and secondary sources) and an optimal solution was achieved, which indicated that the model was well specified and that it was internally consistent. After the optimal solution was found, an experiment featuring free trade was conducted in order to observe the changes in the variables under the new prices. The design of the computational model allows for a comprehensive, comparative static analysis.

14.3.3 Data

A computational general equilibrium model does not require as much data as more detailed econometric models if the elasticities needed in the model are available in the literature. For the present model, the basic data requirement is a social accounting matrix for each region or country, data on employment, regional prices, and some elasticity values.

Some data was collected previously for the ten US regions - Iowa, Illinois, Ohio, Indiana, Michigan, Wisconsin, Pennsylvania, New York, New England region, and the rest of the US (RUS) - on value of output, inputs, value added, employment, etc. The data were aggregated to eight sectors - agriculture, construction, nondurable manufacturing, durable manufacturing, trade, transportation, finance and real state, and services. For those sectors for which the data were not available, an estimation procedure was used, essentially a modification of the Kendrick and Jaycox (1965) technique.

The interregional input-output tables embedded within the model were estimated using a technique developed by Hulu and Hewings (1993) that itself was based on some prior methodology introduced by Round (1978, 1979, 1983, 1988). More details are provided in Hewings and Gazel (1993).

Regional prices were another source of problems due to a lack of data for many of these variables. For the prototype model, specific units were chosen for quantities in order to normalize prices to one. This becomes a problem because the model has more than one trading block, and data have been collected to correct this problem. Differences between wholesale and retail prices would serve as a base for the regional prices differentials.

14.4 Feedbacks and Interregional Linkages[2]

One of the advantages of the CGE modeling framework is the manner in which it captures so many of the operations of sets of markets and their interactions; however, this detail is obtained at some price, namely the apparent loss of model tractability. Feedback loop analysis was conducted to focus attention on the nature of the interregional linkages and associated feedback effects to assist in interpreting the findings from the sensitivity analyses.

14.4.1 Closed Feedback Loops

Consider a system of *n* region and *m* sectors; assume that the transactions between the regions can be presented by an *nm x nm* block matrix *X*, of transactions flows:

$$X = \begin{bmatrix} X_{11} & X_{12} & \cdots & X_{1n} \\ X_{21} & X_{22} & \cdots & X_{2n} \\ \vdots & \vdots & \cdots & \vdots \\ X_{n1} & X_{n2} & \cdots & X_{nn} \end{bmatrix} \qquad (14.1)$$

where each block

$$X_{rs} = \| x_{ij}^{rs} \| \qquad (14.2)$$

represents the transactions flows from sectors in region *r* to those in region *s*. Define:

$$t_{rs} = \sum_{ij} x_{ij}^{rs} \qquad (14.3)$$

namely, the sum of flows between all sectors within each submatrix, X_{rs}. Hence, an aggregated *nxn* matrix may be defined as:

$$T = \| t_{rs} \| \qquad (14.4)$$

The major focus of this section will be the identification of multi-regional feedback loops that reveal the economic self-influence of each region. The economic interpretation of a feedback loop is straightforward. It indicates how strongly (at each hierarchical level) each region is tied to all other regions included in that loop. By focusing on complete loops, one can evaluate the place and position of each and every region vis à vis *all* other regions.

---

[2]This section draws, in part, from Sonis, Hewings and Gazel (1993)

With a three-region system:

$$X = \begin{bmatrix} X_{11} & X_{12} & X_{13} \\ X_{21} & X_{22} & X_{23} \\ X_{31} & X_{32} & X_{33} \end{bmatrix}; \quad T = \begin{bmatrix} t_{11} & t_{12} & t_{13} \\ t_{21} & t_{22} & t_{23} \\ t_{31} & t_{32} & t_{33} \end{bmatrix} \tag{14.5}$$

there are six feedback loops:

$$T_1 = \begin{bmatrix} t_{11} & 0 & 0 \\ 0 & t_{22} & 0 \\ 0 & 0 & t_{33} \end{bmatrix} \quad T_2 = \begin{bmatrix} 0 & 0 & t_{13} \\ t_{21} & 0 & 0 \\ 0 & t_{32} & 0 \end{bmatrix} \quad T_3 = \begin{bmatrix} 0 & t_{12} & 0 \\ 0 & 0 & t_{23} \\ t_{31} & 0 & 0 \end{bmatrix}$$

$$T_4 = \begin{bmatrix} t_{11} & 0 & 0 \\ 0 & 0 & t_{13} \\ 0 & t_{32} & 0 \end{bmatrix} \quad T_5 = \begin{bmatrix} 0 & t_{12} & 0 \\ t_{21} & 0 & 0 \\ 0 & 0 & t_{33} \end{bmatrix} \quad T_6 = \begin{bmatrix} 0 & 0 & t_{13} \\ 0 & t_{22} & 0 \\ t_{31} & 0 & 0 \end{bmatrix} \tag{14.6}$$

It is also possible to combine the interregional and intersectoral interdependencies in which the aggregated table is replaced by a detailed table describing the interplay between the intersectoral and interregional interdependencies. The procedure operates as successive levels on the system, but the approach at each stage is similar. This 'top-down' decomposition may be considered analogously to the exfoliation process in the removal of layers of an onion or, more appropriately, to the construction of 'Matrioshka' dolls in which successively smaller dolls of exactly the same shape and style are nested within the larger dolls. Hence, the *Matrioshka* approach examines the intra- and inter-regional transactions in terms of the hierarchical structure of feedback effects (see Sonis, Hewings and Gazel, 1993 for a full exposition and illustration of this technique).

14.4.2 Regional Feedback Loops for the US Four-Region Aggregation

The input-output tables that formed part of the IRCGE provided the basic data. Eight sectors were originally specified for the ten-region system; in the presentation in this section, the regions were reduced to four in number (with Canada as the fifth region) and the sectors aggregated to three. These are shown in Table 14.2.

The choice of the particular regional groupings of states does not reflect any *a priori* theories about the nature of variations in regional structure; the original ten-region scheme was designed to focus on those states in the Midwest and the Northeast that were the major trading partners with Canada. The sectoral aggregation was used to facilitate the exposition of the methodology and thus provide the basis for a simple cartographic exposition.

Table 14.2. Regional and sector definitions

	Regions		Sectors
1.	Illinois, Iowa, Indiana, Wisconsin	1.	Resources [Agriculture, Forestry, Fishing, Mining]
2.	New England, Pennsylvania, New York	2.	Construction and Manufacturing
3.	Ohio, Michigan	3.	All other activities
4.	Rest of the US		

Table 14.3 presents the *4x4* matrix of aggregate intermediate transactions for the four regions in 1982. The number of complete feedback loops for all these regions would be equal to 4! or 24. Thus, the feedback loop *hierarchical structure* will include only four feedback loops at the regional level.

Table 14.3. Aggregate intermediate transactions for four regions, 1982 [$1982 billion]

FROM/TO	Region 1	Region 2	Region 3	Region 4
Region 1	242.8	6.8	7.8	39.2
Region 2	4.6	145.4	9.4	50.9
Region 3	7.4	13.9	318.8	72.5
Region 4	29.2	41.2	50.7	1605.2

Source: Hewings and Gazel (1993)

Inspection of Table 14.3 reveals that the 'domestic' (i.e., intra-regional) transactions dominate the flow patterns. Hence, the choice of the feedback loop with the largest 'weight' in the system will be a diagonal quasi-permutation matrix:

$$T_1 = \begin{bmatrix} 242.8 & 0 & 0 & 0 \\ 0 & 145.4 & 0 & 0 \\ 0 & 0 & 318.8 & 0 \\ 0 & 0 & 0 & 1605.2 \end{bmatrix}$$

This matrix corresponds to the following permutation, one that focuses entirely on the intra-regional circulations:

$$\begin{array}{cccc} r_1 & r_2 & r_3 & r_4 \\ \downarrow & \downarrow & \downarrow & \downarrow \\ r_1 & r_2 & r_3 & r_4 \end{array}$$

or

$$p_1 = (\tau_1,\tau_2)(\tau_3,\tau_4)$$

The flow intensity of this complete feedback loop is equal to V_l = \$2312.2*b* and accounts for 87.4% of the \$2645.8*b* total intermediate transactions in the US. The remaining percentage of the total transactions, 12.6%, are the inter-regional flows. The inter-regional transactions can be shown to be decomposable into a set of feedback loops. The first will be:

$$T_2 = \begin{bmatrix} 0 & 6.8 & 0 & 0 \\ 4.6 & 0 & 0 & 0 \\ 0 & 0 & 0 & 72.5 \\ 0 & 0 & 50.7 & 0 \end{bmatrix}$$

which corresponds to the permutation:

$$\begin{array}{cccc} r_1 & r_2 & r_3 & r_4 \\ \downarrow & \downarrow & \downarrow & \downarrow \\ r_2 & r_1 & r_4 & r_3 \end{array}$$

or

$$p_2 = (r_1,r_2)(r_3,r_4)$$

This, the most important of the interregional feedback loops, accounts for 40.4% of the inter regional transactions. This feedback loop can be broken down into two independent sub-loops; the sub-loop represents the most intensive pair-wise exchange between region 3 (Ohio, Michigan) and region 4 (Rest of the US). It accounts for 36.9% of the 40.4% that is represented in the complete loop. The remaining 3.5% is reflected in flows between region 1 (Central) and region 2 (New York, Pennsylvania and New England).

The structure of the next two largest inter-regional feedback loops may also be decomposed into two pair-wise sub-loops:

$$p_3 = (r_1,r_3)(r_2,r_4)$$
$$p_4 = (r_1,r_4)(r_2,r_3)$$

and this corresponds to the following matrices:

$$T_3 = \begin{bmatrix} 0 & 0 & 7.8 & 0 \\ 0 & 0 & 0 & 50.9 \\ 7.4 & 0 & 0 & 0 \\ 0 & 41.2 & 0 & 0 \end{bmatrix}; \quad T_4 = \begin{bmatrix} 0 & 0 & 0 & 39.2 \\ 0 & 0 & 9.4 & 0 \\ 0 & 13.9 & 0 & 0 \\ 29.2 & 0 & 0 & 0 \end{bmatrix}$$

The complete scheme of feedback loops is presented in Figure 14.1. Thus, the major self-influence forces connected with the domestic transactions are very heavily concentrated within the region of origin/destination. The interregional transactions reveal a set of pair-wise interactions reflecting an hierarchy of bilateral trade flows in which region 4 assumes a dominant position.

14.4.3 The Feedback Loop System: US and Canada

Figure 14.2 shows the feedback loops when Canada is added to the four-region US system. With five regions, the number of complete feedback loops would be equal to 5!= 120; however, the feedback loop hierarchical structure includes only 5 feedback loops. The results confirm the essence of the data presented in Table 1, namely that Canada's strongest trading relationships are with the first three regions. Together, these account for 86.9% of the interregional transactions. Thus, Canada's trade with the rest of the US involves a filtering process through the dominant northern tier of regions. These findings will be reinforced in the examination and interpretation of the full IRCGE model.

However, a question might be raised at this juncture about the reliability of these estimates, especially given the nature of the modelling exercise and the liberal use of nonsurvey data and estimation methods. Without a benchmark, it is hard to make a direct comparison; instead, limited sensitivity analysis was conducted on some of the parameter estimates and the interregional trade flows to ascertain the degree to which the model results are a function of the data inputs as opposed to the form of the model itself.

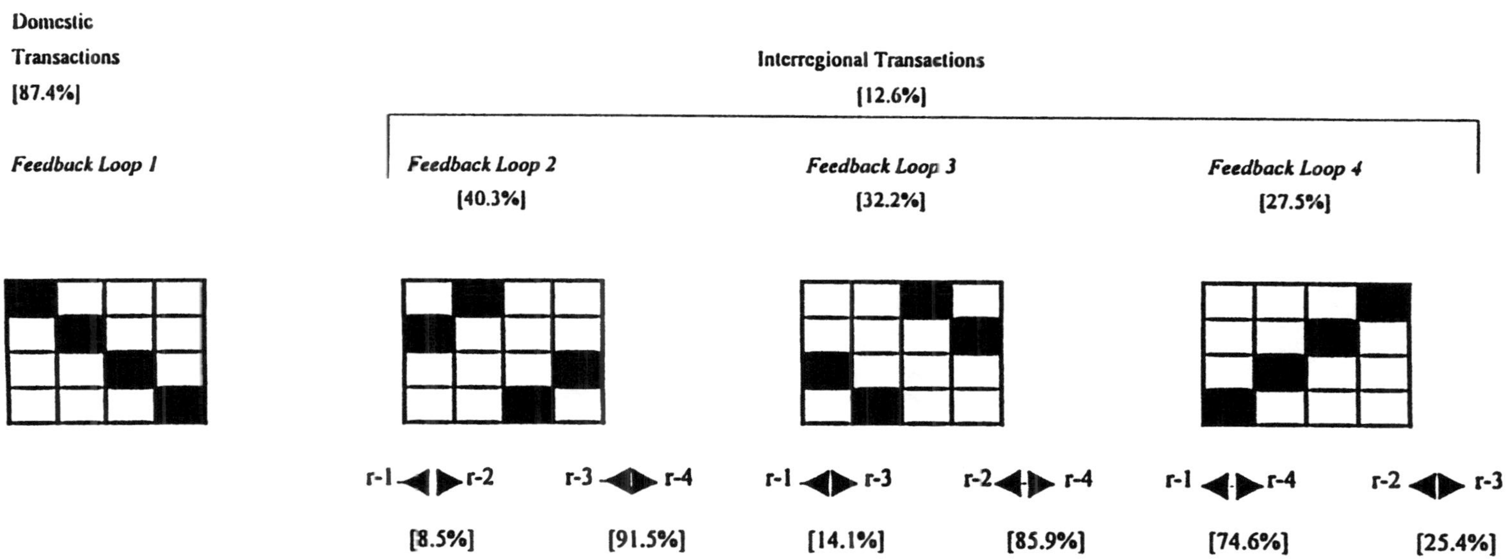

Fig. 14.1. Decomposition of transactions flows into feedback loops: one-sector tables

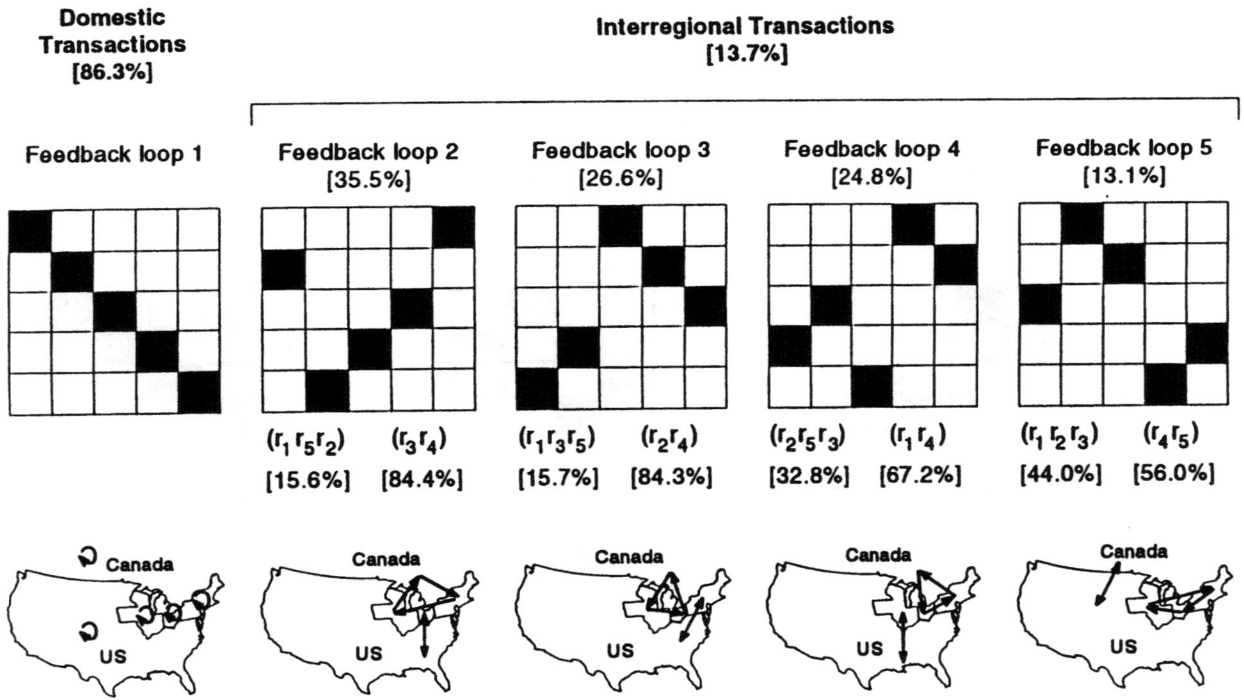

Fig. 14.2. Decomposition of transactions flows into feedback loops: one-sector tables for US and Canada

14.5 Sensitivity Analysis

The analysis in which all tariffs between the US and Canada were eliminated will be referred to as the *Base Experiment*. In this section, the four-region division of the US described in Section 4 will be used, with Canada occupying the fifth and the Rest of the World the sixth region[3]. Several other simulations were performed to test the sensitivity of the model for changes in the benchmark year trade data and some elasticity parameters. These experiments are described below:

Experiment I. The interregional trade flows were modified in the benchmark year data base. Imports among regions within the US were increased by 10 % with equal value added to export of the same commodity in order to keep margins unchanged. In this case the only changes were the increased imports from other regions within the US, increased exports to the same regions and reduced demand for home produced goods in each of those regions.

Experiment II. All imports were increased by 10% in the benchmark year data base. The difference between this and the previous experiment is the increased imports from Canada and the ROW as well as increased imports in Canada. Due to the existence of *ad valorem* tariffs, government revenues increased in all regions. In order not to change government demand for goods and services, the change in revenues were used to decrease government budget deficits instead of increasing spending. As a result of this healthy governmental behavior, more savings were available for investment, resulting in a higher volume of domestic output. All these were small changes due to low tariff rates. Exports demands were higher in order to compensate for lower domestic demand for domestically produced goods.

Experiment III. This experiment uses the same data base as the Base Experiment with 50% higher elasticities of substitution for demand for goods (domestically produced or imported) as well as for allocation of domestically produced goods (domestic sales and exports).

Experiment IV. This experiment is the same as Experiment I with 50% higher elasticities of substitution (same as above). Two changes are associated here, higher import demands and import elasticities.

Experiment V. Same as Experiment III with 100% higher elasticities.

Experiment VI. Same as Experiment IV with 100% higher elasticities.

14.5.1 The Base Experiment

Although exports from region 1 represents almost 15% of the total American exports to Canada, this region captures less than 6% of the total income gains from the FTA (see Tables 14.4 and 14.5). Regions 2 and 3 are the largest relative

[3]The US regions are A [Michigan and Ohio]; B [New England, Pennsylvania and New York]; C [Illinois, Iowa, Indiana and Wisconsin] and D [Rest of the US]

gainers in terms of income compared to their relative size as exporters to Canada. The rest of the country (Region 4) exports more than it gains in percentage terms from the FTA. The discrepancy between share of exports and share of income gains is not surprising given that gains or losses are due to variables other than the direct link through exports between each US region and Canada. These findings reinforce some of the interpretations found from the evaluation of the feedback loop analysis. The concentration of impacts in regions 2 and 3 reflects the high degree of intermediation in industrial production that serves not only markets in those regions but also in the remaining regions and Canada. Nevertheless the low income gains region 1 faces results in negative gains in terms of real GNP due to losses in tariff revenues (assumed to be collected by the region itself).

Table 14.4. Share of exports to Canada and share of income gains by region

Region	Exports	Income
1	14.48	5.65
2	27.35	32.28
3	17.06	25.35
4	41.12	36.71
US	100.0	100.0

Table 14.5. Base Case: Percentage change in major variables

	Region 1	Region 2	Region 3	Region 4	Canada
Capital Inc.	-0.006	0.021	0.039	0.029	0.536
Labour Inc.	0.055	0.374	0.107	0.034	1.040
Total Inc.	0.031	0.250	0.080	0.032	0.841
GNP	-0.012	0.003	0.001	0.005	0.048

Tables 14.6 and 14.7 reveal that the different experiments resulted in different economic effects (measured as change in total income) for the regions in the model. As expected, the economic impacts in all regions of all experiments are very small due to the low level of protectionism between the US and Canada. None of the experiments shows any total income gains greater than one percent. Canada is the largest gainer in all experiments and, within the US, regions 2 and 3 are the largest relative gainers with income increasing around one fourth of one percent in all experiments for 2 and almost one tenth of one percent for 3.

Table 14.6. Perecentage changes in total income by experiment

	Region 1	Region 2	Region 3	Region 4	Canada
BASE	0.0315	0.2496	0.0805	0.0318	0.8410
I	0.0311	0.2499	0.0805	0.0318	0.8410
II	0.0330	0.2751	0.0885	0.0340	0.9254
III	0.0661	0.2538	0.0818	0.0323	0.8570
IV	0.0652	0.2540	0.0818	0.0323	0.8570
V	0.0727	0.2579	0.0832	0.0328	0.8731
VI	0.1005	0.2581	0.0832	0.0328	0.8731

The important point to note is that the changes are very small even when the experiments' values are compared to the base case, except for region 1 (see Table 14.7). For changes in the interregional trade flows in the benchmark year (increased imports within the US), the effects were so small for regions 3, 4, and Canada that they did not differ from zero. Nevertheless, they did matter for Regions 1 and 2 which are relatively more dependent on exports than the other regions. Changes in the value of the elasticities of substitution between home produced and imported goods and between imported goods from different sources have a much stronger impact than changes in the interregional trade flows.

Table 14.7. Percentage change from the base case

	Region 1	Region 2	Region 3	Region 4	Canada
I	-1.270	0.120	0.000	0.000	0.000
II	4.816	10.199	9.946	6.970	10.031
III	109.977	1.685	1.676	1.557	1.899
IV	107.088	1.776	1.666	1.545	1.899
V	130.889	3.339	3.391	3.193	3.821
VI	219.046	3.417	3.370	3.179	3.821

For the presentation of the remaining results from the experiments, the only tabular information provided will be the comparison with the base case; in all cases, the percentage change from the FTA was very small.

14.5.2 Experiment I: Increase in Interregional Imports

As with the Base Case, this experiment results in very small economic effects due to the FTA. Labour gains are positive in all regions, while capital experiences a loss in region 1. Total income increases in all regions and real GNP increases as well, except in region 1 where the result can be explained by the losses in tariff revenues that are not totally compensated by gains in income.

When compared to the Base Case, the results do not differ substantially in magnitude for total income; however, the changes are very important when income is divided between labour and capital. Table 8 shows how changes differ in both cases in percentage terms. As an example, the change in capital income is 26.5 per cent higher under the second experiment than under the Base case, i.e., if capital increased 10% in the first case, it would increase 12.65% in the second case.

Table 14.8. Percentage change from base case

	Region 1	Region 2	Region 3	Region 4	Canada
Capital Inc.	26.488	-78.475	0.660	1.050	-0.006
Labour Inc.	0.709	2.504	-0.154	-0.511	0.003
Total Inc.	-1.181	0.101	0.034	0.090	-0.001
GNP	1.783	-6.873	1.234	1.020	0.005

In region 1, capital gains much more than labour when the interregional trade flows are larger; the opposite occurs in region 2. Changes in the other regions are too small to be taken into consideration.

14.5.3 Experiment II: Import Increase from All Sources and in All US Regions, Canada and the ROW

The results in this experiment parallel those of the others to date, namely, that the percentage changes record small values but some of the changes in comparison to the Base Case reveal rather larger movements, in this case both negatively and positively. Capital income seems to be the major variable of interest with large percentage changes in region 2 and a correspondingly large positive change in region 4. On the other hand, labour income in this later region experiences a strong negative decline. In all regions, the impact on total income and GNP is positive.

Table 14.9. Percentage change from base case

	Region 1	Region 2	Region 3	Region 4	Canada
Capital Inc.	60.859	-106.007	11.630	255.701	8.549
Labour Inc.	8.823	13.715	9.597	-126.935	10.527
Total Inc.	4.910	10.178	9.983	7.066	10.030
GNP	20.469	10.624	11.288	50.607	9.925

14.5.4 Experiment III: Base Case with 50% Higher Elasticities

One of the major problems in the development of CGE models centers around the estimation of some of the elasticities; in this experiment, these elasticities were changed. However, recall that the changes in the Base Case are small to begin with, so even modest changes generated by this experiment may yield rather large percentage changes. This is certainly the case in region 1 (see Table 14.10) where capital income decreases significantly; while a similar result occurs in all regions, the impact on GNP is only negative in region 1.

Table 14.10. Percentage change from base case

	Region 1	Region 2	Region 3	Region 4	Canada
Capital Inc.	-1007.623	-207.442	-53.888	-14.784	-25.205
Labour Inc.	31.993	8.031	14.723	10.505	11.006
Total Inc.	110.167	1.666	1.710	1.649	1.898
GNP	-266.780	79.984	62.838	15.520	55.836

14.5.5 Experiment IV: Same as III with Increased Interregional Imports Within the US

This experiment attempted to examine the sensitivity of the results to the interregional flow estimates; the way the experiment was conducted did not provide a full analysis (in this case, the Field of Influence concept that Sonis and Hewings, 1992 have introduced would be more appropriate) but it did provide some measure of the impacts. Table 14.11 reveals that with the exception of capital in region 2, the changes were modest, suggesting that the elasticities may be more important parameters than the interregional flows. However, a full evaluation of the flows has not been made.

Table 14.11. Percentage change from Experiment III

	Region 1	Region 2	Region 3	Region 4	Canada
Capital Inc.	-1.049	61.394	4.442	1.563	0.000
Labour Inc.	-1.533	1.945	-0.428	-0.667	0.000
Total Inc.	-1.376	0.089	-0.010	-0.012	0.000
GNP	-4.222	-2.484	0.543	1.625	-0.001

14.5.6 Experiment V: Same as Base with 100% Higher Elasticities

To capture the nonlinearities inherent in the model, the elasticities were increased by 100% and comparison made to experiment 3 in which the change was half of that amount. Note (Table 14.12) that the changes are not monotonically related to the earlier changes; there exists here some significant opportunities to explore just exactly which elasticities seem to generate the most important perturbations in the results.

Table 14.12. Percentage change from Experiment III

	Region 1	Region 2	Region 3	Region 4	Canada
Capital Inc.	32.358	162.327	-111.409	-13.317	-32.672
Labour Inc.	-0.813	6.492	12.326	7.815	9.712
Total Inc.	9.959	1.626	1.687	1.611	1.886
GNP	32.915	48.293	43.611	13.983	36.537

14.5.7 Experiment VI: Same as V with Increased Interregional Imports

In the final experiment, the impact of increased interregional imports seems to negatively impact region 2 (in terms of GNP) while having a differential impact on factor incomes in the other regions (Table 14.13).

Table 14.13. Percentage change from Experiment V

	Region 1	Region 2	Region 3	Region 4	Canada
Capital Inc.	-40.726	18.357	-74.032	2.187	0.000
Labour Inc.	88.819	1.439	-0.727	-0.750	0.000
Total Inc.	38.182	0.076	-0.020	-0.014	0.000
GNP	78.286	-1.036	0.796	1.769	-0.001

14.5.8 Evaluation

The sensitivity analysis provides two general assessments; in comparison to the Base Case, the overall results appear reasonably stable. However, when analysis is focused on specific variables and comparison is made in percentage terms, the results appear to be rather sensitive to some of the parameter estimates (especially the elasticities). In cases in which the initial changes might be larger (for example, with the adoption of the North American Free Trade Agreement to add Mexico to the US-Canada FTA), then a careful search of the parameter space and a detailed sensitivity analysis would be warranted. Unfortunately, the density functions for many of the parameters are not known and, as yet, techniques have not been developed to provide a convenient way to assess synergetic changes in parameter estimates (as in the case for the field of influence concept).

14.6 Conclusions

This paper has attempted to report and evaluate the development and application of an interregional computable general equilibrium model to assess the state-level impacts of national free trade agreements. Two features of the work, interpretation of interregional flows via feedback loop analysis and some limited sensitivity experiments, provide the basis for the development of procedures that will enable analysts to 'get inside' CGE models in an attempt to make them more transparent. However, it is clear that some important methodological developments will be required to accomplish this task in an efficient manner. The 'mixed' nature of CGE models (with some econometric estimates, some input-output formalism and a great deal of 'best guesses') creates difficulties for the applications of standard methods for evaluation of parameters and data. In this regard, some of the recent advances in network analysis may yield important insights and opportunities.

From a policy perspective, the analysis suggests that the dimensions of change observed at the national level, when very small, are likely to be repeated at the regional level. However, the sensitivity analysis suggests that there is a significant possibility for some of the sub-national results to depart from those observed nationally, especially if the initial changes are larger. The comments relating to sensitivity analysis and policy interpretation both point to a need for more creativity in the evaluation of CGE models and in the presentation and interpretation of the results.

Acknowledgements

The support provided by the Economic Development Administration, US Department of Commerce and the Canadian Embassy is gratefully acknowledged though neither institution is responsible for the views expressed herein. Professor David Roland-Holst has been particularly generous in providing access to the original US model, on which the interregional version has been crafted. The comments of Roland-Holst, Earl Grinols, John Fieser, Harry Campbell and the participants of the Mallocoota seminar are very much appreciated.

References

Armington PS (1965) A Theory of Demand for Products Distinguished by Place of Production. International Monetary Fund Staff Papers

Brown DK, Stern RM (1988) Computational General Equilibrium Estimates on the Gains from US - Canadian Trade Liberalization. Unpublished Manuscript, University of Michigan

Hamilton R, Whalley J (1985) Geographically Discriminatory Trade Arrangements. Review of Economics and Statistics 67:446-455

Hewings GJD, Gazel R (1993) Regional and Interregional Impacts of the US-Canada Free Trade Agreement. Working Paper 93-P-1. Regional Economics Applications Laboratory, University of Illinois, Urbana

Hulu E, Hewings GJD (1991) The Development and Use of Interregional Input-Output Models Under Conditions of Limited Information. Working Paper 93-P-2, Regional Economics Applications Laboratory, Urbana, Illinois

Kendrick JW, Jaycox MC (1965) The Concept and Estimation of Gross State Product. Southern Economic Journal 32:153-168

Roland-Holst DW, Tokarick S (1990) General Equilibrium Modeling for Trade Policy. An Overview. Unpublished Manuscript. Mills College, California

Round JI (1978) An Interregional Input-Output Approach to the Evaluation of Nonsurvey Methods. Journal of Regional Science 18:179-194

Round JI (1979) Compensating Feedback Effects in Interregional Input-Output Models. Journal of Regional Science 19:145-155

Round JI (1983) Nonsurvey Techniques. A Critical Review of the Theory and the Evidence. International Regional Science Review 8:189-212

Round JI (1988) Incorporating the International, Regional and Spatial Dimension into a SAM. Some Methods and Applications. In: Harrigan FJ, McGregor PG (Eds) Recent Advances in Regional Economic Modeling. London, Pion

Sonis M, Hewings GJD (1992) Coefficient Change in Input-Output Models. Theory and Application. Economic Systems Research, 4:143-57

Sonis M, Hewings GJD, Gazel R (1993) An Examination of Multi-Regional Structure. Hierarchy, Feedbacks and Spatial Linkages. Working Paper 93-T-11. Regional Economics Applications Labouratory, University of Illinois, Urbana

15 The Impact of the Conservation Reserve Program on the US Economy: A General Equilibrium Analysis

Noel D. Uri
U.S. Department of Agriculture

Roy Boyd
Ohio University

15.1 Introduction

While it is often expedient to assess the impact of an agricultural program on just the agricultural sectors of the economy, any inferences drawn run the risk of being inaccurate because of the interrelationships between these sectors and the rest of the US economy. For example, the interactions between supply and demand, both within the markets for agricultural commodities as well as between these markets and the rest of the economy, are quite significant (Harrington, et al. 1986). Given these interrelationships, the use of a computable general equilibrium approach to modelling the effects of an agricultural program is a logical decision. In what follows, the aggregate impacts of the Conservation Reserve Program will be assessed using such an approach.

The Conservation Reserve Program (CRP) is a voluntary cropland retirement program established in the Conservation Title (XII) of the Food Security Act of 1985 and continued in the Food, Agriculture, Conservation, and Trade Act of 1990 (Title XIV). In exchange for placing cropland that is highly erodible into the CRP for ten years, the federal government through the US Department of Agriculture pays participating farmers an annual per acre rent plus one half of the cost associated with establishing conservation practices and a permanent land cover.

Given the design of the CRP program, the expected impact of the original program as detailed in the Food Security Act of 1985 as well as its extension as detailed in the Food, Agriculture, Conservation, and Trade Act of 1990 can be determined. Thus, simple analytics suggest that taking cropland out of production will result in higher agricultural commodity prices. This, in turn, might or might not lead to an increase in net farm income. Whether net farm income will change is a function of the elasticity of demand for the various agricultural commodities together with the changes in the absolute and relative prices of the factor inputs and the quantities used as well as the changes in the absolute and relative prices and mix of outputs.

Most of the empirical efforts[1] that evaluate the impact of the CRP have been rather ad hoc, focusing just on the direct effects of removing cropland from production while ignoring any substitution effects of, e.g. program crops for non-program crops,[2] secondary price effects on both inputs and output, resource use, and import and export response. Given the various interrelationships, however, the use of a general equilibrium approach in conjunction with an agricultural sector specific model to assessing the impact of the CRP would seem to be a more reasonable choice.[3]

The use of a general equilibrium model to assess the impact of government programs on the economy in general and agriculture in particular is not unique to this study. Earlier efforts in this direction include those by Adelman and Robinson (1986) and Dervis, et al. (1982). While it is not the purpose to critique these and other earlier efforts, it should be noted that each one is subject to a variety of shortcomings. One of these limitations is that agriculture is only considered as a component of some larger sector. Thus, for example, Dervis, et al. have only a single agriculture, forestry and fisheries producing sector. Another limitation is that factor substitution (either in the production or consumption of goods and services) is not adequately dealt with. Adelman and Robinson, for example, use only a standard input-output model devoid of any factor substitution.[4]

In the spirit of these earlier general equilibrium efforts, the model developed here attempts to capture the interactions between the CRP and economic activity in general while at the same time endeavoring to overcome some of their limitations. In doing so, the linkages between sectors of the economy are explicitly taken into account and the price responsiveness of producers and consumers both to absolute and relative changes in the prices of the various goods and services is considered. Further, the computable general equilibrium model is linked to an agricultural sector model in order to provide an assessment of sector specific impacts that otherwise would be obfuscated. Thus, the analytical approach used will be a computable general equilibrium model that has been disaggregated into fourteen producing sectors, fourteen consuming sector, six household (income) categories and the government. This level of disaggregation allows for an assessment of the direct effects as well as the indirect effects of the CRP. By measuring these effects, it will be possible to identify the extent to which the agricultural sectors and the other producing and consuming sectors and household groups stand to gain or lose. Hence, equity considerations as well as efficiency considerations can be

---

[1]There are a large number of such efforts including Huang, et al (1990); Ogg, et al (1989); Osborn and Konyar (1990); Reichelderfer and Boggess (1988); Schaible (1989); Shoemaker (1989); Taff (1990) and Young and Osborn (1990).

[2]Note that there is no implicit suggestion in this discussion that the CRP is targeted at cropland on which program crops (or non-program crops, for that matter) are produced.

[3]General equilibrium models in general are not going to be reviewed here. Rather, the interested reader is referred to, e.g. Adelman and Robinson (1978); Ballard (1985), et al; Harberger (1962, 1974) and Shoven and Whalley (1992).

[4]Hertel and Tsigas (1987) details additional limitations.

addressed. Linked to the computable general equilibrium model is a spatial equilibrium model that measures the direct agricultural production impacts. The spatial equilibrium model of agricultural production is of the conventional sort and hence is not discussed in detail. The computable general equilibrium model, however, is somewhat different. Consequently, before conducting the analysis, a brief overview of the computable general equilibrium model is provided.

15.2 A General Equilibrium Model

15.2.1 Introduction

The model used follows in the tradition of the Shoven and Whalley (1972) tax analysis research and incorporates some of the methodological enhancements of the general equilibrium work of Hudson and Jorgenson (1974a, 1974b). For example, it recognizes the differences in preferences of consumers as a function of their incomes and specifies a distinct demand system for each group of households. Additionally, a neoclassical microeconomic model of producer behaviour is employed. The model of consumer behaviour is integrated with the model of producer behaviour (which contains a price-responsive input-output component) to provide a comprehensive framework for policy simulations.

15.2.2 The Producing Sectors

The production sector of the model consists of an input-output matrix with some flexibility with regard to the substitution of the factor inputs (capital, labour, and land). Technologies are represented by production functions that exhibit constant elasticities of substitution. Technological progress (both embodied and disembodied (Uri, 1984)) is assumed not to occur over the period of investigation. Each sector as defined in Table 15.1 is assumed to have a constant elasticity of substitution (CES) production function (Arrow, et al. 1961) where the value added by the specific sector is a function of labour and capital.[5]

For four sectors (the three agricultural sectors and the forestry sector), however, a third factor of production - land - is included. This is done because of the special importance of this input to these sectors. (See, e.g. Heady and Dillon (1961) for a discussion of this issue.) The incorporation into the production function of this factor is accomplished by nesting the CES production function. In particular, an input is defined which is solely a function (in CES form) of land and capital

[5]There is a transformation matrix whereby raw inputs in the producing sectors are transformed into consumption goods and services. Thus, the fact that agricultural goods are combined with, say, manufacturing goods, food processing, and transportation to produce food for consumption is accounted for.

which, in turn, takes the place of capital in the original production function specification.[6,7]

Table 15.1. Classification of producing sectors and consumer goods and services

Industries	Consumer Goods
1. Manufacturing	1. Food
2. Coal Mining	2. Alcohol and Tobacco
3. Other Mining	3. Utilities
4. Service	4. Furnishings and Appliances
5. Chemicals and Plastics	5. Housing
6. Food and Tobacco Products	6. Clothing and Jewelry
7. Petroleum Refining	7. Transportation
8. Financial	8. Financial Services
9. Forestry	9. Other Services
10. Wood Products	10. Motor Vehicles
11. Crude Oil and Natural Gas	11. Gasoline and Other Fuels
12. Agriculture 1 - Program Crops	12. Reading and Recreation
13. Agriculture 2 - Livestock	13. Nondurable Household Items
14. Agriculture 3 - All Other Agriculture	14. Savings

Table 15.2. Household categories based on income

Category	Income Range
I	$ 0 - 9,999
II	$ 10,000 - 14,999
III	$ 15,000 - 19,999
IV	$ 20,000 - 29,999
V	$ 30,000 - 39,999
VI	$ 40,000 and over

15.2.3 The Consuming Sectors

On the demand side, the model captures the behaviour of consumers (who can also serve as investors), the government, and foreigners. Consumers are grouped

[6]While it would be possible to simply add land as an explicit input in the production function, this would implicitly assume that the elasticity of substitution between all pairs of inputs are the same. By nesting, however, the substitution elasticities are permitted to be different between different inputs.

[7]Hertel and Tsigas (1987) have a discussion of how substitutable land is in various uses. The substitution elasticities used here are derived from the Hertel and Tsigas analysis.

according to income (indicated in Table 15.2) and a demand system is specified for each group. Each income group has an endowment of labour and capital and, given the vector of prices, decides the amount to save and invest and the amount of each good and service to consume (purchase). Investment, consequently, is determined by savings.

The output of the 14 producing sectors accrues to the owners of the factors of production (i.e. land, labour and capital) which they sell. With the receipts from sales, these individuals either consume domestic or foreign goods and services, save, or pay taxes to the government. The savings are used for investment and the taxes are ultimately returned to these individuals.

The demand for final goods and services comes from three primary sources: (1) final goods and services which are directly consumed by individuals, (2) investment (which is equal to savings), and (3) foreign demand.

A review of Table 15.1 will show that the composition of the consumer goods and services sectors does not match that of the producing sectors because the final goods and services produced by the producing sectors must go through various channels (i.e. transportation and distribution) before they can be consumed. To address this problem, a transformation matrix is introduced that defines the contribution of each producing sector to the composition of each of the final (consumer) goods and services.

For each category of households (Table 15.2), utility is assumed to be a weighted constant elasticity of substitution (CES) function of the 14 consumer goods and services. The weights on these goods and services (which are household category specific) are computed as the share of total purchases going to a specific consumer good or service. The nature of the CES utility function implies that the elasticity of substitution is the same between any pair of goods and/or services. Because reliable estimates of the respective substitution elasticities across pairs of goods and/or services is difficult to obtain, they are assumed to equal one for all of the combinations. Finally, consumers obtain utility from the consumption of all goods and services including leisure (consumer good and service sector number 12). Hence, it is necessary to determine a weight for this factor in the utility function. For the purpose of the current analysis, this value is assumed to be 0.5 times labour income. The net effect of adding leisure is to incorporate explicitly the fact that consumers not only derive utility from the act of consuming goods and services (which comes through owning the factors of production) but that they also derive utility from leisure. Thus, an increase in leisure can lead to an enhancement of individual well being in the model.

A household's budget constraint is defined such that expenditures on goods and services must be less than or equal to its income, which is defined to equal its portion of the returns to labour plus the returns to capital plus the returns to land. That is, expenditure by a household must be less than or equal to the total factor payments it receives. Maximizing utility subject to this expenditure constraint gives the demand for the various goods and services by household categories (see, e.g. Mixon and Uri (1985) for a discussion of this). Observe that since savings are considered as one of the items in an individual's utility function, the choice between consumption and savings is made explicit. That is, intertemporal tradeoffs are an integral part of the model.

The second component of the demand for goods and services is investment. Like the final demand by individuals, total investment is disaggregated (though a transformation matrix) by the sector of the economy that produces it. For the purpose of constructing the general equilibrium model and calibrating it, investment is taken directly from the national income and product accounts (as compiled by the Bureau of Economic Analysis of the US Department of Commerce) and, since savings are assumed to exactly equal investment, personal savings are scaled to equal the gross investment observed (measured) for each of the 14 producing sectors.

The final component of demand for goods and services is the demand by foreign consumers. The foreign sector produces imports and consumes exports. Trade balance is assumed (that is, the nominal value of exports is assumed to equal the nominal value of imports in equilibrium) but the exchange rate is not explicitly incorporated into the model specification. Exports are scaled to match imports. As a result, foreigners can be regarded as consumers who purchase United States exports with income from the sale of imports to the United States.

In the model exports (i.e. foreign demand) are delineated by producing sector. That is, a transformation matrix analogous to that used for the consumption of final goods and services is employed. A similar delineation is utilized for imports (i.e. foreign supply). The exports and imports are then scaled so that the total foreign account is balanced. By employing elasticity estimates (both demand and supply) found in the literature, export and import demand relationships are constructed for each producing sector.

15.2.4 The Government Sector

The government levies taxes on both production and consumption. That is, there are taxes on factors of production, on output, on income and on consumption. Revenues are used to distribute income back to consumers and to purchase goods and services, as well as capital and labour.

First, there is a question of how to treat the government in a general equilibrium model. For the purpose at hand, it is treated as a separate sector with a constant elasticity of substitution utility function. The elasticity of substitution is assumed to be one. This means that the CES production function collapses to a Cobb-Douglas-type production function. The government collects tax revenue in various forms. The explicitly considered taxes include personal income tax, labour taxes (e.g. a social security tax), capital taxes (e.g. a corporate income tax), property taxes, and sales and excise taxes. All these are treated as *ad valorem* taxes and a marginal rate is used for each household category, consumer good and service sector, producing sector and factor input. In this respect, the model is a distinct improvement over earlier general equilibrium models (Shoven and Whalley, 1972) which simply employed lump sum transfer schemes or used average tax rates.

15.2.5 A Mathematical Statement of the Model

Given these foregoing considerations, it is useful to state precisely the conditions that the model being used here must satisfy for a general equilibrium to exist. First, there cannot be positive excess quantities demanded. That is,

$$\sum_{j=1}^{m} a_{ij} M_j - E_i (p,Y) \geq 0 \; for \; p_i \geq 0 \qquad (15.1)$$

and where i $(i = 1, 2, ..., n)$ denotes the consumer goods and services, M_j $(j = 1, 2, ... m)$ denotes the activity levels, a_{ij} denotes the ij th element in the activity analysis matrix, Y denotes a vector of incomes for the k consumers, p denotes a vector of prices for the n consumer goods and services and E_i denotes the excess demand for good or service i.

The second requirement for general equilibrium is that the profits associated with a given activity are not positive. That is,

$$-\sum_{i=1}^{n} a_{ij} p_i \geq 0 \; for \; M_j \geq 0. \qquad (15.2)$$

Finally, all prices and activity levels must be non-negative. That is,

$$p_i \geq 0, \; i = 1, 2, ..., n \qquad (15.3a)$$

and

$$M_j \geq 0, \; j = 1, 2, ..., m. \qquad (15.3b)$$

The model is solved for a general equilibrium using the iterative algorithm nominally referred to as the Sequence of Linear Complementary Problems (SLCP) developed by Mathiesen (1985a, 1985b). This algorithm is based on the fixed point theorem proved by Scarf (1967).

A complete listing of the equilibrium conditions together with relevant definitions is presented in the Appendix.

15.3 Data for the 1988 Base Year

The general equilibrium model is calibrated for 1988. For each of the 14 producing sectors, data on capital receipts and taxes are computed from reports of the Bureau of Economic Analysis of the US Department of Commerce, the US Department of Agriculture, the US Department of Energy, and from Hertel and Tsigas (1987). The various elasticities of substitution employed in the analysis were obtained from Boyd (1988).

Capital income (earnings) and labour income were obtained from the Bureau of Economic Analysis of the US Department of Commerce. Land income was estimated using factor shares obtained from the Economic Research Service of the US Department of Agriculture and applied to the capital income component noted above.

Data on expenditures on each of the 14 goods and services by each of the 6 household categories were obtained from the *Consumer Expenditure Survey: Interview Survey, 1984* (Bureau of Labor Statistics, 1986). By combining this information with the number of households in each household (income) category (these data come from the Bureau of Economic Analysis), the aggregate expenditures on each category of consumer goods and services by each household category were computed.

The various tax rates used in the analysis were obtained from a variety of sources including the Internal Revenue Service, the Economic Research Service of the Department of Agriculture, Hertel and Tsigas (1987), and Ballard, et al. (1985). These rates, as noted previously, are marginal rates.

The value of exports and imports in 1988 were taken from the *Survey of Current Business* (various issues) with the exception of the energy data which were obtained from the Energy Information Administration of the US Department of Energy and the agriculture data which were obtained from the Economic Research Service of the US Department of Agriculture.

15.4 General Equilibrium Results

Before discussing the results of the general equilibrium model, a couple of items need to be mentioned. First, in order to assess the impact of the CRP it is necessary to determine how the mix of agricultural commodities produced will change in response to taking acreage out of production. The general equilibrium model is too aggregated to make this determination accurately between program crops and between non-program crops. Consequently, a spatial (partial) equilibrium model is used to get a first order approximation of the production impacts. These results are then integrated with the general equilibrium model to determine the aggregate effects. The partial equilibrium model is detailed in Konyar, McCormick, and Osborn (1994). The model is regional (23 regions for the contiguous United States), it considers both program and non-program crops (ten crops), it divides crops by irrigated and non-irrigated acreage, and each crop is produced by seven factor inputs (including labour, fertilizer, agricultural chemicals, energy, etc.). Linear demand and supply functions are assumed making the objective function quadratic. The objective is the maximization of net farm income. The model is run with no CRP acreage and then with the CRP acres taken out of production. The resulting changes in crop mix and total production for both program and non-program crops are used as inputs into the general equilibrium model. The equilibrium levels of production for the agricultural sectors indicated by the spatial equilibrium model in general will differ from those levels indicated by the general equilibrium model. Consequently, it is necessary to iterate between the two models until a consensus is reached. As it turned out for the specific issue under consideration, the solutions from the first iteration were sufficiently close (within 1 percent) to make a second iteration unnecessary.

Second, as observed in a preceding section, the model is solved by the SLCP algorithm of Mathiesen. This algorithm is based on the fixed point theorem proved by Scarf (1967).

Third, the magnitude of the effect that the CRP will have on the substitution of other goods and services consumed for agricultural commodities is an important consideration. Consequently, median values of 1.0 for the elasticities of substitution between the various agricultural commodities and other goods and services incorporated in the model are used. These values are consistent with the values reported by Uri (1982). Because of the potential overall importance of these values to the results of the analysis, however, a sensitivity analysis will be performed whereby the values are assumed to vary around the point estimates.

15.4.1 Reference Case

The Reference Case results (both quantities and normalized prices) are presented in Table 15.3, Table 15.4 and Table 15.5 for the producing sectors, the consuming sectors, and the various household (income) categories, respectively. Note that the nominal values of the quantities are in hundreds of billions of 1988 dollars. The sector numbers and category numbers correspond to those used in Table 15.1 and Table 15.2. By themselves, the values found in Table 15.3 through Table 15.5 provide little useful information beyond showing how the model is calibrated. Rather, the significance of the general equilibrium model and of the equilibrium values is in how these values change in response to the policy initiative that perturbs the general equilibrium.

15.4.2 CRP Implemented at the 1993 Level

By December 1993 there were 36.5 million acres enrolled in the CRP (Economic Research Service, 1994). (This represents approximately 3 percent of the cropland available for production.) Table 15.3, Table 15.4 and Table 15.5 present the general equilibrium values for prices and quantities for the producing sectors, consuming sectors and households, respectively as a result of this level of enrollment. Also indicated on those tables are the percentage changes in the equilibrium quantities in the producing sectors, consuming sectors and households in response to the CRP enrollment.

The reduction in cropland used for agricultural commodity production as a result of the CRP will have several effects. Consider the producing sectors first. In response to the reduced acreage, total output in the producing sectors will fall by 0.002 percent or by about $113 million.[8] This fall, however, is not uniformly spread across producing sectors. For example, the output of food and tobacco

---

[8]Note that these and other effects are in terms of the annual impacts. That is, they indicate what will occur each year.

products sector will fall by 0.020 percent ($69 million).[9] This is the result of a decrease in agricultural production (see below). For the service sector, output falls by 0.002 percent ($60 million). Output in the manufacturing sector will increase by 0.002 percent ($40 million) while output in the chemicals and plastics sector will expand by 0.003 percent ($6 million).

Table 15.3. Equilibrium prices (normalized) and quantities (in hundreds of billions of dollars) or the producing sectors

	Reference Case		CRP Initiative		Percent Change[1]
Sector	Price	Quantity	Price	Quantity	
1 Manufacturing	1.00000	17.1773	1.00000	17.1777	0.002
2 Coal Mining	1.00000	0.24964	1.00002	0.24965	0.004
3 Other Mining	1.00000	0.21266	1.00002	0.21267	0.005
4 Service	1.00000	23.7818	1.00002	23.7812	-0.002
5 Chemicals	1.00000	2.27375	1.00004	2.27381	0.003
6 Food and Tobacco	1.00000	3.50574	1.00002	3.50505	-0.019
7 Petroleum Refining	1.00000	1.61241	1.00006	1.61250	0.005
8 Financial	1.00000	5.54883	1.00006	5.54886	0.001
9 Forestry	1.00000	0.10592	1.00052	0.10589	-0.028
10 Wood Products	1.00000	1.69890	1.00002	1.69893	0.001
11 Crude Oil	1.00000	1.29060	1.00007	1.29068	0.006
12 Agriculture - PC[2]	1.00000	0.45210	1.00507	0.45267	0.126
13 Agriculture - L	1.00000	1.09921	1.00180	1.09899	-0.020
14 Agriculture - O	1.00000	0.61286	1.00143	0.61199	-0.142
Total	1.00000	59.62172	1.00107	59.62059	-0.002

[1] The percent change represents the percentage change in the equilibrium quantities between the CRP initiative and the reference case.

[2] For the agriculture sectors, PC denotes program crops, L denotes livestock, and O denotes all other agricultural activities. Also, some of the other titles have been abbreviated. The complete titles are given in Table 15.1.

What will happen in the three agriculture sectors plus the forestry sector? Output in the program crops sector will rise by 0.126 percent ($57 million), output in the livestock sector will decline by 0.020 percent ($22 million) and output in the all other agriculture commodities sector will be reduced by 0.142 percent ($87 million). Output in the forestry sector will fall by 0.028 percent ($3 million). Thus, the CRP in the aggregate imposes some costs, in terms of reduced output, on the agriculture sectors consisting of the three agriculture sectors plus the forestry sector of about 0.001 percent (or $55 million) in the aggregate. The rise in output in the program crops sector is caused by taking land out of production that, in turn, results in an increase in the price of program crops. This increase

[9] In order to limit the number of tables, some of the equilibrium prices and quantities will not be explicitly presented although selected values will be discussed. Such is the case with the prices and quantities of imported goods and services.

in the price of program crops results in land being used more intensively and consequently, more land being used to grow program crops at the expense of non-program crops and livestock. The total returns to agriculture go up because demand for agricultural commodities is relatively inelastic. This means that the price of land increases. Less is produced because of factor adjustments. Output, however, does not go down by as much as a partial equilibrium analysis would suggest.

Accompanying the changes in agricultural output are changes in the prices of the agricultural commodities. Thus, for example, the price of the output of program crops will rise by 0.507 percent attributable to taking land out of production, the price of the output of the livestock sector will rise by 0.180 percent, the price of the output of all other agricultural commodities will increase by 0.143 percent and the price of the output of forestry products will rise by 0.052 percent. These latter price increases are the result of the land previously devoted to production in these sectors now being used to grow program crops in response to the higher price for program crops.

Finally, with regard to the agricultural sectors, imports of program crops, livestock, and non-program crops will expand by 0.381 percent ($2.03 million), 0.048 percent ($271 thousand), and 0.025 percent ($270 thousand), respectively. These increases in imports are the result of the higher prices for domestically produced agricultural commodities. As the result of an increase in the imports of agricultural commodities, crude oil imports are expected to fall by 3.701 percent ($1.9 million). Other producing sectors will also be impacted but the effect is the largest in the crude oil sector. This result is observed because of the requirement that trade must be balanced (a condition in the general equilibrium model).

With regard to the consuming sectors, the implementation of the CRP results in a slight decrease in the consumption of goods and services by about 0.001 percent ($38 million). The most adversely impacted sector is the food sector which experiences a 0.008 percent ($37 million) fall in consumption. The second most significantly impacted sector is the alcohol and tobacco sector which realizes a 0.002 percent ($10 million) reduction in output. Most other sectors experience minimal changes in consumption attributable to the indirect effects of the CRP. These indirect effects include a lower real income brought about by an increase in the price of the agricultural commodities and a change in relative prices which leads to substitution of relatively less expensive goods and services for relatively more expensive.

Utility falls for all six of the household categories. The aggregate reduction in utility is 0.003 percent ($104 million) for all household categories. The reduction does fall fairly evenly across households, however. Category VI households (i.e. those with incomes in excess of $40,000) experience a reduction in utility of 0.003 percent ($40 million) while Category V households (i.e. those with incomes ranging between $30,000 and $39,999) suffer a 0.003 percent ($17 million) reduction in utility. The remaining household categories incur percentage reductions in utility of about the same order of magnitude. Additionally, when all of the effects of the CRP - both the direct and the indirect effects - are considered, it is not in general regressive. That is, the consequences of the CRP do not fall most heavily on the lowest household (income) category and progressively less heavily on households with larger incomes. Rather, the impacts are approximately equal in relative terms across income categories.

Table 15.4. Equilibrium prices (normalized) and quantities (in hundreds of billions of dollars) for the consuming sectors

Sector	Reference Case Price	Reference Case Quantity	CRP Initiative Price	CRP Initiative Quantity	Percent Change[1]
1 Food[2]	1.00000	4.52066	1.00045	4.52029	-0.008
2 Alcohol and Tobacco	1.00000	0.83300	1.00040	0.83328	-0.001
3 Utilities	1.00000	1.17793	1.00002	1.17793	0.000
4 Furnishings	1.00000	1.46137	1.00000	1.46137	0.000
5 Housing	1.00000	3.74071	1.00006	3.74069	-0.001
6 Clothing	1.00000	1.83323	1.00002	1.83323	0.000
7 Transportation	1.00000	0.28041	1.00002	0.28041	0.000
8 Financial Services	1.00000	1.06161	1.00002	1.06160	-0.001
9 Other Services	1.00000	4.52579	1.00002	4.52578	0.000
10 Motor Vehicles	1.00000	1.46337	1.00001	1.46337	0.000
11 Gasoline	1.00000	0.91156	1.00004	0.91156	0.000
12 R and R	1.00000	1.66132	1.00006	1.66130	-0.001
13 Nondurable Goods	1.00000	0.67238	1.00002	0.67238	0.000
14 Savings	1.00000	3.03333	1.00002	3.03336	0.002
Total	1.00000	27.17694	1.00008	27.17655	-0.001

[1] The percent change represents the percentage change in the equilibrium quantities between the CRP initiative and the reference case.
[2] Note that some of the sector titles in the table have been abbreviated. The complete designations are given in Table 15.1.

Table 15.5. Equilibrium utility levels (in hundreds of billions of dollars) by household categories

Category	Reference Case Utility Level	CRP Initiative Utility Level	Percent Change[1]
I[2]	2.23826	2.23812	-0.006
II	2.10802	2.10795	-0.003
III	2.42417	2.42410	-0.003
IV	6.01311	6.01292	-0.003
V	5.49734	5.49717	-0.003
VI	13.7363	13.7359	-0.003
Total	32.01720	32.01616	0.003
Government	7.45753	7.45786	0.004

[1] The percent change represents the percentage change in the equilibrium quantities between the CRP initiative and the reference case.
[2] Note that the household categories correspond to those defined in Table 14.2.

The government realizes a reduction in expenditures. Net expenditures fall as a result of cropland being idled and because the direct costs of the CRP are less than the reduction in farm program outlays. The aggregate effect is a decline in net government expenditures of 0.004 percent or about $33 million.

15.4.3 An Increase in the CRP to 45 Million Acres

As part of the deliberations for the 1995 farm bill, some considerations are being given to expanding the CRP to 45 million acres (General Accounting Office, 1993). Consequently, consider the situation where 45 million acres of cropland are retired under the CRP. In order to conserve space, complete results are not presented. Just the highlights are given.

The reduction in agricultural commodity production as a result of the CRP increasing to 45 million acres will have several effects. Consider the producing sectors first. In response to fewer acres in production, total output in the producing sectors will fall by 0.004 percent or by about $216 million. This fall is not uniformly spread across producing sectors. For example, the output of food and tobacco products sector will fall by 0.040 percent ($140 million). As before, this is the result of a decrease in agricultural production. For the service sector, output falls by 0.005 percent ($110 million). Output in the manufacturing sector will increase by 0.005 percent ($100 million) while output in the chemicals and plastics sector will expand by 0.006 percent ($13 million).

Output in the three agriculture sectors plus the forestry sector will be affected by the CRP. For example, output in the program crops sector will rise by 0.253 percent ($114 million), output in the livestock sector will decline by 0.041 percent ($45 million) and output in the all other agriculture commodities sector will be reduced by 0.309 percent ($189 million). Output in the forestry sector will fall by 0.037 percent ($3 million). Thus, the CRP in the aggregate stands to potentially impose some costs in terms of reduced output on the agriculture sectors consisting of the three agriculture sectors plus the forestry sector of about 0.001 percent (or $123 million) in the aggregate.

Accompanying the changes in agricultural output are changes in the prices of the agricultural commodities. Thus, for example, the price of the output of program crops will rise by 0.904 percent attributable to taking land out of production, the price of the output of the livestock sector will rise by 0.240 percent, the price of the output of all other agricultural commodities will increase by 0.179 percent and the price of the output of forestry products will rise by 0.052 percent. These latter price increases are the result of the land previously devoted to production in these sectors now being used to grow program crops in response to the higher price for program crops.

Finally, with regard to the agricultural sectors, imports of program crops, livestock, and non-program crops will expand by 0.760 percent ($4.05 million), 0.095 percent ($540 thousand), and 0.049 percent ($536 thousand), respectively. These increases in imports are the result of the higher prices for domestically produced agricultural commodities.

With regard to the consuming sectors, the increase in the number of acres in the CRP to 45 million acres results in a decrease in the consumption of goods and services by about 0.007 percent ($197 million). The most heavily affected is the food sector which experiences a 0.038 percent ($171 million) fall in consumption. The second most significantly impacted sector is the alcohol and tobacco sector which encounters a 0.031 percent ($25 million) reduction in output. Most other sectors experience minimal changes in consumption attributable to the indirect effects of the CRP. These indirect effects include a lower real income brought about by an increase in the price of agricultural commodities and changing relative prices which leads to substitution of relatively less expensive goods and services for relatively more expensive.

Utility falls for all six of the household categories. The aggregate reduction in utility is 0.006 percent ($190 million) for all household categories. The reduction does fall fairly evenly across households, however. Category I households (i.e. those with incomes less than $9,999) experience a reduction in utility of 0.005 percent ($11 million) while Category III households (i.e. those with incomes ranging between $15,000 and $19,999) suffer a 0.006 percent ($13 million) reduction in utility. The remaining household categories incur percentage reductions in utility of about the same order of magnitude.

Unlike the 36.5 million acre CRP case where net government expenditures fell, the government incurs additional costs as the result of a 45 million acre CRP. While program costs fall as the result of program cropland being idled, the direct costs of the CRP more than make up for reduced program outlays. The aggregate effect is a rise in net expenditures of 0.006 percent or about $45 million. The higher cost to the government with the expanded CRP acreage follows because, while CRP payments are higher with the added acreage, program costs are reduced proportionately less.

15.5 A Comparison

How do the results obtained here compare to those obtained by others? Two studies will be used for comparative purposes. Osborn and Konyar (1990) conclude that taking 35 million acres of cropland out of production will result in a fall in agricultural production of 2 percent, a rise in food prices of 0.005 percent, and an increase total government outlays by $63 million per year. Comparable estimates obtained here for 36.5 million acres indicate that agricultural production will fall by 0.001 percent, food prices will rise by 0.045 percent, and government expenditures will fall by $33 million. A portion of the difference in the results is attributable to the partial equilibrium framework used for the Osborn and Konyar analysis. Young and Osborn (1990) find that a CRP of 45 million acres will result in a reduction in agricultural production of 3.2 percent, a rise in food prices of 1 percent. This compares to a reduction in production estimated here to be 0.001 percent and a rise in food prices of 0.065 percent. As with the Osborn and Konyar analysis, some of the difference is the consequence of using a partial equilibrium analysis instead of a general equilibrium approach.

15.6 Sensitivity Analysis

No analysis is complete without an examination of the sensitivity of the results to key assumptions. In the foregoing discussion, many assumptions were made with regard to model structure and parameter estimates. A full examination and discussion of these assumptions would be virtually impossible. Consequently, only the results from the sensitivity analysis of one crucial assumption will be discussed. Namely, what are the effects on the vector of equilibrium prices and quantities of the assumption concerning the elasticity of substitution? Recall that the original point estimates of these elasticities used in the model were 1.0 (i.e. the Cobb-Douglas case). In subsequent simulations, however, these were lowered for all goods and services to 0.5 and then raised them to 1.5. In general, the effect of raising the elasticity of substitution is to magnify the influence of the CRP. The quantitative magnitude of the effect, however, on the results is minimal. Under both a CRP with 36.5 million acres removed from production and one where 45 million acres are remqved from production, neither output nor consumption is affected by more than $150 million and in no case is there any change in the qualitative results discussed above.

These sensitivity results suggest that the values of the substitution elasticities, while important in the determination of the vectors of general equilibrium prices and quantities and significant in determining the implications of a policy initiative affecting the agricultural sectors, are not so pivotal to the model that errors in their values lead to misleading and nonsensical results.

15.7 Conclusion

The foregoing analysis has examined the impact of the Conservation Reserve Program on the United States economy in general and the agricultural sectors in particular. The analytical approach used in the analysis consisted of a general equilibrium model composed of fourteen producing sectors, fourteen consuming sectors, six household categories classified by income and a government. The effects of removing 36.5 million acres and 45 million acres of cropland from agricultural production on prices and quantities are examined. The results are revealing. For example, keeping the CRP at 36.5 million acres will result in lower output by the producing sectors (by about $113 million), a decrease in the consumption of goods and services (by about $38 million), and a reduction in aggregate utility (by about $104 million). The government would realize a net reduction in expenditures of about $33 million. In the case of removing 45 million acres of cropland from production, there would be lower output by the producing sectors (by about \$123 million), a contraction in the consumption of goods and services (by about \$197 million), and a reduction in aggregate utility (by about \$190 million). The government would realize an increase in outlays of \$45 million.

Next, when subjected to a sensitivity analysis, the results are reasonably robust with regard to the assumption of the values of the substitution elasticities. That is, while the model's equilibrium values do vary in response to different assumptions of the values of these elasticities, the fluctuations are not so enormous to suggest that the model is unrealistically sensitive to these parameters.

As a consequence of this analysis, the implications the imposition of the CRP are clear. Namely, besides the reduction in net federal government expenditures, the various producing sectors (in the aggregate) will be adversely impacted in terms of a reduction in output while the various consuming sectors will experience a cumulative fall in the consumption of goods and services. Moreover, social welfare (measured as aggregate utility in the model) will decline. All of these changes, however, are relatively modest.

Acknowledgements

The views expressed are those of the authors and do not necessarily represent the policies of the organizations with which they are affiliated.

References

Adelman I, Robinson S (1986) US Agriculture in a General Equilibrium Framework. American Journal of Agricultural Economics 68:1196-1207

Arrow KJ, Chenery HB, Minhas BS, Solow RM (1961) Capital-Labor Substitution and Economic Efficiency. Review of Economics and Statistics 43:225-235

Ballard CL, Fullerton D, Shoven JB, Whalley J (1985) A General Equilibrium Model for Tax Policy Evaluation. The University of Chicago Press, Chicago

Boyd R (Feb 1988) The Direct and Indirect Effects of Tax Reform on Agriculture. Technical Bulletin Number 1743, Economic Research Service, US Department of Agriculture, Washington, D.C.

Bureau of Labor Statistics (1986) Consumer Expenditure Survey. Interview Survey 1984. US Government Printing Office, Washington

Dervis K, DeMelo J, Robinson S (1982) General Equilibrium Models for Development Policy. Cambridge University Press, Cambridge

Economic Research Service (May 1994) Agricultural Resources. Cropland, Water, and Conservation, Economic Research Service. US Department of Agriculture, Washington

Friedman LS, (1984) Microeconomic Policy Analysis. McGraw-Hill Book Company, New York
General Accounting Office (March 1993) Conservation Reserve Program. Cost Effectiveness is Uncertain. US Government Printing Office, Washington, D.C.
Glaser L (April 1986) Provisions of the Food Security Act of 1985. AIB-498. Economic Research Service. US Department of Agriculture, Washington D.C.
Harberger A (1962) The Incidence of Corporate Taxation. Journal of Political Economy 70:215-240
Harberger A (1974) Taxation and Welfare. University of Chicago Press. Chicago
Harrington D, Schluter G, O'Brien P (Oct 1986) Agriculture's Link to the Macroeconomy. AIB 504. Economic Research Service. US Department of Agriculture, Washington, D.C.
Heady EO, Dillon JL (1961) Agricultural Production Functions. Iowa State University Press, Ames, Iowa
Hertel TW, Tsigas ME (Feb 1987) Tax Policy and US Agriculture. A General Equilibrium Analysis. Staff Paper #87-2. Department of Agricultural Economics. Purdue University, West Lafayette, Indiana
Huang W, Algozin K, Ervin D, Hickenbotham T (1980) Using the Conservation Reserve Program to Protect Groundwater Quality. Soil and Water Conservation 45:341-346
Hudson EA, Jorgenson DW (1974) Energy Policy and Economic Growth. Bell Journal of Economics and Management Science 5:461-514
Hudson EA, Jorgenson DW (1976) Tax Policy and Energy Conservation. In: Econometric Studies of US Energy Policy. North-Holland Publishing Company, Amsterdam
Jorgenson DW (1984) Econometric Methods for Applied General Equilibrium Analysis. In: Applied General Equilibrium Analysis. Scarf HE, Shoven JB (Eds). Cambridge University Press, Cambridge
Konyar K, McCormick I, Osborn T (Jan 1994) The US Agricultural Resource Model (USARM) Documentation. Economic Research Service. US Department of Agriculture, Washington D.C.
MacKinnon J (1984) Comments. In Applied General Equilibrium Analysis. Scarf HE, Shoven JB (Eds). Cambridge University Press, Cambridge
Mathiesen L (1985a) Computational Experience in Solving Equilibrium Models by a Sequence of Linear Complementary Problems. Operations Research 33:1225-1250
Mathiesen L (1985b) Computation of Economic Equilibrium by a Sequence of Linear Complementary Problems. Mathematical Programming Study 23
Mixon JW, Uri ND (1985) Managerial Economics. Macmillan Publishing Company, New York
Ogg C, Aillery M, Ribaudo M (Oct 1989) Implementing the Conservation Reserve Program. AER-618. Economic Research Service. US Department of Agriculture, Washington D.C.
Osborn T, Konyar K (Aug 1990) A Fresh Look at the CRP. Agricultural Outlook. AO-166 pp 33-37
Reichelderfer K, Boggess W (1988) Government Decision Making and the Conservation Reserve Program. American Journal of Agricultural Economics 70:1-11

Scarf HE (1967) The Approximation of Fixed Points of a Continuous Mapping. SIAM Journal of Applied Mathematics 15:1328-1343

Schaible G (July 1989) Irrigated Acreage in the Conservation Reserve Program. AER-610. Economic Research Service. US Department of Agriculture, Washington D.C.

Shoemaker R (Jan 1989) The Conservation Reserve Program and Its Effects on Land Values. AIB-554. Economic Research Service. US Department of Agriculture, Washington D.C.

Shoven JB, Whalley J (1972) A General Equilibrium Calculation of the Effects of Differential Taxation of Income from Capital in the US Journal of Public Economics 1:281-322

Shoven JB, Whalley J (1992) Applying General Equilibrium. Cambridge University Press, Cambridge

Taff S (1990) Using the Conservation Reserve Program to Reduce Program Crop Plantings. North Central Journal of Agricultural Economics 12:89-97

Uri ND (1982) The Demand for Energy and Conservation in the United States. JAI Press Inc, Greenwich

Uri ND (1984) The Impact of Technical Change on the Aggregate Production Function. Applied Economics 16:555-567

Young E, Osborn T (Feb 1990) The Conservation Reserve Program. An Economic Assessment. AER-626. Economic Research Service. US Department of Agriculture, Washington D.C.

Appendix 15.1

Empirical Model

I. Overall Equilibrium by Sector

(1) $Y_j + GE_j + UM_j = \Sigma_L \, RAS_{jL} + GD_j + CD_j + UX_j + INV_j$

(2) $\Sigma_c \, SL_c = \Sigma_j \, DL_j + GDL$

(3) $\Sigma_c \, SK_c = \Sigma_j \, DK_j + GDK$

(4) $\Sigma_c \, SD_c = \Sigma_j \, DD_j + GDD$

where

(5) $GDL = \Sigma_j \, TL_j$

(6) $GDK = \Sigma_j \, TK_j$

(7) $GDD = \Sigma_j \, TD_j$

II. Consumer Goods and Services

(8) $CD_j = \Sigma_i \, Z_{ji} \, (GCE_j - TC_j)$

(9) $\Sigma_c \, RCS_{ic} = GCE_i$

(10) $\Sigma_i \, RCS_{ic} = SL_c + SK_c + SD_c + TRN_c - PIT_c$

(11) $GC_c = \Sigma_i \, RCS_{ic} - SAV_c + (1 - TAU_c) \, (ZTA_c - 1) \, SL_c$

(12) $GC_c = SL_c + SK_c + SD_c + TRN_c \quad PIT_c$

$+ (1 - TAU_c) \, (ZTA_c - 1) \, SL_c.$

(13) $TE = \Sigma_c \, (SL_c \; ZTA_c \; TAU_c + SK_c \, TAU_c + SD_c \, TAU_c$

$- (\Phi_c + TRN))$

where $\Phi_c = SL_c \, TAU_c + SK_c \, TAU_c + SD_c \, TAU_c - PIT_c$

III. Foreign Sector Balance

(14) $\Sigma_j (UM_j (EM_j /(1 + EM_j)) + UM_j / (1 + EM_j))$

$= \Sigma_j (UX_j + FE_j)$

IV. Consistency

(15) $\Sigma_c (SL_c + SK_c + SD_c + TRN_c - PIT_c - TC_c) = \Sigma_c CG_c$

(Net household income equals household expenditures)

(16) $\Sigma_j (GSK_j + GE_j + TL_j + TK_j + TD_j + TXO_j) + GTL$

$= \Sigma_c TRN + \Sigma_j (GDK_j + GD_j) + GD_c)$

(Government income plus endowments equals government outlays)

(17) $\Sigma_j (UM_j - UX_j) = 0$

(Net exports equal zero)

(18) $\Sigma_j (CD_j + GD_j + UX_j - GE_j - UM_j)$

$= \Sigma_j (DL_j + DK_j + TL_j + TK_j + TXO_j)$

(The value of demand equals value added plus taxes)

Y_j - Total production in sector j (j = 1,2,...,14)

CD_j - Consumer demand for product j

GE_j - Government endowment of product j

UM_j - Imports of product j

$\Sigma_L RAS_{jL}$ - RAS balanced input - output intermediate demands

GD_j - Government demand for product j

INV_j - Investment in sector j

UX_j - Exports of product j

SL_c - Supply of labor by household c (c = 1,2,...,6)

SK_c - Supply of capital by household c

SD_c - Supply of land by household c

DL_j - Demand for labor in the industry j

DK_j - Demand for capital in the industry j

DD_j - Demand for land in industry j

GDL - Government demand for labor

GDD - Government demand for land

TL_j - Tax on labor in industry j

TK_j - Tax on capital in industry j

TD_j - Tax on land in industry j

GCE_i - Consumer demand for consumer product i (i = 1,2,...,14)

Z_{ji} - A 14 by 14 transformation matrix

RCS_{ic} - RAS balanced matrix of each houschold's demand for each consumer good

TC_j - Excise tax on consumer good j

TRN_c - Transfer payment to household c

PIT_c - Personal income tax payment for household c

TAU_c - Marginal income tax rate for household c

SAV_c - Savings in household c

GC_c - Gross consumption of household c

ZTA - Consumption plus leisure coefficient

TE - Total government endowments

EM_j - Demand elasticity of export demand

FE_j - Endowment/Demand sector of adjusted elasticity of export demand

GSK_j - Government endowment of capital in industry j

GDK_j - Government demand for capital in industry j

GTL - Government wage taxes on its own employees

TXO_j - Government output tax on industry j

TC_c - Consumption taxes on household c

CG_c - Total government consumption by household c

V

Dynamic Models of Spatial Economies

16 Transport Infrastructure and Technology: Investment, Externalities, and General Equilibrium Effects

Jeroen C.J.M. van den Bergh and Peter Nijkamp
Vrije Universiteit

16.1 Introduction

Transportation fulfils a double role in an economic system; it is an economic sector which, just like any other sector, aims to maximize its performance, while it is also a necessary production factor allowing for commodity trade and human mobility. Transport in the latter capacity - in terms of both transportation services and infrastructure provision - needs to be produced by means of scarce inputs, and hence can be cast in the framework of a generalized production function approach. It is somewhat surprising that transportation has received insufficient attention from the perspective of general equilibrium analysis in economics, as only such an approach can consistently address the economy-wide character of the impacts of transportation as well as its negative external effects. The latter include both system-wide environmental and health effects and transportation-specific congestion and safety effects. In contrast to analyses of a partial nature, it seems clear that an economy-wide perspective cannot easily be regarded as flawless when space is assumed away, as this would essentially mean that one assumes that movement of goods can be realized without transportation costs. In most of the analytical literature however, even less attention has been devoted to the fact that these costs are, in turn, affected by various factors, such as input prices of transportation, spatial demand for transportation services, and the presence of transportation-induced externalities such as congestion.

In order to deal appropriately with these factors, the simultaneous nature of transportation and economy-wide production and consumption of commodities should explicitly be dealt with (see, e.g. Roson 1993). Therefore, it is necessary to recognize that transportation uses scarce economic resources that produce transport services by using inputs, while actors in this field participate as buyers and sellers in input and output markets. It is therefore important to allow for an endogenous price formation of transportation services.

In this paper we will consider in particular an externality cost imposed by the transportation sector onto the society at large. It may be interpreted as a case of negative environmental externalities arising from transportation. The problem to be solved is establishing the direction and order of magnitude of changes in optimal and competitive equilibrium quantities and prices of commodity production and consumption, intermediate inputs, primary inputs and transport with respect to - especially transportation-specific - effects, externalities and policies.

The issue of technology regime switches will also be dealt with in this paper, as it has both economy-wide implications, and may lead to redressing negative externalities of the transport sector. Transportation technology policy can in this context be a critical instrument for achieving a more sustainable transport system in an economy-wide equilibrium. Transport takes clearly place in a force field of conflicting objectives: economic efficiency and performance, spatial access to services, environmental and resource preservation, safety, and so forth.

In light of the previous remarks the paper is organized as follows. Section 16.2 introduces the multi-sector factor allocation problem, as well as the socially optimum combination of production, transport and infrastructure. A dynamic extension of the allocation model, where infrastructure investment is included, is discussed in the third section. In Section 16.4 we offer a competitive equilibrium formulation of the problem of multisectoral production supported by freight transportation, and we examine the inclusion of environmental and other externalities which are imposed by the transportation activity on households. In Section 16.5 we discuss the emergence of new technologies and substitutes for transportation, followed by some suggestions about unresolved issues in Section 16.6.

16.2 Transport and Infrastructure: Optimal Allocation of Production Factors in the Presence of Negative Externalities

In order to offer a systematic framework for analyzing externalities in the transport sector, we will in this section first cast the basic division between production, transport and infrastructure in a planning formulation. One may take for granted here the construct of a government or a public decision-maker concerned with establishing an optimal social welfare level by choosing adequate and consistent input and commodity allocation schemes and activity levels in the production of goods and services (C), transportation (T), and maintenance of infrastructure (B). This is formalized in the model (16.1) to (16.5) by introducing a primary factor (e.g. labour) allocation problem based on a fixed amount of supply (L_S), which can be used in the production sector (L_P), the transportation sector (L_T) and the infrastructure sector (L_B).

$$\underset{\{L_P, L_T, L_B \geq 0\}}{maximize}\, U(C) \tag{16.1}$$

s.t.

$$C = F_P(L_P, T) \tag{16.2}$$

$$T = F_T(L_T, B_T) \tag{16.3}$$

$$B = F_B(L_B) \tag{16.4}$$

$$L_P + L_T + L_B = L_S. \tag{16.5}$$

All functions satisfy positive first-order derivatives. Model (16.1) to (16.5) is a constrained optimization problem. Solving the first-order conditions of the associated Langrangean problem we obtain:

$$\lambda_1 * \frac{\partial F_P}{\partial L_P} = \lambda_2 * \frac{\partial F_T}{\partial L_T} = \lambda_3 * \frac{dF_B}{dL_B}, \tag{16.6}$$

where λ_i (i=1,2,3) are Lagrangean multipliers associated with equations (16.2) to (16.4), respectively. This result can easily be interpreted in standard economic terms, in that the marginal social value of labour is in the optimum equal in all sectors of production, transportation and infrastructure (maintenance).

After substituting the multipliers out, one obtains the following relationships:

$$\frac{\partial F_P}{\partial L_P} = \frac{\partial F_P}{\partial T} * \frac{\partial F_T}{\partial L_T} \tag{16.7}$$

and

$$\frac{\partial F_T}{\partial L_T} = \frac{\partial F_T}{\partial B} * \frac{\partial F_B}{\partial L_B}. \tag{16.8}$$

These relations are to be interpreted as an equality of the direct marginal productivity of labour in a sector (in (16.7) and (16.8) the production and transportation sector, respectively) and the indirect marginal productivity (in terms of the output of the respective sector) of labour in other sectors which render their services to the respective sector (in (16.7) and (16.8) the transportation and infrastructure sector, respectively). This is further illustrated by substitution of equation (16.8) in (16.7), which gives:

$$\frac{\partial F_P}{\partial L_P} = \frac{\partial F_P}{\partial T} * \frac{\partial F_T}{\partial B} * \frac{dF_B}{dL_B}. \tag{16.9}$$

Having derived now some standard results, we will pay attention to the social costs of transport. Inclusion of these can be done as follows. The social welfare function is extended with two arguments, namely the level of transportation activity (T) and gross infrastructure investment (notably maintenance): $U \equiv U(C,T,B)$. These may be used as proxy indicators of negative externalities, so that they negatively contribute to social welfare. Therefore, U has negative first-order partial derivatives with respect to T and B, which can be interpreted by writing them as $\frac{\partial U}{\partial T} = \frac{\partial U}{\partial E} * \frac{\partial E}{\partial T}$ and $\frac{\partial U}{\partial B} = \frac{\partial U}{\partial E} * \frac{\partial E}{\partial B}$; here E stands for the environment (quality and/or quantity), so that the external effect is separated into a physical effect of transport and infrastructure on the environment, and the impact of the resulting environmental change on social welfare. Of course, it may be difficult in practice to distinguish between the effect of transport and infrastructure, which is to some extent due to the fact that some physical effects are the result of the combination of transport and infrastructure. However, one may try to use valuation methods in this case which would aim to estimate the external costs of infrastructure if there were no traffic on it.

In the optimum situation with externalities, the conditions given in (16.6) are again satisfied. However, the relations in (16.7) to (16.9) are now replaced by:

$$\frac{\partial U}{\partial C} * \frac{\partial F_P}{\partial L_P} = \left[\frac{\partial U}{\partial C} * \frac{\partial F_P}{\partial T} - \frac{\partial U}{\partial T}\right] * \frac{\partial F_T}{\partial L_T} \tag{16.10}$$

and

$$\frac{\partial U}{\partial C} * \frac{\partial F_P}{\partial L_P} = \left[\left[\frac{\partial U}{\partial C} * \frac{\partial F_P}{\partial T} - \frac{\partial U}{\partial T}\right] * \frac{\partial F_T}{\partial B} - \frac{\partial U}{\partial B}\right] * \frac{dF_B}{dL_B}. \tag{16.11}$$

This result can be interpreted as follows: all direct and indirect positive and negative contributions to social welfare have to be accounted for when deciding about a change in the allocation structure of the production factor concerned. Positive contributions occur indirectly via increased consumption (in both (16.10) and (16.11)); negative contributions occur via the direct effect of increased transportation on utility (in (16.10)); increased infrastructure affects U negatively directly and indirectly via increased transportation (in (16.11)). It should be noted that (16.10) and (16.11) are generalized formulations of (16.7) and (16.9), respectively. In conclusion, the above model leads to some interesting extensions of conventional economic equilibrium results, even though this is cast in a static framework. In the next section, the limitations of a comparative static model will be alleviated.

16.3 A Dynamic Model: Transport and Infrastructure Investment

A continuous dynamic extension of the model analyzed in the previous section should include investment in infrastructure (I) which is required for transportation, which is in turn necessary for production activities. Let t denote time and r the discount rate. Then we may formulate an appropriate model with a planning horizon s as follows:

$$\underset{\{L_P(t),L_T(t),L_B(t)\geq 0\}}{maximize} \int_0^s e^{-r*t} * U(C(t))dt \tag{16.12}$$

s.t.

$$\frac{dI(t)}{dt} = F_B(L_B(t)) - \delta * I(t) \tag{16.13}$$

$$C(t) = F_P(L_P(t), T(t)) \tag{16.14}$$

$$T(t) = F_T(L_T(t), I(t)) \tag{16.15}$$

$$L_P(t) + L_T(t) + L_B(t) = L_S \tag{16.16}$$

$$I(0) = I_0, \tag{16.17}$$

for all t in the interval $[0,s]$.

The important new equation is (16.13), which reflects that the change in infrastructure I depends on gross investment in it. This in turn is based on a primary input L_B, and depreciation of the existing infrastructure at a rate δ. Gross investment includes maintenance of old and construction of new infrastructure. It should be noted that in the model of the previous section we looked into the case where dI/dt was equal to zero, so that then $B=F_B(L_B)=\delta*I_0$, i.e. 'investment' exclusively includes maintenance of the initial infrastructure. Furthermore, in this model only investment in infrastructure is regarded as a reflection of a long term horizon in which other capital stocks, such as transportation vehicles, and the primary factor input supply (L_S) are kept fixed. Extensions then of this model might amongst others focus on different types of factor inputs and substitution between them. However, our focus is not so much on such standard issues, but rather on the sectoral division between consumption goods, transportation services and infrastructure services.

It is possible to show that for the above optimal control problem condition (16.7) is again satisfied, at each time in the relevant interval. Condition (16.9) is now replaced by

$$\frac{dU}{dC} * \frac{\partial F_P}{\partial L_P} * e^{-r \cdot T} = \lambda * \frac{dF_B}{dL_B}. \tag{16.18}$$

where λ captures the total effect in present utility/value terms of an increase in net infrastructure investment.

The solution procedure to this dynamic multisector allocation problem is somewhat complicated, since there are three control variables which are subject to the condition given in (16.17). Using a standard Cobb-Douglas function to specify the various relations, one may obtain somewhat less general, but more explicit results. The following specifications are chosen here:

$$\left.\begin{aligned} U &= C^{\gamma} \\ F_P &= A * L_P^{\alpha_P} * T^{\beta_P} \\ F_T &= B * L_T^{\alpha_T} * I^{\beta_T} \\ F_B &= D * L_B^{\alpha_B}. \end{aligned}\right\} \tag{16.19}$$

Then, on the basis of (16.7) we can derive that

$$L_P = \frac{\alpha_P}{\beta_P * \alpha_T} * L_T, \tag{16.20}$$

which gives in combination with (16.16):

$$L_B = L_S - \left(1 + \frac{\alpha_P}{\beta_P * \alpha_T}\right) * L_T. \tag{16.21}$$

Substituting now (16.15), (16.20) and (16.21) into (16.13), an expression is obtained in the control variable L_T, the state variable I, and the costate variable λ:

$$\zeta^{\frac{1}{1-\alpha_B}} * I^{\frac{\gamma \cdot \beta_P \cdot \beta_T}{1-\alpha_B}} * \left[L_S - \left(1 + \frac{\alpha_P}{\beta_P * \alpha_T}\right) * L_T\right] - \lambda^{(1-\alpha_B)} * L_T^{\frac{1-\gamma(\alpha_P+\beta_P \cdot \alpha_T)}{1-\alpha_B}} * e^{-rt} = 0. \tag{16.22}$$

ζ denotes some constant which is a function of Cobb-Douglas parameters. Solving (16.22) for L_T in terms of I and λ - by using (16.15), (16.20) and (16.21) - one can by means of substitution into (16.13) and the costate equation:
obtain two interactive differential equations which may be investigated in more detail. If we assume that the nonlinear term with λ and L_T in equation (16.22) is

$$\frac{d\lambda}{dt} = -\frac{dU}{dC}\left(F_P(L_P,F_T(L_T,I))\right) * \frac{\partial F_P}{\partial T} * \frac{\partial F_T}{\partial I} * e^{-rt} + \lambda * \delta \tag{16.23}$$

approximately linear[1], we can derive the following differential state-costate variable system:

$$\frac{dI}{dt} = D * \left[\frac{a * I^b - L_S * \lambda^{\frac{1}{1-\alpha_B}}}{c * I^b + \lambda^{\frac{1}{1-\alpha_B}}}\right]^{\alpha_B} - \delta * I \tag{16.24}$$

and

$$\frac{d\lambda}{dt} = \frac{d * I^{(b * \gamma * (\alpha_P+\beta_P)-1)}}{\left(c * I^b + \lambda^{\frac{1}{1-\alpha_B}}\right)^{\gamma * (\alpha_P+\beta_P)}} * e^{-rt} + \delta * \lambda \tag{16.25}$$

where the new parameters *a, b, c* and *d* are introduced to replace more complex expressions of the Cobb-Douglas parameters; for instance, the parameter *b* equals $\frac{\gamma * \beta_P * \beta_T}{1-\alpha_B}$.

Simplifying the differential system by the following approximation shows that a saddlepoint equilibrium may result:

$$\frac{dI}{dt} = (m-(L_S-1) * \lambda)^{\alpha_B} - \delta * I \tag{16.27}$$

and

$$\frac{d\lambda}{dt} = \frac{l * e^{-rt}}{I} + \delta * \lambda \tag{16.28}$$

where *l* and *m* replace more complex terms in the Cobb-Douglas parameters.

Two different cases of this system are illustrated in Figures 16.1 and 16.2. These depend on the relative positions of the curves on which λ and *I* do not change. In the first situation a saddlepoint results, while in the second situation a divergent unstable system is depicted.

Dependent on the technology of the economy, including production, transportation and infrastructure elements, one may find a pathway of the transport system in the economy towards a stable optimum. However, even when faced with the potentiality to reach a stable optimum, it may be necessary that adequate mechanisms and institutions are available for attaining the precise 'saddlepoint' path. For it is not evident that an in-built mechanism takes care of leading the economic-transport system towards a path converging into the equilibrium. Having

---

[1] For instance, the combination of values $\gamma=0.75$, $\alpha_P=\alpha_B=\beta_B=\alpha_T=0.5$ is close to linearity.

discussed now the general a-spatial situation, we will in the next section introduce the element of space more explicitly.

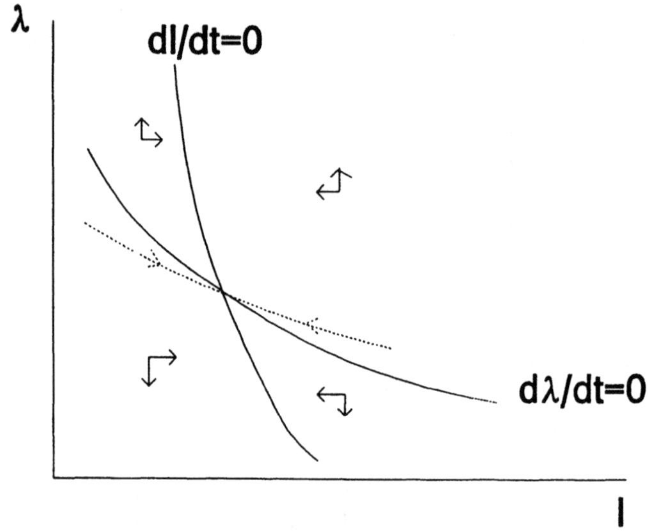

Fig. 16.1. Convergence of a transport-infrastructure-economy system to a saddlepoint.

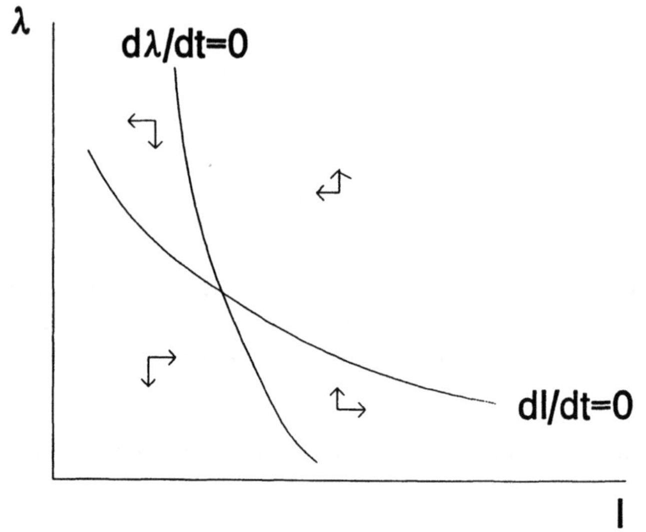

Fig. 16.2. Divergent transport-infrastructure-economy system.

16.4 The Spatial Economic Context of Transportation

In this section we deal with transport in a general equilibrium setting. For this reason the spatial equilibrium character of freight transport is modelled. By choosing a multi-product system we are able to look at spatial flows of intermediate and final commodities. We hope to illustrate the link between interactions among production sectors (and between production sectors and households) on the one hand, and the pricing, production and allocation of freight transportation services on the other hand.[2] An important assumption is that transportation services are required whenever commodities are moved between different locations. The resulting transportation costs depend on a (positive) price for these services, which is in turn dependent on the interaction of supply and demand. Production of the transportation services (T) is assumed to require the input of a primary factor (L_T) - the overall supply of which is fixed - as well as a stock of transportation capital (B_T) (both rolling material and infrastructure). We assume a single household sector with a welfare or utility function ($U(.)$) that consumes two commodity categories (C_1 and C_2). We disaggregate commodity production also into two producing sectors ($i=1,2$), each with a single and unique output. In both sectors the production, represented by a production function ($F_i(.)$), uses as inputs the above mentioned primary factor (L_i) and the transportation capital (K_T), as well as an amount of the commodity produced in the other production sector to serve as an intermediate material input (M_i).

In regard to spatial structure we assume that the first commodity production sector is located in the same place as the household sector. This implies essentially that no transportation is required to move a final good from sector 1 to the household sector. The second production sector however, is located elsewhere. As a result, transportation services are required for trading intermediate inputs and for trading final good 2. This asymmetric treatment of commodities 1 and 2 is chosen to allow for interesting differential interpretations later on.

The overall structure of the model includes constrained optimization formulations as representations of single behaviour of each of the aggregate actors. These include thus one household sector, two production sectors and one transportation sector. All actors are price-takers, all markets are characterized by pure competition, and the primary factor is completely mobile, so that there is a single market and price for it.

[2] Transportation services are measured here in terms of the amount of commodities they can transport from one to another location. In order to prevent an unneccessary complex model it is assumed that an equal amount of transportation services is required to move any commodity over any origin-destination link in any direction. This assumption is plausible without loss of generality with respect to commodity differences, since any commodity measurement may be adapted to the amount of transportation services required to move it over a fixed distance. However, it reduces somewhat the generality with respect to different distances that are likely to exist between different links. Furthermore, we do not address topics of density economies and congestion externalities at this point.

The (aggregate) household sector optimizes its utility by choosing consumption levels for each of the two commodities, subject to an aggregate budget constraint:

$$\textit{maximize} \quad U(C_1, C_2) \tag{16.29}$$

subject to

$$p_{C_1} * C_1 + p_{C_2} * C_2 = Y. \tag{16.30}$$

Here p_{Ci} denotes the CIF price of final good i paid after transportation, i.e. the price paid for commodity i available at the location of the household sector.

The producer of commodity i maximizes his profit by choosing levels of primary and intermediate inputs (L_i and M_i) and output (X_i), consistent with his technological constraint (i=1,2):

$$\textit{maximize} \quad \Pi_i = q_i * X_i - w * L_i - p_{M_i} * M_i \tag{16.31}$$

s.t.:

$$X_i = F_i(L_i, M_i). \tag{16.32}$$

Here q_i is the producer price, w the wage rate, and p_{Mi} the CIF price upon delivery of intermediate inputs. Since we have only two production sectors, the intermediate inputs are always the complement of the commodity produced by the sector concerned.

Next, we state the problem of the transportation sector, which also aims to maximize profits subject to a technological constraint based on only one input. It chooses the level of this primary input (L_T) (which fixes the output of transportation services T) as follows:

$$\textit{maximize} \quad \Pi_T = p_T * T - w * L_T \tag{16.33}$$

subject to

$$T = F_T(L_T). \tag{16.34}$$

The market clearing conditions for commodities 1 and 2 are the following:

$$C_1 + M_2 = X_1 \tag{16.35}$$

and

$$C_2 + M_1 = X_2. \tag{16.36}$$

The market for transportation services is also assumed to clear. Since we assumed transportation services as amounts of moved goods from an origin to a destination, clearing is formulated as follows:

$$M_1 + M_2 + C_2 = T. \tag{16.37}$$

All earnings are flowing to the single household sector. Its income thus originates from 6 sources, namely from each of the three sectors payment for the provision of primary factor services, and the respective earnings out of profits:

$$Y = w * (L_1 + L_2 + L_T) + \Pi_1 + \Pi_2 + \Pi_T. \tag{16.38}$$

Finally, we have four price equations that specify delivered prices of final goods and intermediate inputs in terms of producer and possibly transportation prices[3]:

$$p_{C_1} = q_1, \tag{16.39}$$

$$p_{C_2} = q_2 + p_T, \tag{16.40}$$

$$p_{M_1} = q_2 + p_T, \tag{16.41}$$

and

$$p_{M_2} = q_1 + p_T. \tag{16.42}$$

Notice that the delivered price of final good 1 equals its producer price, since the production of this commodity and the household are located in the same spot. In all other cases transportation costs are incurred.

The Lagrangean for the commodity producer problem i is $\mathcal{L}_i = \Pi_i - \lambda_i[X_i - F_i(L_i, M_i)]$, so that the constrained profit maximization gives, in the case of decreasing returns to scale, $q_i = \lambda_i$, $w = \lambda_i * F_{i,Li}$, and $p_{Mi} = \lambda_i * F_{i,Mi}$, and hence each input is paid the market value of its marginal product. Similarly for the transportation sector, we can derive that $w = p_T * F_{T,LT}$ under decreasing returns to scale.

Households' behaviour results in $U_{C1} = \mu * p_{C1}$ and $U_{C2} = \mu * p_{C2}$, where μ is the shadow price of the budget restriction, interpreted as the marginal utility of income. Since a general equilibrium model can only generate relative prices we may choose a numéraire. One may for instance set $p_{C1} = U_{C1}$, which in fact means that we choose a marginal utility of income equal to 1, and therefore $p_{C2} = U_{C2}$. In

[3] In fact, these equations may be derived by introducing a shipper, which minimizes the transportation cost to move a specific commodity from an origin to a destination by choosing transportation services supplied by carriers. This will then lead to spatial price equilibrium conditions which are stated as delivered prices being equal to the sum of producer prices and transportation costs (see for a very clear discussion and the original statements Samuelson, 1952). Since at this stage we lack such a subdivision of transportation activities, we have to impose these conditions upon some of the general equilibrium prices.

this case delivered final goods' prices reflect marginal utilities of the associated commodities.

Using these results, and substituting the price equations (16.39) to (16.42), we obtain a set of equations the solution of which presents exactly the equilibrium values and prices of the model given in equations (16.29) to (16.42). At this point it also worthwhile to use the well-known result in welfare economics of equivalence between a competitive equilibrium and a Pareto-optimal social outcome. To put this more in a more appropriate transportation-related perspective, one may regard the planning formulation as the traditional approach in transportation-allocation problems, such as the traditional optimal transportation problems (starting with Hitchcock 1941). An extension of this approach is the spatial price equilibrium (SPE) approach, which can be regarded as a generalized transportation problem. A comprehensive treatment of this approach is given in Takayama and Judge (1971), and an interesting discussion of different perspectives is given in Florian and Los (1982). But whereas the planning and SPE approaches are generally limited to one product and do not close the economic system - i.e. no interlinkage of different commodities, inputs and outputs of production processes; no income formation -, our framework is more flexible and consistent. Furthermore, these approaches do certainly not consider endogenous Walrasian prices. At the most they may include variable transportation costs as a result of congestion externalities and density economies.

A difference in calculating the solutions of a planning versus a general equilibrium model is that the general equilibrium model treats prices and quantities in a similar and integrated manner, while prices are a by-product of the planning model. However, more crucial is of course the possibility to deal with transportation policy instruments, imperfect competition, externalities, and inefficient (second-best) outcomes. Of specific interest for our subject is the possibility to disaggregate the transportation subsystem into various (types) activities, for the same level (e.g., different modes of transport) or for separate levels (e.g., organizers and actual transporters, 'shippers' and 'carriers'), and to deal with transportation network characteristics. Before we look into some of these extensions, we consider the system of equations whose solution is the competitive equilibrium for our spatial economy with transportation.

The reduced set of equations representing first-order conditions of the model (16.29) to (16.42) is as follows (11 equations in 11 unkowns):

$$w = q_1 * F_{1,L_1}, \tag{16.43}$$

$$w = q_2 * F_{2,L_2}, \tag{16.44}$$

$$w = p_T * F_{T,L_T}, \tag{16.45}$$

$$q_2 + p_T = q_1 * F_{1,M_1}, \tag{16.46}$$

$$q_1 + p_T = q_2 * F_{2,M_2}, \tag{16.47}$$

$$q_1 = U_{C_1}, \tag{16.48}$$

$$q_2 + p_T = U_{C_2}, \tag{16.49}$$

$$C_1 + M_2 = F_1(L_1, M_1), \tag{16.50}$$

$$C_2 + M_1 = F_2(L_2, M_2), \tag{16.51}$$

$$M_1 + M_2 + C_2 = F_T(L_T), \tag{16.52}$$

$$L_1 + L_2 + L_T = L_S. \tag{16.53}$$

We can solve the price variables in terms of the quantity variables by using equations (16.43), (16.46), (16.47) and (16.48):

$$q_2 = \frac{F_{1,L_1}}{F_{2,L_2}} * U_{C_1}, \tag{16.54}$$

$$p_T = \frac{F_{1,L_1}}{F_{T,L_T}} * U_{C_1}, \tag{16.55}$$

$$w = F_{1,L_1} * U_{C_1}. \tag{16.56}$$

The way to find a solution for this system is to solve a block of only quantity variables and then substitute these in (16.48) and (16.54) to (16.56) to obtain the prices. This 'quantity block' consists of 7 equations in the 7 quantity variables, namely (16.50) to (16.53) plus the following three equations derived by substitution of (16.48) and (16.54) to (16.56) in (16.46), (16.47) and (16.49):

$$\left(1+F_{2,M_2}\right) * F_{1,L_1} = \left(1+F_{1,M_1}\right) * F_{2,L_2}, \tag{16.57}$$

$$F_{T,L_T}+F_{2,L_2} = \frac{F_{1,M_1} * F_{2,L_2} * F_{T,L_T}}{F_{1,L_1}}, \tag{16.58}$$

$$F_{1,M_1} * U_{C_1} = U_{C_2}. \tag{16.59}$$

From (16.57) it follows that in a competitive equilibrium the primary factor is allocated such that the total productivity is equal in each direction, taking both direct and indirect - based on intermediate inputs - productivity effects into account. Equation (16.58) can be interpreted in that primary inputs should be allocated such that the sum of direct productivity effects in sector 1 and the transport sector should equal the indirect productivity of intermediate relative to primary inputs in sector 1. The indirect marginal intermediate productivity is the product of the marginal increases of: output of sector 2 as a result of extra labour; produced transport services; and the direct intermediate input effect in sector 1 resulting from the extra transported intermediate commodities from sector 2 to 1. Relation (16.59) states that in a competitive equilibrium one is indifferent in using commodity two directly for consumption or as in intermediate input in the production of commodity one.

Now we will address the question of how to optimally alleviate the negative environmental consequences of freight transportation. One may think of fuel taxes, car taxes, and road-pricing as examples of specific policy actions.

Two additions compared to the framework of the previous two paragraphs are therefore of importance here. One is the inclusion of external effects from freight transportation impacting on general welfare, i.e. welfare of households. Furthermore, a government (or public decision-maker) has to be included that levies appropriate charges and balances its income with its outlays[4]. In view of the still rudimentary structure of our framework and especially the transportation sector, the environmental policy charge may be either directed at transportation

[4] This means we do not pose any optimization structure upon the government, unlike the case of, for instance, the optimal provision of public transportation. It is interesting to recognize that implicitly the government is a social welfare optimizer when it chooses optimal externality charges. However, the charges are set or calculated by the model analyst, and not by the government in our framework.

inputs (e.g. gasoline reduction) or transport systems use (i.e. the amount of traffic, e.g. through road-pricing).

Because the production of transportation services takes place with only one input, we will choose the latter option here. Then we can be certain about the impact being directly and strictly aimed at transportation. The input aggregation of our framework does in its present form not allow one to distinguish between, for instance, gasoline and other petroleum products. So a charge on inputs would also have unrealistic consequences for other industries using the same input, since primary inputs are assumed homogeneous and mobile. In other words, the choice for charges on the supply of transportation serves allows us to keep the framework simple, which we believe is important to obtain interpretable results in this exploratory stage.

In the externality-extended model, the supply of transportation services T enters the utility function of the aggregate household sector with a negative marginal utility for all values of C_1, C_2 and T, so as to reflect the negative external cost. The social welfare function U is formulated as $U_n(C_1, C_2, T)$. In the general equilibrium formulation the new transportation services' price $p_{n,T}$ is equal to the sum of the marginal value of an additional output (the 'old' price p_T) plus the charge v per unit of transportation service. This implies that we can replace the 'old' price by $p_{n,T}-v$. An alternative way to interpret this is to regard $v*T$ as an additional cost factor for the transportation sector.

The price p_T of transportation services in the revised planning model with externalities is equal to the producer price $p_{n,T}$ of transportation services in the competitive equilibrium solution with optimal transportation charges. Since p_T can be expressed in terms of the externality measure $U_{n,T}$ [5], and $p_{n,T}$ in terms of the charge level v, one can establish the relationship between v and $U_{n,T}$. The optimal charge level is exactly the marginal social externality cost, i.e. $v= -U_{n,T}$. Since $U_{n,T}<0$, the price of transportation services is higher than the producer cost. From (16.55) one can see by substituting $p_{n,T}-v$ for p_T that the new $p_{n,T}$ satisfies:

$$p_{n,T} = \frac{F_{1,L_1}}{F_{T,L_T}} * U_{n,C_1} + U_{n,T}. \tag{16.60}$$

The general equilibrium implications of a positive charge v, compared with the non-externality situation of the previous sections, can be categorized in the following order:

(i) the initially higher transportation price makes the demand for transportation services decline;

(ii) supply adjusts to demand to realize market clearing for transportation services;

(iii) supply for transportation services declines;

(iv) $\partial F_T/\partial L_T$ goes up in the case of decreasing returns to scale (convexity) of $F_T(.)$), and equals zero for constant returns to scale;

(v) in case of decreasing returns to scale both the externality charge and the

[5] The notation $U_{\cdot,k}$ replaces $\partial U_{\cdot}(C_1,C_2,T)/\partial k$, with $k=C_1,C_2,T$.

indirect effect through $\partial F_T/\partial L_T$ push the transportation price p_T further up;

(vi) using the relations for the prices q_1, q_2 and w in equations (16.48), (16.54) and (16.56), which do not alter[6], and - compared to p_T in (16.55) - the increased $p_{n,T}$ as shown in (16.60), one can see that the two commodity and primary factor prices decrease relative to the transportation price; therefore, the relevant 'quantity block' (see the previous section) solution will generally change;

(vii) the delivered prices will increase, as a possible second-order declining effect (i.e. not only relative but also absolutely) on producer prices through relationships (16.48), (16.54) and (16.56) can be expected of a lower order of magnitude than the increase in their transportation cost part through (16.60);

(vi) finally, lower volumes of consumption of both commodities will result; this may be regarded as reflecting the trade-off between final consumption and the negative transportation externalities that they directly and indirectly are dependent on.

We have not yet addressed the question what the government will do with the optimal taxes in the simple model. The above general reasoning is however independent of various different choices of redistribution of the total charge receipts. One way is to immediately return these to the consumer. Alternatively, one may include a fixed or variable payment for supplied factor services by households, necessary to take care of administration and control of the charges (since we do not assume any other task for the government here). The first option has the advantage of having only second-order effects on the primary factor market, so that comparisons with the non-government situation are easier.

Finally, a further type of analysis that can easily be done within the present framework is to charge only specific transportation flows, concerning either final consumption (to households) or intermediate inputs (between firms). It will not give the social optimal outcome that we have studied above. But this approach is interesting for the reason of comparing its outcome with the full externality cost and the socially optimal ones, as well as the differential implications for individual actors. It may, for instance, be the case that compared to the social optimum some agents benefit, while others lose. This would need further investigation.

16.5 New Transportation Technology Regimes

Having discussed now the macro-economic and allocative aspects of the transportation sector as well as its environmental implications in a general equilibrium framework, it is now worthwhile discussing qualitative changes at the supply side of transport systems which aim to alleviate some of the distributional frictions or environmentally disruptive impacts. Such 'jumps' in transport systems

---

[6] More accurately, the three mentioned price formula's in quantity variables do not change, but their values may of course change when the quantity variables have different values.

are caused by the indivisibility and lumpiness of network infrastructure, which cause the implementation of a new system to be of a 0-1 nature. This holds for both expansion of existing transport modes (e.g. the construction of a new motorway) and realization of competing modes next to the existing ones (e.g. a high speed railway system). Such additions to the stock of transport systems capital may also emerge from another source, viz. the implementation of a new technology which serves to make the existing transport system more efficient or environment friendly. The introduction of telematics, aiming at suppressing congestion, lack of safety and external costs of environmental damage, is a good example of another type of qualitative shift which has similar impacts on transport systems performance. Such qualitative 'jumps' may be the result of economic forces (e.g. market contestability on different transport markets), policy interventions or outdated infrastructures in the final stage of a system's life cycle.

Transport and communication systems are thus never static, but always in a state of flux. In general, transport modes - like any other commodity category - exhibit a product life cycle marked by phases of take-off, adoption, market penetration, large-scale use, saturation and declining market shares. Thus transport modes will tend to show a history beginning with a period of low-volume high-cost adoption, followed by further expansion and improvement, and ending with general acceptance and widespread use. Clearly, in way parallel to transport technologies, transport management styles will also exhibit similar life-cycle phenomena. Technical change may intervene here by providing new hardware to facilitate better control or enhance the quality of information which is available to management.

In terms of the model analyzed in Sections 16.2 and 16.3, the following adaptation can be made to allow for technology jumps. In the static model of Section 16.2 one may include an additional transportation technology and/or an additional infrastructure technology. When it is assumed that the one cannot do without the other, a separate infrastructure-transportation system may be added. In terms of variables, this means that one disaggregates T into T_i and B into B_j; i and j refer to distinct types of transportation technology and infrastructure, respectively. The following conditions can then easily be derived for the case where $i=1,2$ and $j=1,2$:

$$\frac{\partial F_{T,1}}{\partial B_1} * \frac{\partial F_{B,1}}{\partial L_{B,1}} = \frac{\partial F_{T,2}}{\partial B_2} * \frac{\partial F_{B,2}}{\partial L_{B,2}} \tag{16.61}$$

where special cases are possible: $F_{T,1}=F_{T,2}$ when a single transportation technology uses the old and the new infrastructure; $B_1=B_2$ (and $F_{B,1}=F_{B,2}$) when the old and new transportation technologies use the same infrastructure. This condition states that in optimum the primary factor is allocated over infrastructure production such that the marginal productivity - measured as the reaction of transportation sector output with regard to a change in primary input in infrastructure production - is equal for all infrastructural and transport technologies.

A second condition that is interesting in this context is as follows:

$$\frac{\partial F_P}{\partial T_1} * \frac{\partial F_{T,1}}{\partial L_{T,1}} = \frac{\partial F_P}{\partial T_2} * \frac{\partial F_{T,2}}{\partial L_{T,2}} \tag{16.62}$$

This says that in optimum marginal changes in the use of each transportat technology will lead to an equal productivity change measured as change in commodity production.

In the static model however only one of the existing technologies may be chosen. Since usually the new technology will be much more efficient than the old one, the conditions in (16.61) and (15.62) imply that no or little use is made of the old technology in the optimum, be it transportation or infrastructure technology. To allow for more flexibility, two options are available. The first one is based on combining the dynamics of investment, as in Section 16.3, with the multiple technology situation, so as to incorporate sluggishness of or delay in technology regime transformations. In that case either a gradual or a sudden shift from one to another system can be modelled. The sudden shift occurs when investment is not effective until a minimum amount of new technology ('capital') is generated, or when a lag period has gone by.

A second option to introduce more flexible model outcomes is the incorporation of negative external effects, especially ones that affect the performance of the own sector, such as congestion. In that case it may be optimal to employ multiple infrastructure-transportation systems alongside using each at its individual optimal level. In order to formalize this however, one should develop a model at the micro-level, or make use of non-monotonic production functions.

Dynamics in transport systems may take various forms: shifts in size and market shares, intermodal substitution or complementarity (e.g. intermodal) transport, emergence of new transport modes, qualitative upgrading of existing modes (e.g. rapid trains). Grübler and Nakicenovic (1991) have made an interesting historical overview of the evolution of many transport systems from the viewpoint of a lifecycle approach. Based on a logistic growth hypothesis related to various technologies, they give a sketch of the expected time trajectory of various transport modes (see Figures 16.3 and 16.4).

It has already been mentioned that qualitative 'jumps' in transport systems may cause major perturbations in the outcomes of such systems. An interesting illustration of such issues can be found in Reggiani et al (1993), where - as part of the EC DRIVE programme on road telematics - an attempt has been made to identify European freight transport corridors which are in particular suitable for a qualitative upgrading by using various types of telematics technologies (e.g. route guidance systems, intelligent vehicle design etc.). Various data on European freight flows have been collected in order to identify those European corridors which would in particular benefit from such a new technology jump. An illustration of such data on a digital map can be found in Figure 16.5.

Apart from flow data, also numerous data on adjacent or crossed regions have been collected. The data can be subdivided into three classes:

- socio-economic attributes of the regions crossed
- transport
- transport demand attributes

Using a typological analysis followed by a multicriteria analysis on the - sometimes qualitative - data, interesting and telematics-relevant corridors and regions could be identified. This has led to the demarcation of a promising European telematics space, denoted by the so-called *Euregg* map (see Figure 16.5), which follows to a large extent the so-called 'blue banana' (see RECLUS, 1989).

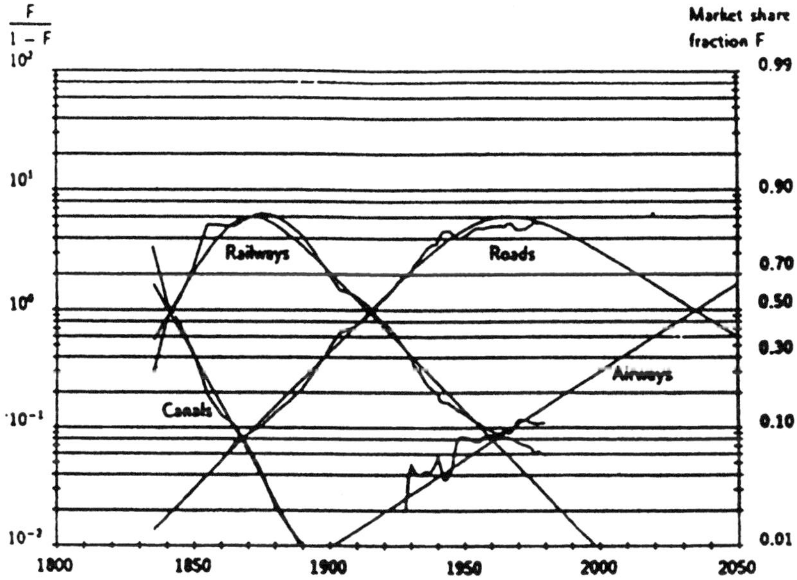

Fig. 16.3. Substitution of transport infrastructures in the USA, shares in length, logit transformation. (Source: Grübler and Nakicenovic 1991).

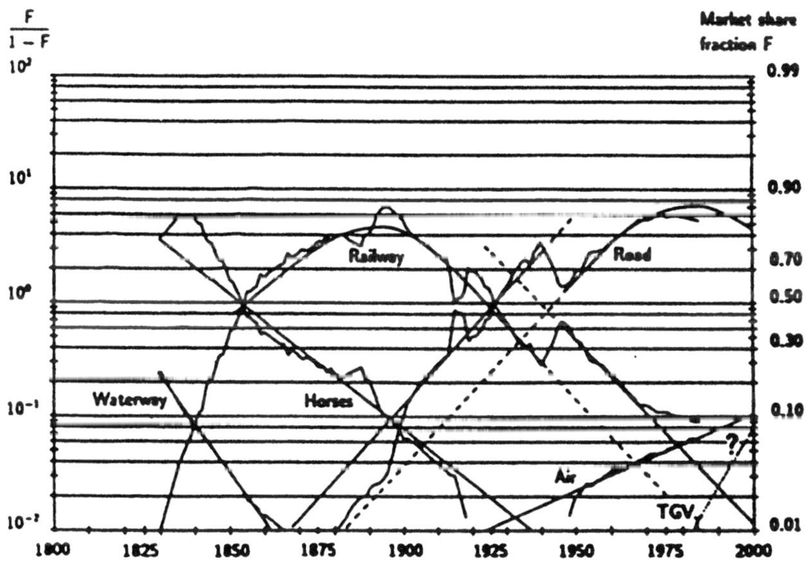

Fig. 16.4. Modal split in passenger (intra-and intercity) transport in France, in fractional share of passenger-km performed, logit transformation. (Source: Grübler and Nakicenivic 1991).

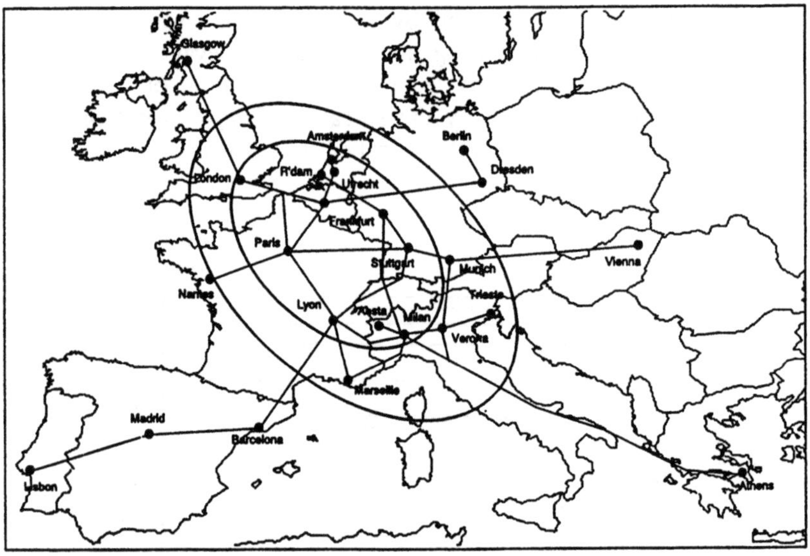

Fig. 16.5. The Euregg model for telematics applications. (Source: Reggiani et al 1993).

16.6 Reflection and Conclusion

In the above discussion an attempt was made to show the usefulness of the static and dynamic general equilibrium framework as a tool for economy-wide analysis of transportation system-related externalities, investments and policies. Two important aspects addressed are the analysis of multiple types of competing transportation activities and infrastructure networks on the one hand, and the analysis of various isolated or combined externalities on the other hand. However, the full complexity of transportation systems as well as the inclusion of more extensive infrastructure networks seems to require the use of numerically solvable (applied) general equilibrium models. This is a logical subsequent step. The aim of this paper was only to offer a conceptual and theoretical background for such more operational studies. It is thus evident that the research agenda in this field is vast. Examples of more fundamental issues to be investigated in more detail are:

- the incorporation of dynamic overlapping generations models for new infrastructure and transportation technologies conditions;
- a linkage to evolutionary economics which would - through dynamic competition models - allow to 'pick the winners';
- a consistent connection between micro behaviour and meso (system's) behaviour, e.g. by using Wardrop's two principles on user equilibrium and system optimum;
- the use of neural networks in order to improve the predictive capability of dynamic transportation models (and applied spatial equilibrium models in general).

Acknowledgement

We are grateful to Piet Rietveld for useful comments.

References

Florian M, Los M (1982) A New Look at Static Spatial Price Equilibrium Models. Regional Science and Urban Economics 12: 579-597
Grübler A, Nakicenovic N (1991) Evolution of Transportation Systems. Research Report RR-91-8. IIASA, Laxenburg
Hitchcock FL (1941) The Distribution of a Product from Several Sources to Numerous Localities. Journal of Mathematics and Physics 20: 224-230
Reggiani A, Lampugniani G, Nijkamp P, Pepping G (1993) European Corridors Characteristics. Paper DRIVE Programme. EC, Brussels
Roson R (1993) Computable Spatial Economic Equilibria and Freight Network Modelling. International Journal of Transport Economics 20: 51-66
Takayama T, Judge GG (1971) Spatial and Temporal Price and Allocation Models. North-Holland, Amsterdam

17 Migration Equilibria in Regional Economies: A Multi-Period CGE Analysis of an Improvement in Local Amenities

Peter G. McGregor, John K. Swales and Ya Ping Yin
University of Strathclyde

17.1 Introduction

In this paper we investigate the possible system-wide impacts on the local economy of an improvement in amenities. The amenity improvement could take a variety of forms, such as upgrading the environment or enhancing local public service provision, and might involve public sector action.[1] However, we here take the change to be entirely exogenous, and concentrate solely on the lasting economic consequences of the amenity improvement (relative to other regions). We are therefore not concerned with any transitory stimulus that expenditures associated with amenity improvement may bring, nor do we attempt to assess 'non-economic' effects on the quality of life in the locality.

The lasting economic effects of an improvement in local amenities are likely to arise through the attraction of firms and households into the region and, possibly, through increased tourism. Here we focus exclusively on the household in-migration effect, using a model which draws on the recent equilibrium, as well as traditional disequilibrium, models of the migration process.[2] In Section 17.2 we provide a theoretical analysis of the impact of an amenity improvement in this context. This generates (perhaps surprisingly) powerful results. In fact, it proves possible to solve for the impact of an amenity improvement on the real wage and the unemployment rate with a simple two equation model, a parameterized version of which allows the derivation of empirical results.

Powerful though the analytical model, and its parameterized empirical counterpart are, where the region is characterized by the presence of persistent nominal inertia, or where interest extends beyond the wage and unemployment characteristics of steady-state equilibria, the simple model is inadequate. Both

---

[1]It may be worth noting that expenditures on environmental improvements are non-trivial. In Scotland, for example, Scottish Enterprise spent some £83 million on such improvements in 1992-3 out of a total budget of £449 million.

[2]For a discussion of these alternative views of the migration process see, for example, Graves (1980), Graves and Mueser (1993), Schacter and Althaus (1989, 1993), Evans (1990, 1993), Harrigan and McGregor (1993) and Hunt (1993).

issues seem to be of importance to regions. First, in a UK context, at least, it has often been argued that the nominal wages of peripheral regions are determined through a centralized bargaining system, giving rise to persistent nominal inertia in such regions. Secondly, the policy-relevance of the analytical results must surely depend on factors other than the properties of steady-state equilibria. For example, the speed of adjustment to equilibria would appear to be critical in governing the policy response: if this were measured in decades rather than months policy-makers responses are likely to be modified considerably. Accordingly, we seek to employ AMOS, our multi-period computable general equilibrium (CGE) model of the Scottish economy, to investigate these (and other) issues of interest which cannot be addressed with the simple two equation model. We outline our simulation model's structure in Section 17.3. In Section 17.4 we present the results of simulating the impact of an improvement in the local environment. We would emphasize that our quantitative analysis is exploratory, since we currently have no way of linking expenditures on environmental improvement to the scales of the amenity differential implied by our net migration function. We present tentative conclusions in Section 17.5, where we also suggest possible avenues for future research.

17.2 Theoretical Analysis of System-Wide Responses to an Improvement in Local Amenities

17.2.1 The Basic Model

We adopt the general form of the net migration function which is presented in Layard, Nickell and Jackman (1991, chapter 6), Greenwood et al. (1991) and Treyz et al. (1993). This relates net migration flows (relative to lagged labour force) to real consumption wages (nominal wages deflated by the relevant consumer price index), employment/unemployment rates and amenity differentials. In the short-run the net migration equation yields the value of net migration flows:

$$M_s = e(W_s, W_r, N_s, N_r, A \;..) \quad e_{W_s}>0;\; e_{W_r}<0;\; e_{N_s}>0;\; e_{N_r}<0;\; e_A>0; \tag{17.1}$$

The net in-migration rate to the target region, Scotland, (M_s) is increasing in both the Scottish real-consumption-wage (W_s) and the Scottish employment rate (N_s) and decreasing in the values of these variables in the rest-of-the-UK (W_r and N_r, both of which are exogenous in our analysis). An improvement in the amenity differential in favour of Scotland (an increase in A) stimulates the rate of net in-migration.

However, these net migration flows continue until a new stock equilibrium is ultimately established. Setting net migration flows equal to zero in (17.1) generates the zero-net-migration condition (17.2). Equation (17.2) incorporates explicit allowance for the endogeneity of the labour force and population.

$$N_s = g(W_s, W_r, N_r, A, ...) \quad g_{W_s} < 0; g_{W_r} > 0; g_{N_r} < 0; g_A < 0; \tag{17.2}$$

An increase in the wage is associated with a reduction in the employment rate in (17.2) (an increase in the unemployment rate), since zero net migration requires that the unemployment rate 'compensates' fully for the increased attractiveness of the region that a higher wage implies. Accordingly, the zero-net-migration locus, ZNM_0, is negatively sloped in Figure 17.1 in real-consumption-wage and employment-rate space, reflecting the fact that a higher real wage has to be compensated by a lower employment rate (higher unemployment rate) to maintain a steady-state equilibrium, characterized by zero net migration. ($\delta M/\delta W = -e_{W_s}/e_{N_s} < 0$)

In the presence of an endogenous labour force, real-consumption-wage and unemployment rate equilibria may be fully characterized by only two equations: one for zero-net-migration, equation (17.2); and one for wage-setting (or labour supply in a competitive context), equation (17.3) below.

$$W_s = f(N_s, ...) \quad f_{N_s} > 0 \tag{17.3}$$

Equation (17.3) is written as a wage-setting equation, with the real-consumption-wage (W_s), which is simply the nominal wage deflated by the consumer price index, increasing in the employment *rate* (N_s). The wage is therefore decreasing in the unemployment rate. Layard, Nickell and Jackman (1991) motivate such a function at the aggregate level in a variety of ways, including union bargaining and efficiency wage considerations, although the simplest interpretation of (17.3) is as an inverted competitive labour supply function. We employ the function here as a generic form which may capture a range of regional wage behaviour.

The wage-setting schedule is upward sloping in real-consumption-wage and employment-rate space, a case captured in Figure 17.1 by the upward sloping competitive labour supply curve, CLS. Under present assumptions equations (17.2) and (17.3) suffice to tie down the real wage - unemployment equilibrium. In Figure 17.1 this equilibrium occurs at point A where the CLS and ZNM_0 schedules intersect. Accordingly, across zero-net-migration equilibria, the real consumption wage and the employment (and unemployment) rate are determined entirely by the arguments of the wage-setting and net migration functions, as Layard, Nickell and Jackman (1991) maintain.

In the short-run, of course, the economy is not tied to the ZNM_0 locus. Short-run equilibrium is in fact determined in the model by the intersection of the wage-setting schedule and the (general equilibrium) demand curve for labour, which is negatively sloped in W-N space[3]. Suppose that point A' represented such a short-

[3]In fact, the general equilibrium demand curve for labour may become positively sloped if commodity demands are very inelastic and substitution possibilities in production are very limited. Indeed, when configured as a conventional I-O system (with endogenous consumption), the general equilibrium demand curve for labour does become positively sloped in AMOS, and this has an important bearing on the system's properties. (See McGregor, Swales and Yin, 1995.) However, with our default parameters, which we employ in this paper, AMOS implies negatively sloped general equilibrium labour demand curves

run equilibrium position. (To avoid cluttering the diagram we do not show the labour demand curve; in any event our purpose is to show that such a position is not sustainable.) At A' the values of W_s and N_s are such as to generate net in-migration. This increase in population has two effects. First, the increase in population stimulates consumption and therefore the demand for labour. However, if the stimulus to demand is outweighed by the increase in the labour force (as seems likely given the presence of any component of demand (such as exports) which is not linked to the population of the region) then the stability of the model is assured. Accordingly, the employment-rate demand curve shifts inwards, leading to a move down the CLS curve. The only sustainable equilibrium is at point A where CLS intersects the ZNM_0 locus.[4]

An improvement in the region's amenities shifts the zero-net-migration curve to the left (to ZNM_1) in Figure 17.1, indicating that the effect of this in attracting in-migrants has to be offset, across zero-net-migration equilibria, by some combination of a lower real wage and higher unemployment rate (given that amenities are here assumed to be exogenous). Point A is now associated with net in-migration and, given the stability assumption, a new equilibrium must ultimately be achieved at point C where CLS intersects the ZNM_1 locus. Of course, although the real wage and the employment rate in the subject region both fall, population, labour supply, employment levels and value-added are all stimulated by the in-migration.

It is straightforward to derive analytical expressions for the induced changes in the equilibrium rate of employment and the wage rate from (17.2) and (17.3):

$$dN_s/dA = \frac{g_A}{(1 - g_{w_s} f_{n_s})} \leq 0 \qquad (17.4)$$

$$dW_s/dA = \frac{f_{N_s} g_A}{(1 - g_{W_s} f_{N_s})} \leq 0; \qquad (17.5)$$

The improvement in local amenities reduces both the equilibrium employment rate and the real wage. It may seem paradoxical that an improvement in the attractiveness of a location in this case results, in general, in some combination of a lower real wage and higher equilibrium ('natural') unemployment rate. However, recall that these results apply across zero-net-migration equilibria: the

[4] If, however, the stimulus imparted by in-migrants to the local economy was sufficiently great, because of their greater wealth, entrepreneurship or whatever, it is conceivable that the employment rate demand curve would shift out to the right from point A': the model would be unstable in this case, and would imply a "cumulative causation" framework. While some regions may periodically have appeared to conform to this case in the presence of rapid expansion or decline, it seems far from typical and we assume that the system is stable in the remainder of the paper. Certainly, AMOS with the default values of key parameters does not exhibit instability.

reduced real wage and higher unemployment rate are required to stem the inflow of new migrants so as to restore steady-state equilibrium.

Clearly the location of the new zero-net-migration equilibrium depends upon the elasticities of the relevant schedules. With a highly elastic labour supply schedule (ie a highly inelastic wage-setting schedule) f_{Ns} tends to zero, increasing the size of the fall in the employment rate (rise in the unemployment rate), and reducing the size of the fall in the real wage. The limiting case of 'real-wage-resistance' is illustrated by curve RWR in Figure 17.1, where only the unemployment rate adjusts to restore a zero net migration equilibrium with wage W_r and the new natural unemployment rate U_r (employment rate N_r). With a very inelastic labour supply schedule (elastic wage-setting schedule), the reverse effects occur. The limiting case here, that of an exogenous employment rate, implies that real wages absorb the full impact of the disturbance, by falling to W_f, and the 'natural rate' of unemployment is unaffected. Various arguments have been advanced to motivate these labour market closures in a regional context (see eg Harrigan et al., 1992), though the exogenous employment rate should simply be regarded as a benchmark (one for which there is little or no empirical evidence to our knowledge).

Ultimately then, the improvement in amenities typically generates some combination of an adverse shift in the real consumption wage and an increase in the natural rate of unemployment. Assessing the need for regional policy intervention on the basis of either real wages or unemployment might therefore seem problematic: 'low-wage, high-unemployment' regions could simply be 'amenity-rich'. This is an idea which has some currency in policy-making circles, manifested most sharply in the notion of the 'Happy Welshman'!

17.2.2 A Specific Example: The Layard, Nickell and Jackman Model

It is instructive to consider a concrete example of the use of the wage-setting/zero-net-migration model: specifically, we take the case of the functional forms which are actually employed in our general equilibrium simulation model, AMOS. We adopt the form of the net migration function employed by Layard, Nickell and Jackman (1991, Chapter 6), which was estimated on UK data.[5]

$$M_s = \alpha_0 - 0.08(u_s - u_r) + 0.06(w_s - w_r) \tag{17.6}$$

where M_s is the net migration rate into Scotland and where lower case letters indicate natural logarithms of the corresponding variables used above. Here we determine the value of α_0 so as to ensure a steady-state equilibrium in the base year of our model, and we interpret this parameter as capturing the amenity

[5]The same functional form is employed by Greenwood *et al* (1991) and Treyz et al (1993) and estimated on US data. The estimated elasticities are roughly double the UK values.

differential in favour of the subject region (if positive, as here).[6] In a zero-net-migration equilibrium, M_s is, of course, zero and imposing this condition on (17.6) yields[7]:

$$u_s = \frac{-\alpha_0 - 0.06(w_s - w_r)}{-0.08} + u_r = -3.26 + 0.75w_s \tag{17.7}$$

This is the zero-net-migration condition that corresponds to the net migration function (17.6). This is a specific form of equation (17.2) and, of course, the partial derivatives take the same signs, although they are here expressed in terms of the unemployment rate (and can be identified here in terms of particular elasticities or combinations thereof).

We introduce the amenity improvement by increasing the value of α_0 by 50%: that is, we increase the implied amenity differential in favour of Scotland by 50%.[8] Of course, it is then possible to re-solve the parameterized version of the two equation system to obtain the estimates of the induced change in the equilibrium wage and unemployment rates. On the assumption of a competitive labour supply, with real wage elasticity of 0.4, the real wage falls by 1.4% and the percentage rate of unemployment rate increases by 13.2% . Under real wage resistance only the unemployment rate adjusts - here by 14.6%. Under the exogenous unemployment rate closure the real wage has to fall by 13.4% to restore a zero-net-migration equilibrium following the disturbance.

17.2.3 Some Complications and Extensions

Other variables. The two equation system identified above may only be solved for the equilibrium values of unemployment and wage rates. If there is interest in other variables such as employment and output levels and their sectoral composition then, in general, these can only be obtained by solving the entire regional general equilibrium system. (The relevant labour demand curve, for example, is one which takes full account of the effects of wage changes on prices and incomes.)

Nominal inertia. The idea of persistent nominal inertia implied by notions of a spatially integrated wage bargaining system is captured by a fixed nominal wage closure, which cannot be handled by means of the simple two equation system:

---

[6]We are currently experimenting with alternative measures of variations in amenity in a function of this type estimated on Scottish / rest-of-UK migration data.

[7]The level of the unemployment rate is expressed as a proportion and u_s is the natural logarithm of this.

[8]Of course, the value of the intercept may, in practice, contain other effects too. For this reason (and others) the numerical analysis must be regarded as indicative. We are currently attempting to obtain direct measures of amenity effects in the estimated net migration function.

again in this case we need to solve the entire system.[9] We consider this case below in the context of our simulation model.

Endogenous capital stocks. First note that the capital stock does not enter the preceding analysis at all. Only the wage-setting and zero-net-migration functions matter for real wage and unemployment rate determination (abstracting from the persistent nominal inertia case), and the level of the capital stock is not relevant to either of these. However, allowing for adjustment of the physical capital stock will influence the equilibrium values of the levels of other variables in the system. (As noted above, these values can only be obtained from a solution of the entire general equilibrium system.)

Adjustment speed. It is appropriate to note the dependence of the significance of our analysis for practical policy-making on the duration of the adjustment period. If static-steady-state equilibria take a very long time to be established, then the properties of such equilibria are of much less interest from a practical policy-making perspective than if such equilibria are very rapidly established. Again it transpires that if there is interest in the speed of adjustment, and in the adjustment path, the full general equilibrium system must typically be solved.

Accordingly, although the analytical results reported in the preceding section are very strong, the simple two equation model of equilibrium unemployment and wage rates is inadequate, and a full general equilibrium analysis is required if: the regional economy is characterized by persistent nominal inertia; there is interest in the equilibrium values of variables other than wage and unemployment rates (and the impact of endogenous capital on these); the speed of adjustment and the adjustment path of the economy is of concern. In our view each of these is sufficient motivation in itself for examining the full system-wide ramifications of the change in amenities. We tackle each of these issues through simulation of our regional CGE model.

17.3 The General Equilibrium Simulation Model

AMOS is a CGE modelling framework parameterized on data from a UK region, Scotland[10]. A very brief description is presented in this section: a full listing of the model is provided in Harrigan et al. (1991). AMOS has four domestic

---

[9]For brevity of exposition, we deal with this case only diagrammatically. Unlike the earlier cases, it cannot be solved analytically using the simple two equation model: the model has to be augmented to include the labour demand curve and consumer price determination, both of which are general equilibrium in character. There is, accordingly, no alternative but to solve the entire system-wide model in this context.

[10]AMOS is an acronym for a macro-micro model of Scotland.

transactor groups, namely households, the non-household personal sector, corporations and government[11]; and three commodities and activities, manufactures, non-manufacturing traded and non-traded[12]. In all the simulations in this paper we impose a single Scottish labour market characterized by perfect sectoral mobility. We do not explicitly model financial flows, our assumption being that Scotland is a price-taker in competitive UK financial markets.

The AMOS framework allows a high degree of flexibility in the choice of key parameter values and model closures. However, a crucial characteristic of the model is the conventional neo-classical nature of the supply side in all sectors. No matter how AMOS is configured, we impose cost minimisation in production with multi-level CES production functions (Leontief and Cobb-Douglas being available as special cases). Additionally, in all the simulations reported here we model all sectors as perfectly competitive.

There are four major components of final demand: consumption, investment, government expenditure and exports. Of these, real government expenditure is exogenous. Consumption is a linear homogeneous function of real disposable income. Exports (and imports) are determined via an Armington link (Armington, 1969) and are therefore relative-price sensitive. However, although we can vary the degree of spatial integration through adjustment of the elasticity of substitution of domestic for other goods we do not here investigate the assumption of perfect spatial integration which characterizes the spatial-price-equilibrium approach developed by Takayama. (See, for example, Takayama and Judge, 1964, 1971). Investment is here treated as exogenous except in the long-run closure (in which capital is optimally adjusted, and which is explained below) wherein investment is just sufficient to maintain the capital stock at its desired level.

We employ three conceptual time intervals here which are defined in terms of our treatment of population and capital stocks. We first consider a medium-run interval over which the zero-net-migration condition for (static) steady-state equilibrium is imposed. Population is endogenous over this interval, and, indeed, is optimally adjusted, but we initially assume that the capital stock is fixed. Secondly, we briefly consider the long-run, which is an interval of sufficient duration to allow for the optimal adjustment of capital as well as population stocks[13]. We do consider the adjustment process in our simulation analysis, but given space constraints we restrict attention to the fixed capital stock case. We do this through a short-run model in which the population stock is exogenous within

---

[11]In AMOS, Scotland is treated as a self-governing economy, in the sense that there is only one consolidated government sector. Central government activity is partitioned to Scotland and combined with local government activity.

[12]Manufacturing comprises sectors 11-65, non-manufacturing traded sectors 1-7 and 67-79 and non-traded sectors 8-10, 66 and 80, 81 and 83 in the 1979 Scottish Input-Output Tables (Industry Department for Scotland, 1984).

[13]To render the analysis comprehensive we would also need to consider the case in which capital is optimally adjusted in the presence of an exogenous population. The motivation of such an interval would presumably depend on a judgment that capital was more mobile than labour. The "mixed" interval of the text implicitly depends upon the opposite presumption.

each interval, but is updated between intervals in response to the last period's net migration flows (but in which the capital stock remains fixed.) By running this model forward we can trace out the adjustment path from one zero-net-migration equilibrium (with fixed capital) to another.

17.4 Simulation Results

Tables 17.1 and 17.2 summarize the impact of an exogenous increase in amenities in Scotland across zero-net-migration equilibria for exogenous and optimally adjusted capital stocks respectively. The results represent the percentage change in variables relative to their base values.

17.4.1 Results Across Zero-Net-Migration Equilibria with Fixed Capital Stocks

The fixed nominal wage closure takes labour to be in infinitely elastic supply at the prevailing nominal wage. This case requires a slight modification to the other 'real wage' cases considered in Figure 17.1. The stimulus to demand pushes up the prices of locally produced goods and thus the cpi. Accordingly, the real consumption wage falls under the fixed nominal wage closure, for example, to W_k in Figure 17.1, where the new equilibrium employment rate is N_k, so that the new equilibrium rate of unemployment is $U_k = 1 - N_k$. (This is not a 'natural' rate of unemployment: the nominal inertia which characterizes the fixed nominal wage case renders unemployment rates sensitive to demand.) In this case even the equilibrium values of unemployment and real wage rates require solution of the general equilibrium system. The demand effects of the in-migration are most apparent in the 0.68% stimulus to consumption in Table 17.1. Since consumption is concentrated in the non-manufacturing sectors, these benefit particularly, but total employment increases only very slightly by 0.16%. Manufacturing value-added and employment actually fall (negligibly) because of the increase in the price of intermediate goods used by the manufacturing sector: this increases manufacturing output prices and reduces the demand for manufactures very marginally (with small reductions in exports). Note that the biggest impact by far is on the percentage rate of unemployment, which increases by over 14.5%, an increase generated by a rise in the population of 1.21%. In this closure it is the unemployment rate which does most of the adjustment: the real consumption wage only falls by 0.10%.

As we would expect from our theoretical analysis, the nominal wage actually rises under real-wage-resistance to compensate for the reduction in the real consumption wage which we observe in the fixed nominal wage case. Accordingly, the stimulus to total employment is even less in this case than under the fixed nominal wage closure (0.07% as opposed to 0.16%). This is because the increase in the nominal wage induces 'crowding out' effects through reductions in net exports, effects which hit the manufacturing sector particularly. Consumption demand in this case increases by 0.66%. Again the major impact is on the unemployment rate which rises by 14.6% through the in-migration that ultimately increases population by 1.13%.

The 'fix-wage' closures, although distinct in ways which are explicable, give very similar results. However, these contrast with the results of the 'flex-wage' closures. Our theoretical analysis implies that the contrast should be greatest for the exogenous labour supply closure and Table 17.1 does confirm this. Of course, the unemployment rate is ultimately unaffected, but the large reduction in the real consumption wage which (as we already know) is required to re-establish a zero-net-migration equilibrium generates a stimulus to employment of nearly 13% and is associated with an increase in population of the same percentage amount. In this closure only the real-consumption wage adjusts to restore equilibrium, and it has to move substantially to achieve this, and thereby imparts a significant stimulus to employment. Under real-wage-resistance, only the unemployment rate adjusts. However, while re-establishment of a zero-net-migration equilibrium does require a significant percentage change in the unemployment rate, this is comparatively painlessly achieved given the sensitivity of unemployment rates to demand changes in fix-wage closures.

The competitive outcome is intermediate to the real-wage-resistance and exogenous-labour-supply closures as our theoretical analysis implies. However, notwithstanding the comparative wage-inelasticity of the (default) 'labour supply' equation, the competitive outcome is still associated with major stimuli to employment in all sectors, especially manufacturing. This sector benefits more from the wage cut than any other, even though it does not benefit directly from the consumption stimulus.

17.4.2 Zero-Net-Migration Equilibria with Optimally Adjusted Capital Stocks

Comparison of the results of Tables 17.1 and 17.2 confirms that equilibrium wage and unemployment rates are unaffected by allowance for optimal capital stock adjustment. In the presence of optimal capital stock adjustment, the 'fix-wage' closures now behave like I-O systems. (This result is implied by Samuleson's non-substitution theorem, given the exogenous interest rate.) In these cases it is 'as if' the supply-side becomes passive across such long-run equilibria. Since prices are unaffected here the distinction between the fixed-nominal and fixed-real-wage cases disappears (see, for example, McGregor and Swales, 1994). In general, irrespective of the assumed nature of regional wage determination, the increased elasticity of supply in the presence of endogenous capital stocks leads us to expect the scale of all effects (other than those on real consumption wage rates and unemployment rates) to be increased relative to the exogenous capital stock case.

The RWR and fixed-nominal-wage closures now converge as can be seen from inspection of the relevant columns of Table 17.2. These results reflect the greater elasticity of supply in the presence of optimally adjusted capital stocks, as compared to the fixed capital stock case of Table 17.1. With a variable capital stock, manufacturing employment actually increases (in contrast to the results of

Table 17.1) and the increase in the population required to restore a steady-state is now greater (1.38% as compared to 1.13%). The results of this case are essentially input-output in character, with employment, capital and value-added in each sector increasing equiproportionately with no changes in prices. However, note that unemployment increases significantly, despite the greater stimulus to demand in the long-run (notice the consumption change). This is in stark contrast to the results which would be generated by a conventional I-O system in which the unemployment rate would typically fall as employment increases. The other model closures also imply greater effects in the presence of more elastic commodity supply curves and here real wages fall in response to net in-migration, so prices fall and stimulate demand, especially in the traded sectors.

17.4.3 The Process of Adjustment to Zero-Net-Migration Equilibria

The properties of zero-net-migration equilibria are of interest, but arguably of little practical relevance from a policy perspective if they are only established after a very prolonged adjustment period. Figure 17.2 plots the adjustment path for the real-wage-resistance closure. Convergence to the steady-state equilibrium is not completed until period 13. However, adjustment of total employment is all but complete within 7 periods[14]. The fixed-nominal-wage closure exhibits more rapid adjustment, with the new steady-state achieved by period 11 and total employment adjustment complete by period 7.

The flex-price closures exhibit much less rapid adjustment. In fact, even by period 40 the exogenous labour supply closure has only achieved some 33% of adjustment to the steady-state level of employment![15] Figure 17.3 shows the adjustment path for the competitive closure. Complete adjustment takes 27 years, but 66% of the adjustment in total employment is completed by year 6. As labour supply expands wages fall and this imparts a stimulus to the traded sectors in general and manufacturing in particular.

17.5 Conclusions

We began this paper by examining analytically the impact of an improvement in local amenities on the local economy across zero-net-migration equilibria. Since this constitutes a 'supply-side' disturbance, the 'equilibrium' unemployment rate of the region is affected. In general, it transpires that the greater attractiveness of the local economy has to be 'compensated for' by some combination of a higher

---

[14]We interpret a period here as approximately one year. This reflects the fact that all econometric estimation and all calibration procedures were implemented on annual data sets.

[15]The graph of the multi-period results of the exogenous labour supply closure is not shown due to space constraints.

equilibrium rate of unemployment and lower real consumption wage, provided the system is stable (and is not, therefore, characterized by cumulative causation processes). These changes are required to offset the induced net in-migration which the improved amenities generate.

The tradition in the UK of placing great emphasis on the spatial distribution of unemployment rates to identify regions in need of assistance seems to be partly challenged by this type of analysis: 'high amenity' regions *may*, *ceteris paribus*, be associated with 'low' wage and 'high' unemployment, although our experiment here is simply with a disturbance in a single region. The idea here is that if Welshmen (or Scotsmen) were unhappy with their lower relative wage and higher relative unemployment rates they would migrate. Since a static steady-state is defined by the absence of such flows then, in equilibrium, we have happy Welshmen!

However, while unemployment and real wage rates typically must fall to restore a steady-state equilibrium following an amenity improvement, employment and output both rise (particularly so when capital stocks are fully adjusted), and in this sense the change constitutes 'good news' for the economy of the region. Furthermore, it is important to keep in mind the premises on which the analysis of this paper is based. First we assume that we start from a position of steady-state equilibrium. Many would interpret the continuing net migration outflows which have characterized much of the past experience of Scotland as reflecting continuing spatial disequilibrium: if this is so then an improvement in amenities could act so as to stem the outflow. A second point, which reinforces the first, is that, with UK parameter estimates at least, our simulation experiments suggest that such spatial disequilibria are likely to be prolonged, especially so if regional wages do respond to labour market excess demands. Thirdly, we have assumed homogeneous labour. In fact, migration is known (at least in a UK context) to be selective in age and skill. It is conceivable that skill disaggregation of the model could call for significant modification of our results. Finally, we have abstracted from the impact of firm movement. Further exploration of the effects of relaxing these assumptions constitutes an agenda for future research.

Acknowledgements

The authors are all members of the Fraser of Allander Institute and International Centre for Macroeconomic Modelling, Department of Economics, University of Strathclyde. The authors are grateful to Andrew Bain, Ron Botham and an anonymous referee for comments on an earlier version and to participants in a presentation of a related paper delivered at the workshop in Recent Advances in Spatial Equilibrium Modelling, held in honour of Tak Takayama, Amsterdam, January 1994. The authors gratefully acknowledge the research assistance of Sarah Le Tissier and Kenneth Low and the financial support of Scottish Enterprise National. However, the authors bear sole responsibility for the views expressed and for any errors and misconceptions.

References

Evans AW (1990) The Assumption of Equilibrium in the Analysis of Migration and Interregional Differences. Journal of Regional Science 30:515-31

Evans AW (1993) Interregional Equilibrium. A Transatlantic View. Journal of Regional Science 33:89-97

Graves PE (1980) Migration and Climate Journal of Regional Science 20:227-37

Graves PE, Mueser PR (1993) The Role of Equilibrium and Disequilibrium in Modelling Regional Growth and Decline. A Critical Reassessment. Journal of Regional Science 33:69-84

Greenwood MJ, Hunt G, Rickman DS, Treyz GI (1991) Migration, Regional Equilibrium, and the Estimation of Compensating Differentials. American Economic Review 81:1382-90

Harrigan F, McGregor PG, Dourmashkin N, Perman R, Swales JK Yin YP (1991) AMOS. A Macro-Micro Model of Scotland. Economic Modelling 8:424-479

Harrigan F, McGregor PG, Swales JK (1992) Imperfect Competition in Regional Labour Markets. A Computable General Equilibrium Analysis. Environment and Planning A 24:1463-1481

Harrigan F, McGregor PG (1993) Equilibrium and Disequilibrium Perspectives on Regional Labor Migration. Journal of Regional Science 33:49-68

Hunt G (1993) Equilibrium and Disequilibrium in Migration Modelling. Regional Studies. (Special Issue on Migration, edited by M.J. Greenwood) 27:341-349

Layard R, Nickell S, Jackman R (1991) Unemployment. Macroeconomic Performance and the Labour Market, Oxford University Press, Oxford

McGregor PG, Swales JK (1994) An Investigation into a Neo-Classical Interpretation of Regional Input-Output. Strathclyde Papers in Economics 94/1

McGregor PG, Swales JK, Yin YP (1995) Input-Output Analysis, Labour Scarcity and Relative Price Endogeneity. Demand Disturbances in a 'Flex-Price' Leontief System. Economics Systems Research, forthcoming

Schacter J, Althaus PG (1989) An Equilibrium Model of Gross Migration. Journal of Regional Science 29:143-59

Schacter J, Althaus PG (1993) The Assumption of Equilibrium in Models of Migration. Journal of Regional Science 33:85-88

Takayama T, Judge GG (1964) Equilibrium Among Spatially Separated Markets. A Reformulation. Econometrica 41:510-514

Takayama T, Judge GG (1971) Spatial and Temporal Price and Allocation Models. North-Holland, Amsterdam

Treyz GI, Rickman DS, Hunt GL, Greenwood MJ (1993) The Dynamics of U.S. Internal Migration. Review of Economics and Statistics ?:209-214.

18 Towards a Science of Complexity in Spatial-Economic Systems

Aura Reggiani
University of Bologna

Peter Nijkamp
Vrije Universiteit

18.1 Introduction

Conventional (spatial) economic equilibrium models have usually taken for granted the existence of well-behaved functional forms of state equations, so that existence conditions for optimal solutions could in principle be established. Irregular behaviour was an exception in the tradition of equilibrium analysis. Recent years have witnessed among economists an increasing popularity of *non-linear dynamic models*. The wide range of theories and reflections on evolutionary dynamic systems reflects the continuously rising interest of economists in 'economics without equilibrium' (Kaldor 1985). Kaldor notes in the Okun Memorial Lectures: "it seems clear that if we are to get out of the present impasse we must begin by constructing a different kind of abstract model, one that recognizes from the beginning that time is a continuing and irreversible process; that it is impossible to assume the constancy of anything over time, such as the supply of labour or capital, the psychological preferences for commodities, the nature and number of commodities, or technical knowledge" (Kaldor 1985 p 61).

It is evident that the question of stable economic processes provokes important research issues on the *specification* of economic relationships, not only in formal mathematical terms but also in a qualitative respect, namely: how to model complex feedback processes of economic behaviour of actors, in which socio-economic interactions and externalities are dominant forces? These notions of interactive economies cast doubt on standard assumptions in many economic models (see Morishima 1991; Okun 1981).

In the present paper a critical review of the recent history of non-linear dynamic modelling will be given. A plea will be made to investigate more rigorously the potential of *chaos* modelling, and in particular *bio-ecological* modelling, for economic research. This new endeavour will be illustrated by referring to various applications. Incorporation of interactive and competitive elements from bio-ecological thinking will generate a new type of economic analysis in which *complexity theory* plays a major role.

18.2 Towards Non-linear Modelling

Despite the strong modelling traditions in various social sciences (inter alia, economics, geography, regional science, transportation science), until a few years back surprisingly little attention has been given to *non-linear* models. Apparently, the large majority of model experiments and applications has taken for granted the existence of linear - and thus regular - growth processes. Admittedly, linear dynamic models are certainly able to generate *unstable* solutions, but the solutions of such models are restricted to the following four regular standard types: oscillatory and stable, oscillatory and explosive, monotonous changing and stable, and monotonous changing and explosive. Such models may provide approximate replications of short- or medium-term changes, but fail to encapsulate long-term developments characterized by structure shifts of an irregular or abrupt nature.

The interest in modelling long-run future development paths dates back to the late sixties/early seventies with the emergence of dynamic systems theory (see among others Forrester 1968; Meadows and Meadows 1973). Such models - mainly simulation models - focused attention on the impact of feedback effects in dynamic systems on stable behaviour. The availability of computer software led to a high degree of popularity of such experiments, although also severe criticism arose. In contrast to linear models, such systems dynamic models were able to generate also a-periodic growth patterns.

The interest in unstable systems behaviour remained in the 1970s and 1980s, and generated new contributions in the field of structure changes (such as tensor analysis, singularity theory, bifurcation theory, etc.). Especially *catastrophe theory*[1] gained much popularity in the late seventies, mainly because it is able to describe turning points and jumps - of an asymmetric and sometimes irreversible nature - in seemingly stable systems (see for various illustrations Poston and Stewart 1978; Zeeman 1977). In such systems the direction of movement and the level of threshold values determines the occurrence of catastrophes, which are essentially characterized by the fact that a dynamic system may have various equilibrium points for the same values of state variables.

A weak element in catastrophe models is the fact that the identification and explanation of turning points is fraught with difficulties, so that the operational basis of catastrophe theory remained feeble. The main problem in analyzing turning points and non-linear growth patterns is the fact that such phenomena occur in a very irregular fashion, so that normal time series are often inadequate in providing sufficient statistical evidence. Thus instability can *ex post* be traced, but is very hard to predict *ex ante* (Peters 1988). Prediction is apparently only meaningful in case of stable evolution.

In light of these observations *chaos theory* has in recent years become a popular tool. The next section will offer a sketch of the road towards chaos.

---

[1]Catastrophe is a term coined by Thom (1975) for explaining biological morphogenesis. Thom also gave a topological classification of the elementary types of catastrophes depending on the number of control variables in a dynamical system (see also footnote 14).

18.3 Towards Chaos Theory

18.3.1 Structural Instability

The interest in the dynamics of real-world systems is not recent, but can already be found in early analytical dynamics. Analytical dynamics aims to describe and explain transformations in complex real-world systems. It emerged about 300 years ago with the concept and calculus of differential equations developed by Newton and Leibniz. Most theories in physics (such as Einstein's relativity theory or Maxwell's electro-magnetic theory) can be represented by means of differential equations.[2] The use of ordinary differential equations in economics has a shorter life span than in physics and dates back to Walras in 1874. However, it was Samuelson (1947) who firmly demonstrated the scientific significance of the use of differential equations in economics. He introduced the new concept of 'dynamic analysis', as a substitute for the commonly accepted concept of 'static analysis', by pointing at Frisch's article "Propagation Problems and Impulse Problems in Dynamic Economics" (1933) (see also Medio 1979). In this context the notion of stability in economics emerged in the late 1950s (see, for example, Allen 1959). It should be noted at this stage that stability concepts are in economics strictly linked to dynamic equilibrium phenomena.

Roughly speaking, following Zhang (1990 p 19) we may define equilibrium "as to the state of a dynamic system at which all variables are invariant with respect to time". Then we may assert that "an equilibrium x is stable if all nearby solutions stay nearby. It is asymptotically stable if all nearby solutions not only stay nearby, but also tend to x" (cfr. Hirsch and Smale 1947 p 180).[3] Inversely "a system is considered to be at non equilibrium if the variables have different rates of change with respect to time. Limit cycles, aperiodic behaviour and chaos (which will be described subsequently) are examples of non-equilibria" (see again Zhang 1990 p 19).

In the past decades differential equations have been extensively employed as powerful analytical tools by economic researchers. However, it soon became clear that conventional differential equations have an intrinsic limitation, namely they do usually only describe phenomena displaying regular and ordered behaviour, since their analytical solutions need differentiability conditions. Thus dis-continuities or sudden and dramatic changes or jumps which often characterize real-world phenomena cannot satisfactorily be analyzed by differential equations.

---

[2]Differential equations can in general be formalized as follows:
$\dot{x}_i = f_i\,(x_1, x_2, \ldots, x_n; \alpha_1, \alpha_2, \ldots, \alpha_m)$ $\quad i = 1, 2, \ldots, n$ where the symbol · indicates *dx/dt* (i.e. the derivative of function *x* with respect to the variable *t*, with *t* being time), and where α are relevant parameters.

[3]For the classification of different kinds of (in)stability related to a general system of two differential equations, see also Annex 6A in Nijkamp and Reggiani (1992a).

A second shortcoming is that it is often impossible to find the explicit solution of non-linear differential equations (i.e. equations where the coefficients of the state variables are not constant), unless one resorts to approximations via linear equations. But this means again that unexpected discontinuities, a multiplicity of (dis)equilibria as well as a multiplicity of bifurcations and catastrophe behaviour are missed out. More modern approaches try to overcome the above mentioned limitations by analyzing not just the properties of particular solutions of differential equations, but *the sensitivity of solutions to perturbations (or their stability)*, in other words, their *structural (in)stability*.

In this framework we can define a system as 'structurally unstable' when even *small perturbations in parameters or functional forms* change the qualitative properties of the system. In other words, structural instability means the possible existence of significant qualitative changes in the behaviour of the system (i.e. in the state variables) which are closely connected with bifurcation and catastrophe phenomena that (can) occur when the parameters values (i.e. the control variables) are changing (see for a brief review on bifurcation theory and catastrophe theory applied to spatial-economic patterns among others Day 1985; Dendrinos and Mullally; 1985; Zhang 1991).

It is well known from the literature that a simple bifurcation can transform the equilibrium points from one state (e.g. stable) to another one (e.g. unstable) (or vice versa). An example of such structural instability is provided by a prey-predator system, since a bifurcation (depending on the parameter values) can transform the equilibrium point from a stable 'focus' into an unstable 'focus' via a 'centre' (for a precise description we refer to Nijkamp and Reggiani 1992a Ch 8).

In conclusion, it is possible to analyze, by means of the concept of structural stability, various types of complex dynamics in real world systems, such as the evolution of cities (see Mees 1975), regional development processes (Andersson and Batten 1988; Batten et al 1987; Wilson 1981), dynamic choices in spatial systems (Fischer et al 1990) or economic equilibrium analysis (Turner 1980).

It should be added that a further analysis of dynamics in complex systems should focus on the time-dependent behaviour of the system (compared to the phase portrait or the bifurcation diagrams plotting the state variable values against the parameter values). In this context the identification of (a)periodic motions (or oscillations in the state variables) results as an interesting outcome.

18.3.2 (A)Periodic Motions in Behaviour

Periodic or cyclic solutions are very common in economics (see, for example, the wealth of business cycle literature reviewed by Lorenz 1989; Kelsey 1988; Puu 1989; Zhang 1991) as well as in ecology (see, e.g. Haken 1983a, 1983b), and recently also in geography and regional science (see, e.g. Dendrinos and Mullally 1985; Nijkamp 1987; Nijkamp and Reggiani 1990).

From a mathematical point of view periodic solutions are strictly connected with the existence of *closed orbits* and *limit cycles*. In the first case states are repeated from one orbit to the next; in the second one this does not hold, but then the orbits are asymptotically close to a closed orbit. In other words, a limit cycle is a closed orbit (emerging as an equilibrium solution from two-dimensional differential equations) which is also an *attractor* (i.e. a bounded region towards which every trajectory of a dynamic system tends to move). Thus a limit cycle is "closed so that a point moving along the curve will return to its starting position at fixed time intervals and thus execute periodic motion" (cf. Zhang 1990 p 207).

It should be noted that there is also the possibility of so-called aperiodic or 'chaotic' oscillations when the amplitude and period vary without any trace of a recognizable pattern (see also next subsections).[4]

A method to find limit cycles in the two-dimensional plane is provided by the Poincaré-Bendixson theorem (see e.g. Haken 1983a). The Poincaré-Bendixson theorem refers in particular to the existence of two kinds of attractors for a two-dimensional flow: (1) stable fixed points (see for a classification of fixed points, Nijkamp and Reggiani 1992a) and (2) periodic solutions or limit cycles. Consequently, as we will show hereafter, the chaotic motion which is associated with the so-called *strange attractor*, is not possible for two-dimensional flows (see also Lichtenberg and Lieberman 1983). A *strange attractor* is defined here as a bounded region of phase space in which initially close trajectories separate exponentially such that the motion becomes chaotic.

Now, it should be noted that the change from a fixed point into a limit cycle (if a single parameter varies) is a phenomenon known as *Hopf bifurcation*. It is interesting to note that in an at least three-dimensional system the Hopf bifurcation can be associated with chaotic behaviour (see Marsden and McCracken 1976). In this framework Newhouse et al (1978) have shown that after three Hopf bifurcations regular motion becomes highly unstable in favour of motion to a *strange attractor*. It is thus clear that central in the analysis of dynamical systems is also the idea of an attractor, i.e. a set of states in the state space of the system which attracts all trajectories emanating from neighbouring points.

18.3.3 (Strange) Attractors

Given the above definitions of attractors, it is clear that the attractor does not describe change over time. Indeed, for many parameter values of a dynamical system its attractors (if they exist) are regular objects described by equilibrium points or closed curves. For other parameter values they may have an extremely complex structure. Such attractors are called "strange", a term coined by Ruelle and Takens (1971) in order to indicate an "exponential separation of orbits" (see Day 1985; Eckmann and Ruelle 1985). Strange attractors are strictly connected to chaotic motion (see Annex 6B in Nijkamp and Reggiani 1992a, for a review of major classes of strange attractors). In particular, in 1975 Li and Yorke christened '*chaotic*' a system with strange

---

[4]An interesting review on illustrations of (a)periodic motions with a special focus on technological change, business cycles and economic growth and competition can be found in Batten et al. (1987).

attractors or a dynamic situation exhibiting aperiodic - though bounded - trajectories.

A more strict definition of a strange attractor is provided by Schuster (1988 pp 105-106) as follows:

a) It is an attractor, i.e. a bounded region of phase space to which all sufficiently close trajectories from the so-called basin of attraction are attracted asymptotically for long enough times. We note that the basin of attraction can have a very complicated structure. Furthermore, the attractor itself should be indecomposable; i.e. the trajectory should visit every point in the attractor in the course of time. A collection of isolated fixed points is no single attractor.
b) The property which makes the attractor strange is the sensitive dependence on the initial conditions; i.e, despite the contraction in volume, lengths need not shrink in all directions, and *points which are arbitrarily close initially, become exponentially separated at the attractor for sufficiently long times.*

Thus strange attractors (together with the other attracting regions in phase such as fixed points and limit cycles) are pertinent to *dissipative*[5] systems for which volume elements in phase space shrink with increasing time.

Strange attractors have often 'fractal' characteristics. The term fractal has been introduced by Mandelbrot (1977) who discovered that similar structures may repeat themselves at higher orders or dimensions. Essentially, a fractal set is a set having the property of being invariant at different scales (self-similarity and irregularity property) and having a non-integer, fractional dimension (for an application of fractal geometry to urban structure see, e.g. Batty and Longley 1986; Longley and Batty 1994; Frankhauser 1991). Therefore, the notion of fractals only refers to the geometry of attractors (see also Mandelbrot 1977; Peitgen and Richter 1986). However, since the notion of strange attractors usually refers to the dynamics of the attractors (and not just to their geometry), strange attractors need not have a fractal structure, while attractors with a fractal structure need not be chaotic (see, e.g. Holden and Muhamad 1986).

The concept of 'chaos' results straightforward from the previous basic concepts. Its significance rests essentially on the possibility of studying a variety of natural and social science based models where non-differentiability is typical.

18.3.4 Chaos

Chaos has in recent years become the new paradigm for studying non-linear dynamic systems. During the 1980s chaos theory "received widespread prominence and some scientists have even placed it alongside two other great

---

[5]However one can find 'chaotic' (in the sense of irregular) regions also in conservative systems, but they are not attractors.

revolutions of physical theory in the twentieth century - relativity and quantum mechanics. While those theories challenged the Newtonian system of dynamics, chaos has questioned traditional beliefs from within the Newtonian framework" (Crilly 1991 p 193).

The interesting characteristic feature of chaos theory is that it addresses essentially the (in)stability of *deterministic, non linear dynamic* systems which are able to produce *complex motions* of such nature that they are sometimes seemingly random. In particular such systems incorporate the property that small uncertainties may grow exponentially (although all time paths are bounded), leading to a broad spectrum of different trajectories in the long run, so that precise or plausible predictions are almost impossible. This phenomenon, which is typical of chaotic dynamics, is known as *sensitivity to initial conditions*. Already at the beginning of this century it was recognized by Poincaré that "it may happen that small differences in the initial conditions produce very great ones in the final phenomena. A small error in the former will produce an enormous error in the latter. Prediction becomes impossible, and we have the fortuitous phenomenon" (Poincaré 1913 p 397).

Particularly in a chaos system, two sets of conditions, which initially are very close together may become exponentially separated[6] giving rise to widely different states in the long run[7]. Thus, it follows from the previous remarks that predictability of long term behaviour may become problematic for those non-linear systems which incorporate chaotic dynamics.

Consequently, the new logic which has emerged in the field of non-linear dynamics by the introduction of the theory of chaos has also an interesting psychological appeal; model builders need not necessarily be blamed any more for false predictions, as errors in predictions may be a result of the system's complexity, as can be demonstrated by examining more carefully the properties of the underlying non-linear dynamic model structure. However, even though from a behavioural point of view chaos is sometimes identified with 'dynamical stochasticity', 'self-generated noise', or 'intrinsic stochasticity' (see, for example, Hao 1984), from a mathematical point of view, there is still some uncertainty regarding the precise definition of chaos: "clearly there is still a lot of 'chaos' in chaos theory" (Rosser 1991 p 30).

On the one hand, chaos is identified by many authors as aperiodic behaviour (see, for example, Guckenheimer 1979; Nusse 1987), starting from Li and Yorke (1975) who, in their well-known theorem 'Period Three Implies Chaos', identify chaos - for any continuous mapping of a one-dimensional interval onto itself - with the existence of cycles of all orders and a scrambled set in which all trajectories are non-periodic and asymptotically unstable. In this spirit we recognize also the 'Feigenbaum route' and the 'intermittency route' to chaos.

---

[6]This exponential separation of adjacent points can be measured by means of a positive Liapunov exponent. For its analytical formulation see Schuster (1988) and for the calculus of algorithms see Wolf et al. (1985).

[7]This is the so-called 'butterfly effect' (Lorenz, 1963): the single flap of a butterfly's wings might theoretically alter the initial conditions of weather systems giving rise to completely different weather scenarios in the long term.

Feigenbaum (1978) in particular discovered the phenomenon of a cascade of bifurcations, each leading to period doubling sequences as a system approaches chaos. Manneville and Pomeau (1979) found that chaos can be intermittent, i.e. chaos can emerge and then be replaced by a new zone of stable equilibrium.

On the other hand, chaos can be identified with the existence of strange attractors, like in the 'Ruelle-Takens-Newhouse route' (Ruelle and Takens 1971; Newhouse et al 1978). These authors showed in particular that after three Hopf bifurcations, it is 'likely' that regular motion becomes highly unstable in favour of motion to a strange attractor.

However, Eckmann and Ruelle (1985) argued that it is the sensitive dependence on initial conditions which is the *true* meaning of chaos, as indicated by the values of Liapunov exponents (see footnote 18.6). It appears to be a more 'rigorous' and computationally easier approach to test chaos by means of a positive K or a positive Liapunov exponent instead of using the Li-Yorke theorem, the Feigenbaum route or the other routes to chaos.

It should be added, however, that in an empirical context (e.g. in a time-series analysis) Liapunov exponents do not seem to be sufficient in providing evidence for chaos (see for more details, Nijkamp and Reggiani 1995; Brock 1986; Frank and Stengos 1988), since there are many independent sources of noise which can affect economic data. It is however possible that some of the observed noise is not 'extrinsic noise', independent of the economic system, but 'intrinsic noise' generated by chaotic dynamics of the system itself. It is interesting to note in this context that it has recently been shown that chaotic behaviour is less sensitive to noise than periodic orbits (see Kelsey 1988). Because of this low sensitivity of chaotic equations to random errors, it turns out to be difficult to identify the nature of random disturbances in time series.

In conclusion, chaos is not regarded any more as a peculiar theoretical possibility, but also as a practical issue for empirical research. Interest in economics is imminent, starting in the '80s and still developing with a rich set of contributions in many economic fields, such as economic growth theory, innovation theory, economic competition, economic trade, urban and regional economics, transport economics, etc.[8].

In recent years chaos and non-linear dynamic principles have been incorporated in evolutionary algorithms which had shown successful applications in the domains of optimization, selection, adaptation and learning. In this context, formalised language and related simulation experiments are an important tool for generalizing "from the specification to the state space, and in principle to suggest states other than those which are intuitively accessible. Thus, the system is able to bootstrap from specification to generalization" (Leydesdorff 1994 p 190).

---

[8]See for a review on the applications of chaos theory in economics, Hommes (1991) and Nijkamp and Reggiani (1995), as well as the recent edited books by Leydesdorff and Van den Besselaar (1994) and Nijkamp and Reggiani (1993).

18.3.5 Modelling Chaos

Already in 1976 May argued in his seminal contribution "Simple Mathematical Models with Very Complicated Dynamics": "Not only in research, but also in the everyday world of politics and economics, we would all be better off if more people realized that simple non linear systems do not necessarily possess simple dynamical properties" (p 157). In this article, May also investigated how apparently erratic fluctuations in census data (e.g. for an animal population) may originate from a simple deterministic population growth relationship such as the following first-order difference equation:

$$x_{t+1} = F(x_t) \tag{18.1}$$

where F(x) is a non-linear function. More specifically, May showed the existence of chaotic motion in simple first-order differential equations of a 'logistic' type[9]. From a mathematical point of view, chaos can also arise in higher-order difference equations, in systems with more than one variable, and in systems of differential equations of dimension of at least three[10]. Two well-known examples are the Hénon map and the Lorenz equations for these two cases, respectively (see, for an illustration, Annex 6 B in Nijkamp and Reggiani 1992a).

Non-linearities of the type of equation (1) including a saturation effect have a structure often found in (spatial and industrial) economics, but have posed a difficult question on the relevance and possibility of using discrete time periods in applied economic research. The hypothesis of a fixed delay is often considered as a shortcoming, since it implies the presence of rigid discontinuities in technology or in the economic agents' memory and expectations. Next, it may be questioned whether the results obtained by applications of (1) capture plausible indigenous features of real economic systems or whether they are only of mathematical interest. An interesting approach in this debate is the one provided by Medio (1989) and Sparrow (1980) who show how to transform a system with a fixed delay into a continuous one such that the dynamic behaviour is presented. It should be noted that there are also many economists who favour a discrete time formulation, since they assert that decisions of economic agents occur only at certain points in time. For example, Eckalbar (1992) mentions six arguments defending the use of discrete time modelling (see Greiner and Kugler 1994).

In conclusion, chaos models essentially support the 'endogenous' viewpoint in (spatial) economic development as identified by Grandmont: "There are traditionally two contrasted viewpoints on the working of dynamic socio-

---

[9]Since this equation is well-known, we will not discuss it here in great detail. The interested reader is referred to Nijkamp and Reggiani (1992a, 1995).

[10]The fact that chaos cannot arise in one- or two-dimensional systems of differential equations can be explained by the Poincaré-Bendixson theorem (see Section 18.3.2) which demonstrates that any attractor of a two-dimensional system of differential equations is either a fixed point or a limit cycle.

economic systems. The 'fundamentalist' viewpoint claims that such systems are inherently stable and the origins of observed fluctuations is to be found exclusively in exogenous variations of the 'fundamentals'. Another ('Keynesian') viewpoint claims that a significant part of observed fluctuations may be due to endogenous factors, in particular to volatile expectations" (Grandmont 1991 p 293).

It is clear that the 'adventure' of chaos theory and models stimulates scientists to focus on a fundamental issue, viz. the character of (regular or irregular) evolution in (spatial) economic systems. In this framework, recent research has started to orient itself towards new tools, often based on the methodology of natural sciences which, in connection with chaos theory, may give new perspectives for evolutionary economics. This will be dealt with in the next sections.

18.4 Towards Bio-ecological Modelling

18.4.1 The Relevance of Niche Theory

A new paradigm which has received a great deal of attention in the modelling literature of the last decade concerns the competition (substitution/-complementarity) among populations in a space-economy network, generally formalized by means of (bio)ecologically-based models. The potential of using the formalism of mathematical ecology in economics has been advocated by an increasing number of scientists. This new orientation is based on the awareness that the modern ecological methodology is often using approaches which are based on lack of precision (witness the often graphical and qualitative solutions) for generality and realism (given the role played by competitive interactions, diversity, adaptivity and niche configurations in both uniform and patchy environments). The above properties make this type of models - in the framework of the three main purposes of a model, viz. *generality, realism* and *precision* - preferable to other types of models which either sacrifice realism for precision and generality, or sacrifice generality for realism and precision (see Levins 1968; Maggioni 1993).

A second important property related to ecologically based models, is their 'methodological' connection with the main classes of non-linear models used in economics. It is noteworthy that non-linear dynamic systems were applied in biology well before they were applied in economics (see e.g. Rosser 1991). In particular, it has been shown (see Nijkamp and Reggiani 1992b, 1995) that many economic growth models which are actually in use derive essentially from two classes of ecological models, viz. the May model (outlined in the previous section), and the Lotka-Volterra (or prey-predator) model (which essentially represents a system of interrelated logistics). These two prototypes of models[11] have also the possibility of giving rise to oscillating and chaotic behaviour.

In this framework, an interesting approach, viz. *niche theory*, which formally embeds both the May model and the Lotka-Volterra model (see Nijkamp and Reggiani 1992b), has been recently come to the fore.

The niche concept[12] refers to the phenomenon of inter-species competition and to dynamic patterns of resource utilization. It seems thus plausible that socio-cultural and economic evolution (where the population dynamics can be extended to urban and regional development, economic activities, diffusion of ideas, trade and transport systems) can be analyzed by means of niche chains, where new 'species' can (partially or totally) replace old ones. Moreover, also the self-organizing paradigm (see Prigogine 1976) can be cast in the framework of niche formulation, where the new species (competitors) can be considered in terms of ecological fluctuations. Consequently, empirical examples - such as the substitution of transport infrastructure and energy systems over time as well the introduction of technological innovations - can be interpreted in the framework of niche theory (see Reggiani and Nijkamp 1995).

A further step is the analysis/application of niche theory with the possibility of irregular or chaotic dynamics in order to evaluate the overall stability of the economic system at hand. This brings us to the more general issue of a network constituted by multi-layer structures where the related interactions/interdependencies - which can be formalized in a 'niche' form - can produce cycles, fluctuations and bifurcations, with the consequence of likely structural changes in the overall economic network.

18.4.2 Modelling a Niche Network

By conceiving of a network as a "space-time (ordered) connectivity structure" (see Reggiani and Nijkamp 1994b) it is clear that its evolution can be considered as the result of the competition between different subsystems/niches within the main system/niche, leading - as noted above - to a self organizing process.

Such a dynamic process can generally be modelled by means of the so-called logistically-defined niche system:

$$\dot{x}_i = x_i \left(K_i - \sum_{j=1}^{N} a_{ij}\, x_j\right) \tag{18.2}$$

where x_I is the population of species i $(i=1,2,..N)$; K_i is the carrying capacity for species i and the coefficients a_{ij} are the interaction/competition coefficients measuring the niche overlapping[13].

[11]For a review on applications of the May and Lotka-Volterra model in economics and regional science, see Diappi and Reggiani (1996), Nijkamp and Reggiani (1992b) and Rosser (1991).

[12]For a formalization of the niche concept see again Nijkamp and Reggiani (1992b).

[13]See, for formal details, Nijkamp and Reggiani (1992b) and Reggiani and Nijkamp (1995b).

In order to take into account, in the above general network (2), dynamic interaction effects among the capacities K_i as well as among the coefficients a_{ij}, it is then necessary to interpret these *parameters* also as *dynamic functions*. Then the network evolution is the result of multilayer niche dynamics, representing the niche species interacting with the niche capacities and/or with the niche growth rates.

18.4.2.1 Multilayer Niches with Dynamic Capacity

An interesting example in this respect can be offered by the dynamic relationships between economic systems and environmental systems, by adhering to the dynamic interdependence illustrated in Figure 18.1, where the time dependent variables K_l, K_m, K_n, represent the related capacity of the three systems *e, m, n,* shown as logistic functions (see Reggiani and Nijkamp 1994).

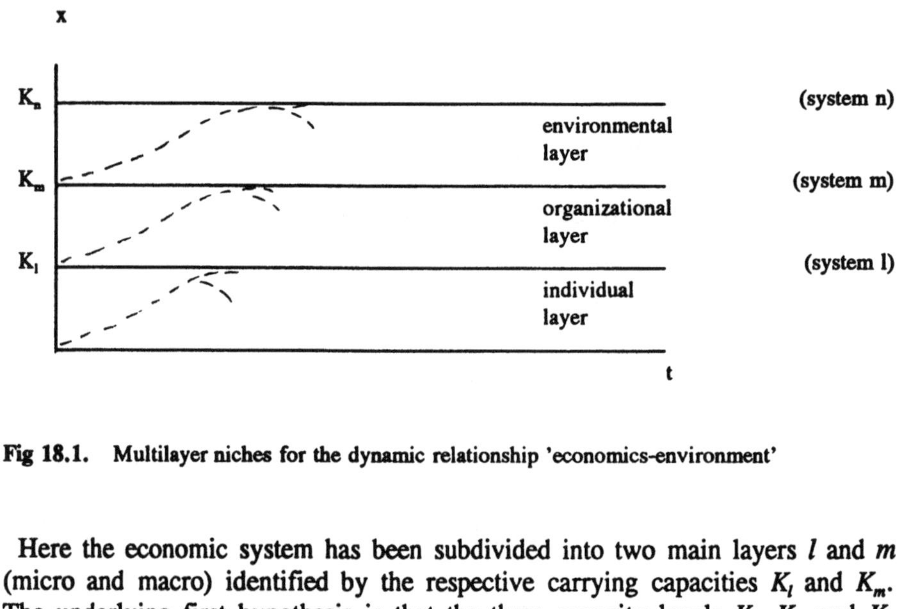

Fig 18.1. Multilayer niches for the dynamic relationship 'economics-environment'

Here the economic system has been subdivided into two main layers *l* and *m* (micro and macro) identified by the respective carrying capacities K_l and K_m. The underlying first hypothesis is that the three capacity levels K_l, K_m and K_n are not fixed, but may vary over time leading to the possibility of both evolutionary and devolutionary processes in the network. The second hypothesis is then the possibility of both continuous and discontinuous niche evolutions for each layer, thus driving feedback patterns to resonance effects.

The above multilayer niche structure has recently been applied - by means of simulation experiments - to the case of three competing modes in a transport-economic system (with variables x_i), nested in a niche form with respect to the competition of the three capacities K_i concerned and limited by a final threshold determined by environmental costs (see Reggiani and Nijkamp 1994, 1995b). This analysis, which can easily be extended to other economic phenomena with competing capacities such as migration and trade developments, shows how some subsystems may additionally drift in terms of their relevant resonances, while certain niches may improve shielding against variations in other parts of the network (see also Leydesdorff 1994).

18.4.2.2 Multilayer Niches with Dynamic Growth Rates

Next, we will deal with time dependent growth rates in a niche system. The relevance of a dynamic growth rate a_{ij} is clear, if we consider the evolution of niche system (2) as characterized by a *morphogenetic*[14] aspect; in other words, by placing emphasis on the form and shape of the network, like for example, in models of innovation diffusion and technological change. It is well-known that in the innovation diffusion literature many models based on logistic functions have been proposed with both a symmetric and a non-symmetric S shape (see, for a review, Mahajan et al 1990; Kapur et al 1992). However, the explanations of the diffusion process of technological innovations - mainly based on socio-attitudinal features of the groups adopting the innovation - can be only captured by the growth rate of the adopters. In other words, by encapsulating in the growth rate of the usual logistic function a dynamic attitudinal function measuring the preference intensity, we may understand why the evolution of innovation is shaped in a certain way in some geographical contexts/groups, while - in the same time period - it is shaped in a completely different way elsewhere. The resulting multilayer structure is displayed in Figure 18.2, where x_i represents the amount of adopters in a certain group (region i, $i=1...N$) and a_{ij} the growth parameter leading to an attitudinal response of population x_i.

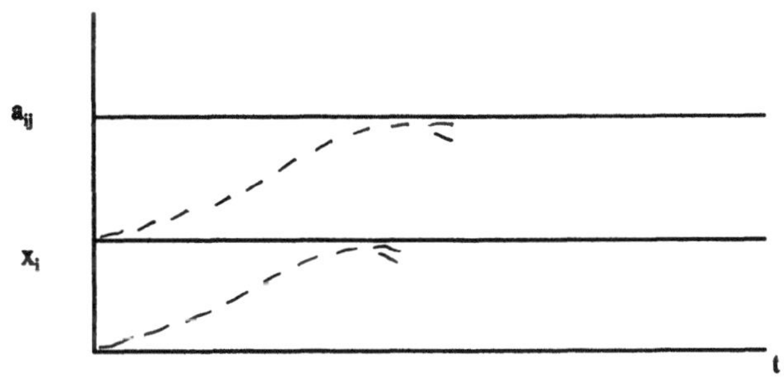

Fig. 18.2. Multilayer niches for the dynamic relationship 'innovation diffusion-preference function'

In consequence, instability and resonance effects can also appear in this second type of multilayer niche network, as demonstrated in simulation experiments carried out for the case of two and four centers adopting the same

[14]Following Thom (1989), a morphogenetic field always implies the appearance of new discontinuities in the developing species (see also Thom, 1975).

innovation (see Nijkamp and Reggiani 1995b). We may thus argue that, for a full integrated network configuration with multiple nodes and links, the general niche models displayed in Figure 18.1 and 18.2 will tend to produce 'complex' phenomena owing to cooperative and interacting/competitive effects. This brings us to the fundamental issue of complexity in a network and its connection with the stability concept; in other words, with the robustness/sustainability of cooperative
phenomena.

18.5 Towards Complexity Theory

Complex systems share the common feature of exhibiting a great variety of behaviour. Different types of complexity can be identified - which clearly can influence the evolution of a network - inherent to both *static and dynamic complexity* (see Reggiani and Nijkamp 1995a); however, the fundamental mechanism underlying complexity is instability, whose principal manifestations are bifurcations towards a multiplicity of states and chaos. This provokes the idea of a 'complex' system, with sensitive trajectories characterized by unavoidable perturbations of an intrinsic or environmental origin present in the real world. It is evident that such phenomena raise the problem of prediction and control of system's behaviour. However, for a long time the idea prevailed that complexity is not an intrinsic property of nature, but rather a temporary drawback arising from the presence of a great number of variables and parameters blurring some other fundamental underlying principle. The most common evidence of complexity comes from phenomena at a microscopic level, even though there is an increasing awareness that complexity appears also at the microscopic level. As pointed out by Davies (1989), complexity theory was hampered in the past by two conceptual obstacles: reductionism and vitalism, while the relevance of the organizational aspects was neglected. "In this post-reductionist era of physics, one looks not to the individual molecules for an explanation of large-scale self-organization, nor even to the action of one molecule upon its neighbour. Instead attention is focused on *global* issues, such as boundary conditions, constraints, distances from equilibrium and so on. True, the properties of individual molecules and their interaction are necessary ingredients in providing a full explanation, but they are not sufficient.." (Davies 1989 p 103).

In this context the *resilience* concept deserves serious attention, since resilience refers to the capacity of a network to retain its organizational structure consequent upon a perturbation of some state variable from a given value (Common and Perrings 1992; Perrings 1994). Thus the network-stability concept can be revisited and reinterpreted in the framework of the sustainability/fragility[15] issue. In other words, if we define *sustainability* as a stable domain in a network, it is clear that complex systems are likely fragile, while simple systems are dynamically robust. As a consequence, given the limited predictability - as noted above - of complex systems, a predictable environment is necessary in order to permit the complex (fragile) system to exist or to continue.

It is now evident that also complex economic processes, need new tools of analysis, as is witnessed by the following quotation: "the only way to analyze economical dynamics is to develop an approach defined as an "analytico-physiological" approach which must take its inspiration from biology" (Georgescu-Roegen 1974; Vivien 1994). In this framework the analysis of relationships - and especially complexity aspects - in a economic network, such as diversity (a measure for multiple dynamic functions), maturity (when the system gradually approaches a more or less stable stage) or hierarchical organization, is necessary, as well as the study of their 'resilience' in the presence of spatio-temporal variations.

In addition to new methodological tools, such as the sustainability apparatus, there is a need for complementary ways of exploring complexity in spatial economics, mainly based an interactive computer simulations of prototype partial differential equations and specially designed experiments to depict qualitative behaviours in both space and time. More insight into such complexity issues will certainly be generated by the development of artificial intelligence and neural network approaches.

Such complex systems, which are capable of supervised and unsupervised inductive learning as well as decision-making based on generalization from experience (without the need for knowledge on the underlying rules), show increasingly relevant applications in spatial economics, where communication, pattern recognition, optimization, adaptivity, quality control and predictability, are necessary - as is the case in complex economic systems.

A blend of complex systems theory and techniques seems to be critical for a better understanding of the mechanisms behind the dynamics of economic processes, in particular the links between the relatively simple behaviour at the individual level and the more complex and adaptive group behaviour at a meso/macro level. In this framework, cross-fertilization between different disciplines, such as economics, biology, ecology, mathematics, physics, sociology and informatics, have proven to be very fruitful and can be expected to continue to bear many more fruits in the future.

---

[15]We define a dynamic fragile system as a system which will persist only for tightly circumscribed values of the environmental parameters and which will tend to collapse under significant perturbations either to environmental parameters or to population values.

Acknowledgement

The present work has been developed under the Bilateral Project CNR n° 940850.CT11.

References

Allen RGD (1959) Mathematical Economics. MacMillan, London (2nd edition)

Andersson ÄE, Batten DF (1988) Creative Nodes. Logistical Networks and the Future of the Metropolis. Transportation 14:281-293

Baker GL, Gollub JP (1990) Chaotic Dynamics. An Introduction. Cambridge University Press, Cambridge

Batten D, Casti J, Johansson B (Eds) (1987) Economic Evolution and Structural Adjustment. Springer-Verlag, Berlin

Batty M, Longley PA (1986) The Fractal Simulation of Urban Structure. Environment and Planning A 18:1143-1179

Brock WA (1986) Distinguishing Random and Deterministic Systems. Abridged Version. Journal of Economic Theory 40:168-195

Common M, Perrings C (1992) Towards on Ecological Economics of Sustainability. Ecological Economics 6:7-34

Crilly T (1991) The Roots of Chaos. A Brief Guide. Fractals and Chaos. Crilly AJ, Earnshaw RA, Jones H (Eds). Springer-Verlag, Berlin, pp 193-209

Davies PCW (1989) The Physics of Complex Organization. Theoretical Biology. Epigenetic and Evolutionary Order from Complex Systems. Goodwin B, Saunders P (Eds) Edinburgh University Press, Edinburgh pp 101-111

Day RH (1985) Dynamical Systems Theory and Complicated Economic Behaviour. Environment and Planning B 2:55-64

Dendrinos DS, Mullally H (1985) Urban Evolution. Studies in the Mathematical Ecology of Cities. Oxford University Press, Oxford

Diappi L, Reggiani A (1996) Evolutionary Models in Spatial Systems. Qualitative vs Quantitative Approaches. Ten Years of Regional Science. Camagni R, Senn L (Eds) Franco Angeli, Milano (forthcoming)

Eckalbar JA (1992) Book Review of 'From Catastrophe to Chaos' by JB Rosser. Journal of Economic Literature 30:2150-2151

Eckmann JP, Ruelle D (1985) Ergodic Theory of Chaos and Strange Attractors. Review of Modern Physics 57:617-656

Feigenbaum HJ (1978) Quantitative Universitality for a Class of Non-Linear Transformations. Journal of Statistical Planning 19:25-52

Fischer MM, Nijkamp P, Papageorgiou YY (1990) Spatial Choices and Processes. North-Holland, Amsterdam

Forrester J (1968) Principles of Systems. Wright-Allen, Press, Cambridge, Mass.

Frank M, Stengos T (1988) Chaotic Dynamics in Economic Time-Series. Journal of Economic Surveys 2:103-133

Frankhauser P (1991) Aspects Fractals des Structures Urbaines. L'Espace Géographique 1:45-69

Frisch R (1933) Propagations Problems and Impulse Problems in Dynamic Economics. Economic Essays in Honour of Gustav Cassel. Allen & Unwin, London

Georgescu-Roegen (1974) Dynamic Models and Economic Growth. Economic Appliquée 27:529-563

Grandmont J-M (1991) Expectations Driven Business Cycles. European Economic Review 35:293-299

Greiner A, Kugler F (1994) A Note on Competition Among Techniques in the Presence of Increasing Returns of Scale. Evolutionary Economics and Chaos Theory. Leydesdorff L, Besselaar P van den (Eds) Pinter, London, pp 44-52

Guckenheimer J (1979) Sensitive Dependence to Initial Condions for One-Dimensional Maps. Communications in Mathematical Physics 70:133-160

Haken H (1983a) Synergetics. Springer-Verlag, Berlin

Haken H (1983b) Advanced Synergetics. Springer-Verlag, Berlin

Hao B-L (Ed) (1984) Chaos. World Scientific Publication Co, Singapore

Hirsh M, Smale S (1974) Differential Equations, Dynamical Systems and Linear Algebra. Academic Press, London

Holden AV, Muhamad MA (1986) A Graphical Zoo of Strange and Peculiar Attractors. Chaos. Holden AV (Ed) Manchester University Press, Manchester pp 15-35

Hommes CH (1991) Chaotic Dynamics in Economic Models. Wolters-Noordhoff, Groningen

Kaldor N (1985) Economics Without Equilibrium. University College Cardiff Press, Cardiff

Kapur JN, Kumar U Kumar V. (1992) Some Possible Models for Technological Innovation Diffusion. Journal of Scientific and Industrial Research 51:202-208

Kelsey D (1988) The Economics of Chaos or the Chaos of the Economics. Oxford Economic Papers 40:1-3

Leydesdorff L, Besselaar P van den (Eds) (1994) Evolutionary Economics and Chaos Theory. Pinter, London

Leydesdorff L (1994) New Models of Technological Change. New Theories for Technology Studies? Evolutionary Economics and Chaos Theory. Leydesdorff L, Besselaar P van den, (Eds) Pinter, London pp 181-192

Levins R (1968) Evolution in Changing Environments. Some Theoretical Explorations. Princeton University Press, Princeton

Li TY, Yorke JA (1975) Period Three Implies Chaos. American Mathematical Monthly 82:985-992

Lichtenberg AJ, Lieberman MA (1983) Regular and Stochastic Motion. Springer-Verlag, Berlin

Longley P, Batty M (1994) Fractal Cities. Academic Press, London

Lorenz EN (1963) Deterministic Non-Periodic Flow. Journal of the Atmospheric Sciences 20:130-141

Lorenz H-W (1989) Non-Linear Dynamical Economics and Chaotic Motion. Lecture Notes in Economics and Mathematical System 334. Springer-Verlag, Berlin

Maggioni MA (1993) Ecological Dynamics and Critical Mass in the Location of High-Tech Firms. Paper presented at the 40th North American RSAI Conference. Houston, TX

Mahajan V, Muller E, Bass FM (1990) New Product Diffusion Models in Marketing. A Review and Directions for Research. Journal of Marketing 54:1-21

Mandelbrot B (1977) The Fractal Geometry of Nature. VH Dreeman and Company, New York

Manneville P, Pomeau Y (1979) Intermittency and the Lorenz Model. Physics Letters A 75:1-2

Marsden JE, McCracken M (1976) The Hopf Bifurcation and its Applications. Springer-Verlag, Berlin

May RM (1976) Simple Mathematical Models with Very Complicated Dynamics. Nature 261:459-467

Meadows DL, Meadows DH (Eds) (1973) Toward Global Equilibrium. Collected Papers. Wright-Allen Press, Cambridge, Mass.

Medio A (1979) Teoria Non Lineare del Ciclo Economico. Il Mulino, Bologna

Medio A (1989) Discrete and Continuous Models of Chaotic Dynamics in Economics. Department of Economics, University of Venice

Mees A (1975) The Revival of Cities in Mediaeval Europe. An Application of Catastrophe Theory. Regional Science and Urban Economics 5:403-425

Morishima M (1991) General Equilibrium Theory in the Twenty-First Century. The Economic Journal 101:69-74

Newhouse S, Ruelle D, Takens F (1978) Occurrence of Strange Axiom-A Attractors near Quasiperiodic Flow or T^{m}, $m \geq 3$. Communications in Mathematical Physics 64:35-40

Nijkamp P (1987) Long Term Economic Fluctuations. A Spatial View. Socio-Economic Planning Sciences 21:189-197

Nijkamp P, Reggiani A (1990) An Evolutionary Approach to the Analysis of Dynamic Systems with Special Reference to Spatial Interaction Models. Sistemi Urbani 1:601-614

Nijkamp P, Reggiani A (1992a) Interaction, Evolution and Chaos in Space. Springer-Verlag, Berlin

Nijkamp P, Reggiani A (1992b) Spatial Competition and Ecologically Based Socio-Economic Models. Socio-Spatial Dynamics 3

Nijkamp P, Reggiani A (Eds) (1993) Nonlinear Evolution of Spatial Economic Systems. Springer-Verlag, Berlin

Nijkamp P, Reggiani A (1995) Nonlinear Evolution of Dynamic Spatial Systems. The Relevance of Chaos Theory and Ecologically Based Models. Regional Science and Urban Economics 25:183-210

Nijkamp P, Reggiani A (1996) Space Time Synergetics in Innovation Diffusion. A Nested Network Simulation Approach. Geographical Analysis (forthcoming)

Nusse HE (1987) Asymptotically Periodic Behaviour in the Dynamics of Chaotic Mapping. SIAM Journal of Applied Mathematics 47:498-515

Okun AM (1981) Prices and Quantities. Basil Blackwell, New York

Peitgen HO, Richter PH (1986) The Beauty of Fractals. Springer-Verlag, Berlin

Perrings C (1994) Ecological Resilience in the Sustainability of Economic Development. Proceedings International Symposium on Models of Sustainable Development. AFCET, Paris pp 27-41

Peters T (1988) Thriving on Chaos. MacMillan, London

Poincaré H (1913) The Foundation of Science. Science and Method. (English Translation: The Science Press, Lancaster, 1946)

Poston T, Stewart I (1978) Catastrophe Theory and its Applications. Pitman, London

Prigogine I (1976) Order through Fluctuation. Self-Organization and Social System. Evolution and Consciousness. Jantsch E, Waddington CH (Eds) Addison-Wesley, Reading, Mass

Puu T (1989) Non-Linear Economic Dynamics. Springer-Verlag, Berlin

Reggiani A, Nijkamp P (1994) Evolutionary Dynamics in Technological Systems. A Multi-layer Niche Approach. Evolutionary Economics and Chaos Theory. Leydesdorff L, Besselaar P van den (Eds) Pinter, London pp 93-108

Reggiani A, Nijkamp P (1995a) Competition and Complexity in Spatially Connected Networks. Analysis and Simulation. Systems Dynamics Review 11:51-66

Reggiani A, Nijkamp P (1995b) Multilayer Networks and Dynamics Transport Systems. Geographical Systems 2:39-57

Rosser JB (1991) From Catastrophe to Chaos: A General Theory of Economic Discontinuities. Kluwer, Dordrecht

Ruelle D, Takens F (1971) On the Nature of Turbulence. Communications in Mathematical Physics 20:167-192

Samuelson PA (1947) Foundations of Economic Analysis. Harvard University Press, Cambridge, Mass.

Schuster HG (1988) Deterministic Chaos. VCH, Weinheim

Sparrow C (1980) Bifurcation and Chaotic Behaviour in Simple Feedback Systems. Journal of Theoretical Biology 83:93-105

Turner J (1980) Non Equilibrium Thermodynamics, Dissipative Structures and Self-Organization. Dissipative Structures and Spatio-Temporal Organization Studies in Biomedical Research. Scott G, MacMillan J (Eds) Iowa State University Press, Iowa pp 13-52

Thom R (1975) Structural Stability and Morphogenesis. Addison-Wesley, Reading

Thom R (1989) An Inventory of Waddingtonian Concepts. Theoretical Biology. Goodwin B, Saunders P (Eds) Edinburgh University Press, Edinburgh pp 1-7

Vivien (1994) Bioeconomics and Sustainable Development. Proceedings International Symposium on Models of Sustainable Development. AFCET, Paris pp 875-886

Wilson AG (1981) Catastrophe Theory and Bifurcation. Croom Helm, London

Wolf A, Swift JB, Swinney HL, Vastano JA (1985) Determining Lyapunov Exponents from a Time Series. Physica D 16:285-317

Zeeman EC (1977) Catastrophe Theory. Selected Papers 1972-1977. Addison-Wesley, Reading
Zhang WB (1990) Economic Dynamics. Springer-Verlag, Berlin
Zhang WB (1991) Synergetic Economics. Springer-Verlag, Berlin

19 Advances in Spatial Equilibrium Modelling: Retrospect and Prospect

Jeroen C.J.M. van den Bergh, Peter Nijkamp and Piet Rietveld
Vrije Universiteit

19.1 Retrospect

The contributions in this book demonstrate the wealth of efforts which have been made to study spatial economic interactions by means of formal analysis, using input-output (IO), spatial price equilibrium (SPE) and general equilibrium (GE) type of models. The SPE framework, developed by Samuelson and Takayama, assumes a special place in the field of spatial economics, as it is focusing on a socially optimal set of spatial commodity flows derived by using a market-based system description. The SPE model has theoretically and methodologically been elaborated in many directions and has also seen many applications (see especially Chapters 1, 2, 3, 4, 6, 9 and 10 of the present volume). In the first chapter, Professor Takayama carefully describes the historical process and the critical researches and publications that have shaped the present field of Spatial Price Equilibrium Modelling. In the second Chapter the SPE model is discussed in a wider context of spatial equilibrium analysis in a multi-regional framework, including also IO, econometric, (C)GE, linear endogenous price, and hybrid sequential models. To illustrate a special and important type of application of such models, this discussion is placed in the context of transport policy issues.

Combining elements of economic structure and spatial organization in one analytical context has given rise to the development of various formal methods for dealing with issues of commodity flows, transport, location choice, trade, etc. The basic neoclassical model that has a spatial disaggregation and uses price-equilibrating processes for clearing markets is the Spatial Price Equilibrium (SPE) model. It adopts an approach of analysing spatially separated markets and transport from an equilibrium point of view. Therefore it may, amongst others, shed light on the integration of economics and space in wider economic models as well. The basic SPE formulation aims to solve an interregional trade and spatial price equilibrium, where demand prices equal the sum of supply costs and transportation costs. Samuelson was in 1952 the first to derive a formal representation of the SPE problem as a linear programming model. The SPE problem can be formulated as a model that is able to identify the spacial competitive equilibrium, given transportation costs, domestic demand and supply curves at regional or spatial markets. This equilibrium embraces the set of market-clearing prices in all spatial markets, the levels of demands and supplies at each location, as well as of all exports and imports.

The representation of the spatial equilibrium problem into primal and dual programming forms was offered by Takayama and Judge (1964) and Takayama and Woodland (1970). Following this work, the most extensive contribution to the field of SPE theory and modelling has been the standard text by Takayama and Judge (1971). It not only covers the basic elements of spatial equilibrium models, but also treats extensions with multicommodity systems and dynamic processes in detail. The historical development of the relevant work on SPE-related questions is discussed by Takayama himself in the first chapter of this book. More details on relevant and recent literature is provided in particular in Chapters 2, 4 and 9. In Chapter 3 the question is addressed of how one should deal with the issue of re-exports in the context of SPE analysis. It is clarified why re-exports cause SPE models not to lead to a spatial equilibrium. In order to solve this basic problem a slight reformulation of the SPE model is then needed. Chapter 4 focuses on the interesting relationship between three types of spatial models, namely with fixed demand and supply (the Hitchcock-Kantorovich-Koopmans LP transportation model), with fixed demand and endogenous supply, and with endogenous demand and supply (the Samuelson-Takayama-Judge SPE model). The sensitivity of these models is compared in the context of the analogy with the thermodynamic 'Le Châtelier Principle'. The main conclusion is that the least constrained spatial systems can absorb exogenous shocks most effectively in terms of the welfare measure via flexible adjustment processes.

The SPE models have been widely applied (an early overview is Takayama and Judge, 1973). The applications fall into two main categories, namely analysis of mineral energy resource markets and agricultural commodity markets. Also in Labys et al (1989) overviews of such SPE applications are extensively given (see also Chapters 1 and 4). However, applications of multi-commodity SPE studies are rare. Interaction between commodities may occur on the transportation network, or in terms of substitution in demand functions. For an interesting application of a world food model in such a context see Takayama et al (1989).

Finally, it is important to note that the mathematical and algorithmic aspects of SPE models have received a great deal of attention. Different approaches have been adopted by various authors. Interested readers may consult the discussion and references provided in Chapters 1, 4, and 9. Especially Chapter 9 is interesting for its combination of SPE and trade policy instrument analysis, which complicates the analytical representation of the associated optimization problem. A solution algorithm is based on the theory of variational inequalities and tested on different machines (normal and parallel processors).

19.2 Prospect

We discuss the prospects of spatial equilibrium modelling on the basis of a number of themes that appear in the contributions in this book.

19.2.1 Networks and Equilibrium

Several steps have in the past years been taken to fill the gaps between the SPE framework and alternative approaches such as network, general equilibrium and dynamic models. They include adoption of variable transportation costs, infrastructure or perceived networks, multi-commodity systems, and various dynamic phenomena related to cumulative processes, market imperfections, price mechanisms, and evolutionary development. For a valuable contribution we refer, amongst others, to Takayama et al (1984). Chapter 7 is devoted to an interesting combination of network and imperfect market elements. It considers the question under which conditions airline firms will tend to adopt a hub-and-spoke network rather than a fully connected network. Chapter 8 discusses an interregional input-output growth model to study allocation of public investment in transportation infrastructure.

Extensions of spatial equilibrium notions to an economy-wide context can be found in a number of chapters in this book, notably in Part IV. A discussion of equilibrium in spatial labour markets is contained in Chapter 5. Here the optimal allocation of labour from dispersed residential locations to competing firms in various other locations is analysed. The optimal set of wages and rents, realized through residential mobility, is then calculated by using a linear programming model.

19.2.2 General and Spatial Equilibrium

Although the multi-commodity SPE problem shares several elements with the general equilibrium (GE) framework, there is one important difference, namely that demand and supply are all mutually independent, and as a result the commodities only 'interact' in the transportation system. From a general equilibrium point of view this means in fact that even the most extended SPE formulation still remains very much in the tradition of partial transportation and interregional trade optimization. The approach of SPE versus GE models is therefore to some extent analogous to partial planning versus general behaviourial models. Important in SPE models is the assumption of separable demand (and supply) functions (i.e., cross-price elasticities zero), which is not required in GE formulations. This assumption underlines the independent treatment of products in the SPE model. In conclusion, one may say that the interesting contribution of SPE is its spatial or interregional price character, which results from linking a multiregional system to a transportation system, whereas the GE model concentrates on intersectoral relationships. It is obvious that combining the two approaches leads to more complicated issues in general. How to circumvent or deal with these questions is discussed in Part IV of this book. In Chapter 12 multiregional general equilibrium modelling is approached from the perspective of linking computable GE models. A discussion is offered on the relationship and the difference between spatial equilibrium theory and the use of the well-known Armington assumption. Chapter 13 is devoted to an analysis of interactions between transport and the general economy. For this purpose a number of model types are linked: a CGE model, a dispersed SPE model, a freight network equilibrium model, and a regional econometric model. In Chapter 14 an interregional computable GE model is investigated which was developed for an analysis of the impacts of the US-Canada Free Trade Agreement. The focus of this chapter is on the sensitivity of the model with respect to feedback loops in the trading relationships among regions. Finally, in Chapter 15 a GE analysis is performed to assess the impact of an agricultural program, the conservation reserve program, on the U.S. economy. The GE model is linked to an agricultural sector model of an SPE type, to allow for a more detailed analysis of the agricultural sector by measuring the direct agricultural production impacts.

It appears that integration between notions of general spatial and general economic equilibrium is well underway. Such a development may be especially useful for investigating the economy-wide effects of transport and for dealing with policy issues in a multi-regional or multi-national setting. It may also allow for a closer relationship between planning and behavioural models.

19.2.3 Externalities

In the field of spatial economics and transport several issues receive ample attention at present, notably those related to externalities, environmental quality, congestion and infrastructure investment. Some of these may be tackled by using spatial equilibrium models. For instance, network oriented models can deal with congestion effects. These may be localized on individual links, or may cause spill-over effects, both in a spatial-network and an economic sense. Environmental impacts of transport may be investigated by spatial equilibrium analysis, especially when the spatial diversity of either environmental sensitivity or transport flow intensities is great. This book addresses such issues, notably in Chapters 6, 10, 11, 13 and 16. In Chapter 6 three types of models are investigated with respect to externality analysis, namely gravity type of models, SPE type of models and general equilibrium models. In Chapter 10 an SPE analysis is performed of optimal and second-best environmental policies, which is based on a distinction between external costs caused by mobile and immobile sources of pollution. The approach is interesting, as it is not only employing the SPE quasi-social-welfare objective, but describes in addition micro level decisions through behavioural rules. In Chapter 11 network congestion problems, focused on road transport, are considered in a general equilibrium framework. Traditional second best taxation theory is used to study the Pigouvian congestion tax as an element of a broader package of government instruments to be optimized simultaneously. In Chapter 13 traffic congestion is investigated in the context of the model for Italy mentioned above. Finally,

Chapter 16 devotes some attention to the combination of infrastructure dynamics, transportation and environmental externalities, using a simple theoretical framework.

Externalities are by definition related to a systems', multi-actor or economy-wide perspective, providing for a link between planned or desired systems on the one hand and market or actual systems on the other hand. For this reason it seems that SPE models may serve a useful role in studying externalities in a spatial context, including congestion as well as environmental effects, and in dealing with price-oriented policy instruments. It is clear that there are many research opportunities to be seized here, related to issues of optimal spatial structure and mobility, optimal policy of immobile and mobile sources of pollution, and optimal investment in infrastructure and the spatial structure of production and consumption. In addition, one can use static SPE-oriented approaches, for instance, to link natural resource availability and quality to employment, in order to be able to provide for an integrated analysis of two main indicators of spatial system performance.

19.2.4 Dynamics of Spatial Economies

A major limitation of many of the models discussed so far is their static nature. Dynamic adjustment processes and evolutionary elements are, however, intriguing topics which would have to be considered as well. This is the subject matter of Part IV. A topic which is dealt with in this last part is of major importance for deciding about large scale investments in spatio-economic systems, for example, those related to physical infrastructure. This is particularly relevant, as most contributions in the book address comparative static analysis. However, the information obtained this way may be insufficient in judging or evaluating system shocks which are in every respect significant. Chapter 16 is devoted to the use of static and dynamic general equilibrium models for analyzing economy-wide impacts of transport technology and transport policy. Chapter 17 presents the results of a study concerned with the lasting economic effects of an improvement in local or regional amenities, which are likely to arise via attraction of firms and households into the region (migration) and via tourism. Both an analytical model and an empirical regional applied GE model are used for dealing with this topic. In Chapter 18 a wider approach is considered for looking at dynamic behaviour of spatial economies. Here the implications of modern topics in dynamic systems for spatial economies are considered, which are based on the important evolutionary concepts, like non-linearity, structural instability, a-periodic motions, strange attractors, chaos, niche and complexity. Dynamic aspects are further discussed in Chapter 3, where SPE models with inventories are discussed, and in Chapter 13, where a multi-period computable GE model is used to study adjustment processes towards a new equilibrium after a disturbance of the regional economic system. Future work on dynamic models can, in addition to the issues discussed here, be aimed at investment issues (see also Chapter 8) and at extending the spatial externality perspective formalized in an SPE model to include future generations and environmental sustainability.

19.3 Conclusions

The foregoing observations have made clear that many topics related to spatial equilibrium models have received ample attention in the past decades. The SPE framework may be regarded as the core and pure case of a wider field of spatio-economic analysis based on equilibrium notions. As is clear from many contributions in this book, the field of SPE analysis and modelling has been influenced by two main forces. In the first place, it has been studied from the perspective of applications, e.g., related to interregional transport and international or interregional trade. In the second place, a methodological perspective, related to mathematical formulation and solution algorithms, has allowed certain models to be developed and applied more fruitfully than others. In both perspectives much concern has been given to the large dimensions of the systems being modelled. This may be clarified by the many applications of the SPE model to complex situations, such as multi-regional agricultural and energy markets. In addition, the SPE method is also sometimes used in combination with other models, to which it can supply an optimal system of commodity or resource flows. Application in a more theoretical sense, such as dealing with specific instruments, externalities, behaviour and extended system characteristics closing in on other economic models (e.g., GE) has received relatively little attention.

All in all, it seems that the tool of SPE modelling may allow for a broader 'application' than has been realized so far. This book shows various new results in this context. Most important seems the methodological level, where there is still much work to be done with regard to integrating concepts and models of general and spatial equilibrium. In terms of methods this means research focused on some type of integration between the GE and the SPE frameworks. With regard to application areas, it seems that the interaction between transport and the economy will require further development of adequate theory and methods, especially focusing on the role of networks, imperfect markets, congestion, environmental externalities and instruments.

References

Takayama T, Judge G (1964) Equilibrium Among Spatially Separated Markets. A Reformulation. Econometrica 32:510-524

Takayama T, Woodland AD (1970) Equivalence of Price and Quantity Formulations of Spatial Equilibrium. Purified Duality in Quadratic and Concave Programming. Econometrica 38:889-906

Takayama T, Judge G (1971) Spatial and Temporal Price and Allocation Models. North-Holland, Amsterdam

Takayama T, Judge G (1973) Studies in Economic Planning over Space and Time. North-Holland, Amsterdam

Takayama T, Hasimoto H, Uri ND (1984) Spatial and Temporal Price and Allocation Models. Some Extensions. Socio-Economic Planning Sciences 18: 227-234.
Labys WC, Takayama T, Uri ND (Eds) (1989) Quantitative Methods for Market-Oriented Economic Analysis over Space and Time. Gower Publ. Co., London
Takayama T, Hashimoto H, Nguyen DH, Whitacre RC (1989) Application of the Spatial and Temporal Price and Allocation Model to the World Food Economy. In: Labys et al

Contributors

M.J. Beckmann
Department of Economics
Brown University
Providence
Rhode Island 02912
USA

J. Berechman
Public Policy and Economics Departments
Tel Aviv University
Tel Aviv
Israel

P.M. Bishop
Department of Agriculture
Resource and Management Economics
Cornell University
Ithaca
New York
USA

R. Boyd
Department of Economics
Ohio University
Athens
Ohio
USA

C. Elbers
Department of Development and Agricultural Economics
Faculty of Economics
Vrije Universiteit
De Boelelaan 1105
1081 HV Amsterdam
The Netherlands

R. Gazel
Department of Geography
University Of Illinois
1-3 Observatory
901 S. Mathews
Urbana Ill 61801-3682
USA

H. Hashimoto
Faculty of Economics
University of Osaka
1-1 Machikaneyama-cho
Toyonaka
Osaka 560
Japan

G.J.D. Hewings
Department of Geography
University of Illinois
1-3 Observatory
901 S. Mathews
Urbana Ill 61801-3682
USA

H. Kohno
Institute of Socio-Economic Planning
University of Tsukuba
Tsukuba
Ibaraki 305
Japan

W.C. Labys
Department of Resource Economics
Division of Resource Management
College of Agriculture and Forestry
West Virginia University
Morgantown WV 26506-6108
USA

I. Mayeres
Centre for Economic Studies
University of Leuven
Naamse Straat 69
B-3000 Leuven
Belgium

P.G. McGregor
The Fraser of Allander Institute and
International Centre for Macroeconomic Modelling
Department of Economics
University of Strathclyde
Curran Building
100 Cathedral Street
Glasgow G4 0LN
United Kingdom

A. Nagurney
Finance and Operations Management
School of Management
University of Massachusetts
Amherst MA 01003
USA

C.F. Nicholson
Department of Agriculture, Resource and Managerial Economics
Cornell University
Ithaca
New York
USA

P. Nijkamp
Department of Spatial Economics
Faculty of Economics
Vrije Universiteit
De Boelelaan 1105
1081 HV Amsterdam
The Netherlands

S. Proost
Centre for Economic Studies
University of Leuven
Naamse Straat 69
B-3000 Leuven
Belgium

A. Reggiani
Department of Economics
University of Bologna
Piazza Scaravilli 2
40126 Bologna
Italy

P. Rietveld
Department of Spatial Economics
Faculty of Economics
Vrije Universiteit
De Boelelaan 1105
1081 HV Amsterdam
The Netherlands

R. Roson
Department of Economics
University of Venice
Ca' Foscari
Dorsoduro 3246
30123 Venice
Italy

O. Shy
Department of Economics
The University of Michigan
Ann Arbor, Michigan 48109 1221
USA

M. Sonis
Department of Geography
University of Illinois
1-3 Observatory
901 S. Mathews
Urbana Ill 61801-3682
USA

and

Bar Ilan University
Ramat Ban
Israel

J.K. Swales
The Fraser of Allander Institute and
International Centre for Macroeconomic Modelling
Department of Economics
University of Strathclyde
Curran Building
100 Cathedral Street
Glasgow G4 0LN
United Kingdom

T. Takayama
The Japanese Studies Unit
Department of Economics
The University of Western Australia, Nedlands
Western Australia 6009
Australia

N.D. Uri
Economic Research Service
US Department of Agriculture
1301 New York Avenue NW
Washington DC 20005-4788
USA

J.C.J.M. van den Bergh
Department of Spatial Economics
Faculty of Economics
Vrije Universiteit
De Boelelaan 1105
1081 HV Amsterdam
The Netherlands

E.T. Verhoef
Department of Spatial Economics
Faculty of Economics
Vrije Universiteit
De Boelelaan 1105
1081 HV Amsterdam
The Netherlands

C.W. Yang
Department of Economics
Clarion University of Pennsylvania
Clarion
Pennsylvania
USA

Ya Ping Yin
The Fraser of Allander Institute and
International Centre for Macroeconomic Modelling
Department of Economics
University of Strathclyde
Curran Building
100 Cathedral Street
Glasgow G4 0LN
United Kingdom

MIX
Papier aus verantwortungsvollen Quellen
Paper from responsible sources
FSC® C105338
FSC
www.fsc.org

If you have any concerns about our products,
you can contact us on
ProductSafety@springernature.com

In case Publisher is established outside the EU,
the EU authorized representative is:
Springer Nature Customer Service Center GmbH
Europaplatz 3, 69115 Heidelberg, Germany

Printed by Libri Plureos GmbH
in Hamburg, Germany